RESEARCH TRENDS IN MODERN BOTANY

RESEARCH TRENDS IN MODERN BOTANY

(MORPHOLOGICAL, ANATOMICAL, EMBRYOLOGICAL AND CHEMOSYSTEMATIC STUDIES IN ASTERACEAE)

Dr. S. Sundara Rajan
Prof. and Head
Dept. of Post Graduate Studies and Research in Botany
St. Joseph's College
Bangalore - 560 025

ANMOL PUBLICATIONS PVT. LTD.
NEW DELHI - 110 002 (INDIA)

ANMOL PUBLICATIONS PVT. LTD.
4374/4B, Ansari Road, Daryaganj
New Delhi - 110 002

Research Trends in Modern Botany

First Edition, 2000

ISBN 81-261-0659-X

PRINTED IN INDIA

Published by J.L. Kumar for Anmol Publications Pvt. Ltd., New Delhi - 110 002 and Printed at Mehra Offset Press, Delhi.

CONTENTS

PREFACE

Asteraceae is one of the most perplexing, enigmatic and fascinating families of flowering plants. The family represents a dynamic biological system having the widest range of adaptive potentiality that can be imagined for a living material. Having occupied almost the pinnacle of evolution amongst the sympetalous dicotyledons, the family offers a very favourable material for the enquiry into various processes that characterize the most successful line of evolution.

Asteraceae exhibits a wide array of diverse characters morphologically, anatomically, ecologically and even phytochemically. Yet it also has a basic unified pattern upon which is built up the widest possible diversity. While the vegetative characrters are highly plastic, the reproductive characters are highly conservative. It is this diversity in unity (which perhaps exemplifies the basic dictum of life itself) that is primarily responsible for the sucess of Asteraceae. No wonder then Asteraceae is regarded as the second largest family of flowering plants next only to Orchidaceae (opinions on this however are divided).

Eversince I started my research career way back in 1966 , Asteraceae has been the centre of my attention whether the investigations are Embryology, Anatomy or Chemosystematics.

The present work is a compilation of research on Asteraceae dealing with Nodal and Foliar Anatomy, Embryology and Chemosystematics of three Tribes viz., Eupatorieae, Heliantheae and Senecioneae. A part of these researches constituted the dessertation for which I was awarded D.Sc (Doctor of Science) degree from Mysore University.

The book has four basic chapters constituting General Introduction, Nodal and Foliar Anatomical studies, Embryology and Chemosystematics. Chapter 5 constitutes Synopsis followed by Chapters 6-9 dealing with tabular results.

The three aspects taken for investigation viz., Anatomy, Embryology and Chemosystematics are not only dealt descriptively but also comparatively emphasizing the evolutionary processes there in.

The book is meant to be a reference work for those working on Asteraceae. Constructive criticisms from the actual users of the book are most welcome.

Dr. S. Sundara Rajan
Professor and Head

1

GENERAL INTRODUCTION

Asteraceae (Compositae) is the second largest family of flowering plants next only to Orchidaceae. The innumerable representatives of the family on this planet have made it (the family) the most versatile and enigmatic among all the flowering plants. The family enjoys almost a universal distribution being present practically in almost all geographical areas of the world. The members have a wide range of adaptation to suit different ecological niche and can thrive well even in the most inhospitable terrain. Mesophytes, xerophytes, marsh plants and even epiphytes have their representatives in thc family. With such diverse adaptive radiation, no wonder then Asteraceae is known to have over 900 genera and 13,000 species (Willis, 1960). One of the recent works on Asteraceae (Heywood *et al.,* 1977) lists nearly 1350 genera and 22,000 species. Of the total number of flowering plants on Earth, the share of Asteraceae is nearly 10%.

Fossil records indicate that the Asteraceae must have had its origin earlier than Oligocene because by then the family ws already well developed. From Miocene onwards, rock deposits show an abundance of Asteraceae fossils. Small (1919), opines that later Cretaceous or early Ecocene must have been the probable time of origin of Asteraceae. The origin of Asteraceae coincides with approximately the first upheaval of Andes (Small, 1919).

Benthem (1873), Samll (1919), Raven and Axelrod (1974), and Turner (1977) agree that Asteraceae must have had its origin in the southern hemisphere, more specifically in the montane northwestern portions of South America. Rzedowski (1972) and Hu (1958) also agree with a montane bias for the family and opine that the current world-wide distribution must have been due to dispersal along the major mountain chains of the land masses. Turner (1977), considers that the world-wide distribution of the family is mainly due to continental drift and not due to "explosive evolution" as suggested by Raven and Axelrod (1974).

Classification

The basic classification of the family recognised today is practically same as established by Bentham in his celebrated essay "Notes on the classification, history and geographical distribution of Compositae" (1873b). Bentham's work was largely based on the works of earlier authors like Cassini, DeCandolle and Lessing. There are 13 tribes in the family as recognised by Bentham (1873). These are - Vernoniaceae, Eupatoriaceae, Asteroideae, Inuloideae, Helianthoideae, Senecionideae, Calendulaceae, Cynaroideae, Mutisaceae, Cichoriaceae, Helenieae, Anthemideae and Arctotideae. (The spellings of tribal names are according to Bentham).

Hoffmann in his account of the family in Engler and Prantl's *Die naturlichen pflazenfamilien* (1889-94), has introduced some modifications. The tribes as currently recognized are same as those listed by Bentham (1873), except for Helenieae which has been disbanded; most of its members being included in the tribes Heliantheae and Senecioneae. Currently 13 tribes are recognised in the family and they are as follows :

Vernonieae, Eupatorieae, Astereae, Inuleae, Heliantheae, Anthemideae, Senecioneae, Tageteae, Arctoteae, Calenduleae, Cynareae, Mutiseae and Lactuceae.

Divisions of the family.

Cronquist (1955), divided the family into two sub families, Asteroideae and Lactucoideae based on the absence of latex in the former and its presence in the latter. He included the tribes Heliantheae, Astereae. Anthemideae, Arctotideae, Inuleae, Senecioneae, Calenduleae, Eupatorieae, Vernoieae, Cynareae, Mutisieae under the sub family Asteroideae and the Lactuceae (Cichorieae) under Lactucoideae.

Wagenitz (1975), after a survey of distribution of several characters identifies the existence of two rather distinct groups of tribes inside the Asteroideae. The two groups are :

Group I : Vernonieae, Liabeae, Mutisieae, Cardueae, Echinopeae, Arctotideae and Lactuceae.
Group II: Eupatorieae, Heliantheae, Helenieae, Senecioneae, Calenduleae, Astereae, Inuleae and Anthemideae.

Jeffrey (1978), splits the family into two sub families and then suggests that the tribes Eupatorieae and Senecioneae might warrant seggregation into a third sub family as they are in some ways intermediate between the other two sub families.

Carlquist (1976), also proposes a biphyletic grouping and includes his 12 tribes in two sub families.

Sub family *Cichorioideae* (Lactuceae) : Mutiseae, Vernonieae, Cardueae, Arctotideae, Cichorieae and Eupatorieae.

Sub family *Asteroideae* : Heliantheae, Astereae, Inuleae, Calenduleae, Senecioneae and Anthemideae.

Intertribal relationships

Bentham (1873), did not suggest a phyletic scheme for the tribes, but arranged them in such a way as to indicate a relationship between them (Fig. 1.0). He considers the Helianthoideae to be the most ancient tribe on account of the foliaceous involucral bracts, aristate pappus and free anthers of some genera. From the ancestral Helianthoideae, by the regular development of a tubular corolla and the substitution of a pappus for a reduced calyx, he suggests the origin of Eupatorieae in America, the Vernoniaceae in the new and old Worlds, the Cynaroideae in the northern and the Mutisaceae in the southern hemispheres.

Bessey (1897), raised the 14 tribes of Asteraceae to the rank of families and regards the Helianthaceae as the primitive group from which two principal phyletic lines have arisen culminating in the Eupatoriaceae on the one hand and the Lactucaceae on the other.

Hutchinson (1916), also regards the Heliantheae as the most primitive tribe and regards the pappus structure as the most important feature in the phylogeny and affinities of the genera.

Small (1919), after a study of the variation in styles and stamens and the elucidation of the function of the appendages of these organs regards Senecioneae as the most primitive tribe. He suggests a direct origin for Cichorieae from Senecioneae.

Cronquist (1955) proposed a more complex interrelationship for the tribes deriving all of them from the tribe Heliantheae (Fig. 1.1). Turner (1977), supports the scheme of intertribal relationship proposed by Cronquist.

Jeffrey (1977), based on corolla forms has suggested a new scheme of intertribal relationship. According to him, the ancestral corolla type in Asteraceae was zygomorphic bilabiate, and other corolla forms are derivatives. On this assumption, he regards Mutisieae to be closest to the ancestral stock with Lactuceae standing near it as a distinct line.

Structural attributes of Asteraceae

The members of the family are mostly annual or perennial herbs, shrubs, and rarely trees. Leaves are generally exstipulate, either simple or compound or even whorled. Inflorescence is a centripetal capitulum which may be solitary or grouped into secondary inflorescences like spikes, racemes, corymbs etc. The aggregation of flowers into capitula is by far the most unique and unifying character of the family. Various modifications in the floral structure have been attempted by the members without however crossing the limit of uniformity of a capitulum. The several names applied to Asteraceae like **Syanthereae, Androtomeae, Nevramphipetalae, Aggregatae** etc., all refer to the unique floral aggregation. Capitula are one to many flowered, homogamous or heterogamous and are of three categories.

i. Discoid with tubular florets
ii. Radiate with only marginal florets ligulate and
iii. Ligulate with all florets ligulate.

Capitula are always subtended by bracts. Receptacle is concave, convex, elongated or flat. Florets are bisexual or unisexual, regular or irregular. Calyx is modified to pappus and exhibits a great range of variety. The successful distribution of Asteraceae owes much to the diversity of its pappus. Corolla is gamopetalous, actinomorphic or zygomorphic. Stamens 5, epipetalous with syngenesious anthers. Anthers may be simple or have basal or apical appendages. Ovary is inferior, bicarpellary syncarpous with a single anatropous ovule on basal placenta. Fruit is an indehiscent achene.

The mechanism of pollination is a unique feature in Asteraceae. The floral parts largely favour cross pollination and later in the event of failure of cross pollination help self fertilization. The anthers are introrse and fuse by their edges to form an anther tube. In the initial stages pollen is shed into the anther tube which is exposed to biological pollinators. Later, the style grows out of the anther tube and carries with it the pollen deposited on the anther tube. The stigmas separate, exposing them to pollinators. In the event of failure of cross pollination, the style arms recurve so that the stigmas make contact with the pollen deposited on the style.

Chemistry of the family

There have been extensive investigations into the chemistry of the family revealing a wide array of chemical compounds. flavonoids are ubiquitous in the family. Most of the members contain volatile oils and triterpenes. Two classes of natural products seem to be very characteristic of the family. These are - sesquiterpene lactones and a variety of different types of acetylenes. The absence of non protein amino acids, iridoids, tannins and major classes of alkaloids (except for pyrrolizidine alkaloids in Senecioneae and to some extent in Eupatorieae) is characteristic of the family. Table 1.1 gives an account of the chemical diversity of the family.

Interrelationships of the family

The origin and interrelationships of the family is much debated and not surprisingly (due to the diversity of characters) there is no unanimity among synantherologists regarding the systematic position of the family, although every one agrees that the family is highly specialized. There is a wide spectrum of views regarding the natural allies of the family ranging from polypetalous families to families of tubiflorae. Unfortunately the diverse views regarding the phylogeny of the family depend on different character sources which appear to be heterogenous and hence the confusion. The following is a brief review of various opinions on the interrelationships of the family.

Batsch (1802) indicated that the family shows relationship with Campanulaceae, Valerianaceae and other families. Lindley (1845) opined that Asteraceae has affinities with Calyceraceae and Dipsacaceae, while Delpino (1871) suggests origin from Campanulaceae through Lobelioideae. Hallier (1905), regards that Asteraceae represents the highest expression of several tendencies shown by the Campanulatae.

Wernham (1913), in his account of floral evolution in sympetalae, supports the view that Asteraceae has arisen from Campanulaceae.

Bessey (1915), placed Asterales (which includes asteraceae) at the peak in the Rubialian line of evolution. The four families included in the order were supposed to have been derived as follows :

Valerianaceae → Dipsacaceae → Calyceraceae → Asteraceae.

Small (1919), traces close affinities between Lobelioideae and Asteraceae on the basis of

i. Protandry,
ii. Pollen presented outside the style
iii. Autogamy by curving of the style
iv. Presence of capitula as in *Jasione* and *Phyteuma.*

Gundersen (1943), has placed Asteraceae with Campanulaceae and Dipsacaceae in one group together with other gamopetalous families having inferior overy. He, later (1950) united them with Calyceraceae to constitute the order Asterales following Rubiales. Philipson (1953), on the basis of the morphology of inflorescence in Rubiales and Campanulaceae agreed that Asteraceae bears close resemblance to Campanulaceae.

Erdtman (1952), based on the characters of the pollen, expressed that Asteraceae is similar to Brunoniaceae, Calyceraceae and Goodeniaceae, but differs from Campanulaceae, Dipsacaceae, Stylidiaceae and Valerianacaeae.

Some of the more recent views also present a divided opinion on the interrelationships of the family.

Turner (1977), is of the opinion that evidence points out to the South American herbaceous family Calyceraceae being closest to Asteraceae. The two families are also similar palynologically (Skavarla *et al* 1977).

Jeffery (1977), based on the analysis of corolla forms suggests that protoasteroids (ancestral Compositae) are likely to be found in racemose and often bilabiate families such as Campanulaceae, Goodeeniaceae and Stylidiaceae rather than in cymose, predominantly actinmorphic ones such as Rubiaceae Caprifoliaceae and Dipasacaceae. According to Jeffrey (1977), the Brunoniaceae and Calyceraceae which resemble Asteraceae, have evolved parallely.

Interrelationships of the family based on chemical evidences are equally puzzling and fail to give a true picture of the phylogeny of the family.

Hegnauer (1977), considering only the chemical data suggess a relationship between Polemoniales (Boraginaceae) and Asterales (Asteraceae). On the other hand, Mabry and Bohlmann (1977), suggest that among the extant families umbelliferae and Campanulaceae are phyletically closest to Asteraceae. Sesquiterpene lactones and acetylenes are found in both Asteraceae and Umbelliferae. Acetylenic chemistry also links the family to Campanulaceae. In addition to lactones and acetylenes, coumarins of similar types are found in Umbelliferae and Asteraceae. Mebry and Bohlmann (1977), also opine that Campanulaceae may be

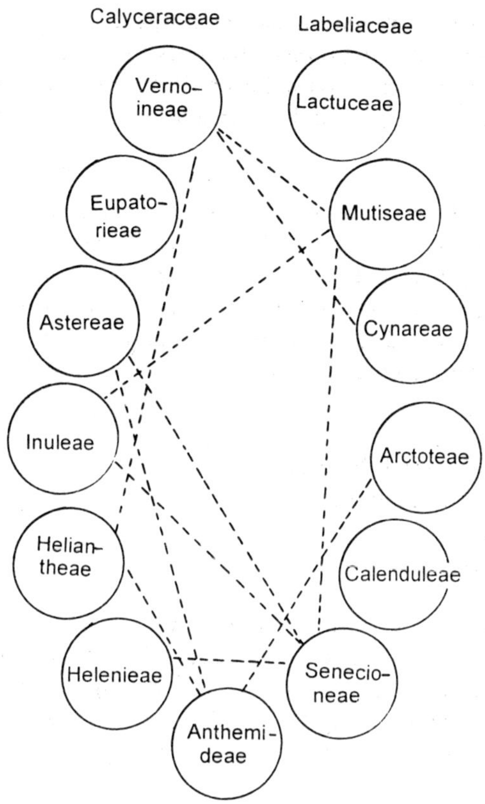

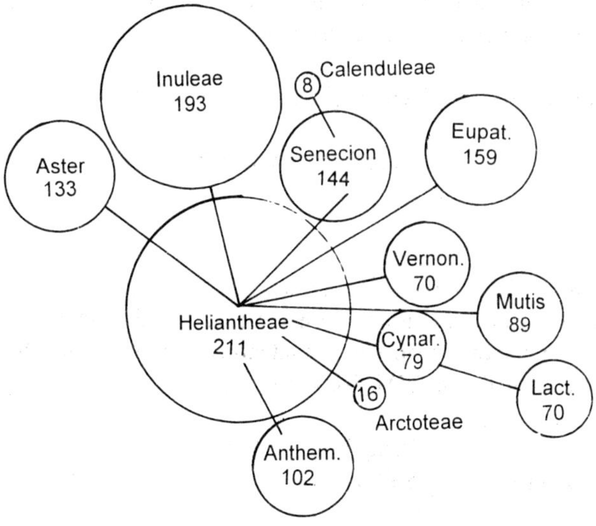

Plate 1.1 Intertribal relationships in Asteraceae

Fig 1.1. Natural relationships among the tribes of Asteraceae as envisaged by Bentham (1873); **Fig 1.2.** Phyletic relationships among the tribes of Asteraceae as envisaged by Cronquist (1955).

further linked to Asteraceae as both of them accumulate inulin type fructans. The Calyceraceae regarded as related to Asteraceae on morphological grounds (Turner 1977) does not seem to go with the family (Asteraceae) as for as chemical data are concerned. Mebry and Bohlmann (1977) observe "Not only do the Calyceraceae examined to date do not produce acetylenes, but they contain seco-iridoids similar to those occurring in Gentianales".

To sum up, phylogeny of the family based on chemical data seems to be at variance with those obtained from morphology. This raises a question as to how far conclusions based on only one source can be relied upon. Stebbins (1977), based on a detailed study of development and comparative anatomy of Asteraceae has come to the conclusion that "The Compositae can not be regarded as descended from or closely related to any other modern family". The cytochrome *c* data also seem to support this view (Boulter *et al* 1972). Thus, the origin of Asteraceae has still remained an open question.

The present work

The present investigations encompass three parts

i. Nodal and Foliar anatomical studies
ii. Embryological studies and
iii. Chemosystematic studies

The scope of the work has been explained individually in the introductory remarks on chapters concerning the aspects mentioned above. A total of 29 taxa are included in Nodal and Foliar anatomical studies, 8 in embryology and 51 in chemosystematic. The names of taxa and the aspects of investigations carried out are given in table 1.1. The taxa investigated belong to tribes Eupatorieae, Senecioneae and Heliantheae. The following is a brief morphological description of the taxa studied (original citations are given for Indigenous plants only).

1. *Ageratum conyzoides* L. Sp. Pl. 839.1753.
   Erect herbs with opposite or upper alternate leaves. Heads in dense corymbs or penicles. Receptacle flat. Anthers have appendages. Achenes 5 angled; pappus of 5 scales.

   The plant is a softly hairy annual weed with white or pale blue flowers.

   Place of collection : Bangalore, South India,
   Herbarium specimen examined : SSR, 1001.

2. *Ageratum houstonianum* Mill. (= *A. mexicanum* Sims).
   Small cultivated herbs; leaves usually heart shaped at base. Heads commonly little more than 0.4 cm across. Flowers blue, involucral bracts lance linear, attennuate, entire, ciliate, the back finely, densley and somewhat viscidly hairy (named after WM Houston).

   Place of collection : Bangalore, South India
   Herbarium specimen examined : SSR, 1002.

3. *Adenostemma lavenia* (L.) Kuntze. Rev. Gen. Pl. 1: 304. 1891.
   An erect sticky herb growing up to 1.5 m. Leaves Usually opposite. Heads homogamous, arranged in terminal corymbs. Bracts in 12 rows, sometimes connate. Receptacle flat. Corolla white, all tubular. Anthers truncate with glandular tip. Achenes 3-5 angled. Pappus made up of 3 hairs.

Place of collection : Kemmannugundi, Karnataka, South India
Herbarium specimen examined : SSR, 1003.

4. *Chromolaena odorata* (L.) King and Robinson. Phytologia. 20 : 204, 1970 (= *Eupatorium odoratum L.*)
Stem and branches glandular and pubescent. Leaves opposite, sub palmately veined; heads many in terminal corymbs, homogamous and discoid. Bracts uni or multiseriate surrounding a flat receptacle. Anthers obtuse below. Achenes 5 angled, truncate.

Common undershrub in open forests of Western ghats, Karnataka.

Place of collection : Sakalespur, Karnataka, South India
Herbarium specimen examined : SSR, 1004.

5. *Acanthospermum hispidum* DC. Prodr. 5:522. 1936.
Annual erect, hairy herbs. Leaves opposite, palmately veined. Heads solitary axillary 2.5 - 3cm across, radiate. Bracts few. Outer female florets with 2 dentate ligules. Achenes compressed; pappus absent.

Place of collection : Bangalore, South India
Herbarium specimen examined : SSR, 1005.

6. *Polymnia grandis* Linn.
Tall shrubs; stem cylindrical and pubescent. Leaves palmately lobed. Head heterogamous; ligules white. Achenes compressed.

Place of collection : Nandi Hills, near Bangalore, South India
Herbarium specimen examined : SSR, 1007.

7. *Siegesbeckia orientalis* L. Sp. Pl. 900. 1753.
Annual herbs with opposite crenate palmately 3 nerved leaves. Heads few, in panicles; bracts biseriate, outer glandular, the inner enclosing ray florets. Receptacle paleaceous. Anthers obtuse below. Achene curved, truncate and glandular; pappus absent.

Place of collection : Nandi Hills, near Bangalore, South India
Herbarium specimen examined : SSR, 1007.

8. *Blainvillea acmella* (L). Philpson. Blumea 6:350. 1950. (= *B . rhomboidea* Cass).
Annual, scabrid, hairy herbs. Leaves opposite or alternate; palmately veined and hairy. Heads usually solitary, axillary. Involucral bracts few. Receptacle flat. Outer florets ligulate; ligule 2-3 lobed. Anthers obtuse at base. Achenes of female florets trigonous, of bisexual florets laterally compressed pappus 0 or of 2-3 bristles.

Place of collection : Bangalore, South India
Herbarium specimen examined : SSR, 1008.

9. *Eclipta prostrata* (L). L. Mant. 2:206. 1771. (= *Eclipta alba* (L). Hassk).
Annual or perennial herbs. Leaves 3-4 cms. long, opposite, Bracts biseriate. Ray florets female and white ligulate. Anthers obtuse below. Achenes in female florets 3 angled, in bisexual 4 angled; pappus absent.

Place of collection : Bangalore, South India
Herbarium specimen examined : SSR. 1009.

10. *Eclipta prostrata* (L.) L. (Small leaf)
This is a small leaf form similar to the normal one in all respects except for the size of the leaves (1-2 cms long) and head and prostrate habit.
Place of collection : Castle rock (along railway track), Northern Karnataka, South India.
Herbarium specimen examined : SSR. 1110.

11. *Synedrella nodiflora* (L.) Gaertin., Fruct. 2:456. 1791.
Annual herbs with opposite leaves growing up to 29 cms tall. Stem green terete. Lamina 2.8-6 cm x 1.4 -4 cm. Heads axillary or terminal, heterogamous, with yellow rays and surrounded by four bracts. Calyx modified into two awns. Anthers hooded. Achenes of ray florets winged; of disc florets winged obscurely.

Place of collection : Bangalore, South India
Herbarium specimen examined : SSR. 111.

12. *Synedrella nodiflora* (L.) Gaertn. (smallo leaf)
Annual erect herbs up to 26cms tall. Stem green below but violet upwards. Leaves 1-3 cm x 0.5-1.8 cm. Same as the normal *S. nodiflora* in all other respects.

This plant has appeared in Bangalore during the past two years and is propagating well; generally found in association with the normal leaf form. The plant is true breeding.

Place of collection : Bangalore, South India
Herbarium specimen examined : SSR. 1112.

13. *Calyptocarpus vialis* Less. (= Synedrella vialis Less A. Gray)
Annual decumbent branching herb rooting at the lower nodes. Stem terete greenish to pale violet, pubescent, up to 1.5mm thick; leaves opposite, petiolate, lamina ovate to ovate lanceolate, attenuate at base, acute at apex, unicostately reticulate veined, pubescent on both the surfaces 2.0 - 5.7 cm x 0.6-3 cm. Heads heterogamous, yellow, axillary or terminal, sessile or up to 25 mm pedicellate, surrounded by four bracts. Outer bracts (2), green and leafy ± 6mm x 2mm, inner two hyaline, ± 5mm x 1.4mm. Ray florets 4-6 in number. Calyx is modified into two awns, pubescent. 1-5mm long. Corolla ligulate with 2-3 lobed ligules. Ovary pubescent; style slender with slightly recurved stigmatic lobes. Disc florets 6-9 in number: corolla 5 lobed. Stamens 5, anthers slightly exerted, hooded and connate. Ovary inferior, bicarpellary syncarpous bilocular with a single basal ovule. Stigma deeply 2 lobed, glandular and hairy. Achenes of ray florets winged obscurely, light brown in colour; of disc florets dark brown in colour, not winged.

This taxon has appeared in Bangalore during the past two years. The identity of the taxon has been determined by Dr.Jeffrey C. of Royal Botanic Gardens, Kew, where a voucher specimen has been deposited.

This is the first report of the taxon for South India. The earlier two records for North India are at Pune, lawns of BSI (Ahuja 1968) and New Delhi, NBPGR Khanna (1980).

Place of collection : Bangalore, South India
Herbarium specimen examined : SSR. 1013.

14. *Lagascea mollis* Cav., Anal. Cienc. Nat. 6:331, t.44. 1903.
Grey pubescent herbs, with alternate or opposite leaves. Heads 1 flowered on a common receptacle supported by leaves which appear like bracts. All flowers in the compound head fertile. Corolla regular. tubular. Anther bases sagittate. Style arms curved and hairy; achenes cuneate, compressed or three angled.

Place of collection : Bangalore, South India
Herbarium specimen examined : SSR. 1014.

15. *Sclerocarpus africanus* Jacq., Ic. Pl. Rar. 17, t. 176. 1781.
Strigose or villous herbs. Leaves alternate or opposite. Heads heterogamous, rayed; outer flowers female but sterile; inner flowers bisexual and fertile. Bracts few, biseriate. Receptacle convex. Corolla of ray flowers yellow with 0-3 lobed ligule.

Anther bases obtuse. Achenes subovoid, compressed; pappus absent or a short circle of scales.

Placc of collection : Bangalore, South India
Herbarium specimen examined : SSR. 1015.

16. *Helianthus rigidus* Desf.
Shrubs growing up to 8 feet high, sparingly branched; leaves opposite, thick oblong to ovate, gradually narrowed into petioles, scabrous on both sides; heads showy, long stalked, involucral bracts ovate, obtuse or acute, ciliate, disc purple or brown (cultivated plant).

Place of collection : Bangalore, South India
Herbarium specimen examined : SSR, 1016.

17. *Rudbeckia nitida* Nutt.
Plants 1-1.5 m in height. Leaves bright green, entire or minutely toothed, the basal and lower cauline, ovate lanceolate to oblong. Heads solitary or few; rays several to numerous; pure yellow in colour. Style branches short and obtuse. Pappus conspicuous (cultivated plant).

Place of collection : Bangalore, South India
Herbarium specimen examined : SSR, 1017.

18. *Bidens pilosa* auct. non L. (=*B. biternata* Merr & Sherff, ex Sherff) (Normal ray)
Perennial herbs. Leaves opposite 1-2 pinnatisect or lobed. Heads heterogamous, in corymbs or subsolitary. Rays uniseriate female; disc bisexual; bracts biseriate. Receptacle flat or convex. Corollas of ray florets white; of discs yellow in colour. Anthers subsagittate or obtuse. Achenes tetragonous.

Place of collection : Nandi Hills, near Bangalore, South India.
Herbarium specimen examined : SSR, 1018.

19 and 20 : The intermediae and rayless forms of *B. pilosa* differ from the normal, only in terms of rays. In the rayless form, the corolla of peripheral florets is equally divided into 5 lobes. In the intermediates the rays are much reduced in size.

Place of collection : Nandi Hills, near Bangalore, South India.
Herbarium specimen examined : SSR, 1019 and 1020.

21. *B. humilis*, H.B. and K.
This taxon is similar to *B. pilosa*. Leaf margin dissected much more than in *B.pilosa*. Heads heterogamous. Receptacle convex. Anther bases obtuse. Style arms of female flowers oblong. Corollas of both rays and discs yellow in colour.
Place of collection : Nandi Hills, near Bangalore, South India.
Herbarium specimen examined : SSR, 1021.

22. *Moonia heterophylla* Arn., Nov. act. Nat. Cur. 18:349. 1936.
Perennial undershrub. Leaves opposite entire or pinnately toothed. Heads heterogamous rayed, solitary axillary or terminal.

Rays deeply cleft in linear lobes, bract 0.3 - 0.5 cm long with ciliate margins.

Place of collection : Kemmannugundi, Karnataka, South India.
Herbarium specimen examined : SSR, 1022.

23. *Galinsoga quadriradiata* Ruiz Lopez and Pavon, Syst. Veg. 1:198. 1798 (=*G.ciliata* Raffin & Blake).
Small hispid hairy herbs. Leaves 2-3.5 cm long, ovate acute at apex cuneate at base; hispid on both surfaces. Capitula 2-5mm across on short hairy peduncle. Involucre biseriate, outer and inner phyllaries 5 each. Ray flowers 5, tube 1mm long, ovate, lobes deeply dissected up to 0.5mm long; corolla of disc yellow; 5 lobed. Achenes 1-2mm long, black.

Place of collection : Bangalore and Kemmannugundi, Karnataka, South India.
Herbarium specimen examined : SSR, 1023, 1024.

24. *Galinsoga parviflora* Cav., Ic. 3:41. t. 281. 1795.
Herbs; stems sparsely hairy or nearly glabrous; leaves ovate to ovate lanceolate, acuminate, cuneate at base, glabrous; petiole 0-3-1.5mm long. Involucre biseriate; 5 each in inner and outer phyllaries. Ray flowers 5 with 1mm long pubescent tube; ligule 1mm or slightly longer, lobes very shallowly dissected. Achenes 1-2mm long and black in colour; pappus scaly and 10 in number.

Place of collection : Bangalore and Kemmannugundi, Karnataka, South India.
Herbarium specimen examined : SSR, 1025, 1026.

25. *Galinsoga* (intermediate)
This taxon is encountered in two populations, in Kemmannugundi (Chikmaglur dist) and Bangalore. The characters are some what intermediate between *G.quadriradiata* and *G.parviflora*

Small herbs; stem sparsley hairy; leaves like in *G.quadriradiata* but not hispid; sparsely hairy; capitula heterogamous; ray flowers tubular, tube up to 1mm long, ligule 0.8 - 1.4mm long; lobes dissected *to the middle of their length* (upto 0.4 - 0.7 mm); corollas of disc flowers as in *G.quadriradiata* Achenes also as in *G.quadriradiata*.

Place of collection : Kemmannugundi and Bangalore, Karnataka, South India.
Herbarium specimen examined : SSR, 1027, 1028.

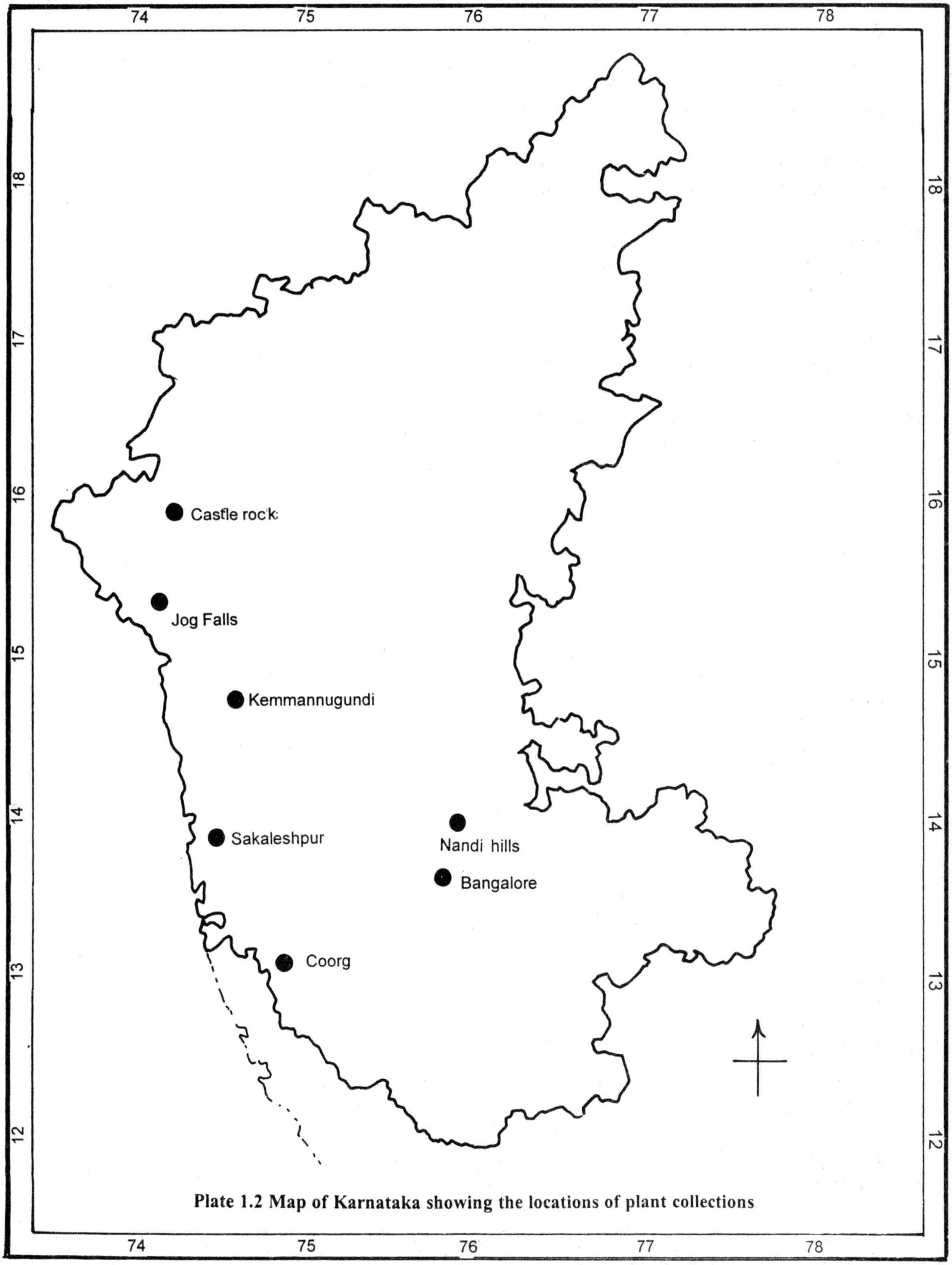

Plate 1.2 Map of Karnataka showing the locations of plant collections

26. *Spilanthes calva* Dc., in Wt. Contrib. Bot. Ind. 19.1834. (=*S.acmella* Murr). (small head)
Creeping, ascending herbs. Leaves 2-3 x 1-2 cm, ovate, acute subcrenate. Heads usually solitary, ovoid - 1.1 x 0.7 cm in size. Ray and disc florets yellow in colour. Bracts ovate acute Achenes dorsally compressed and glabrous; pappus absent.

Place of collection : Bangalore, South India.
Herbarium specimen examined : SSR, 1029.

27. *Spilanthes calva* DC. (large head)
This taxon is similar to *S. calva* in all respects except in size of the head and to some extent in leaves. Heads 1.5 x 2.5 cms in size.

Place of collection : Kemmannugundi and Bangalore, Karnataka, South India.
Herbarium specimen examined : SSR, 1030.

28. *Spilanthes paniculate* Wall. ex. Dc., Prodr. 5:625. 1836.
Tall herbs, leaves 3-6 cm long, broadly ovate, acute with sub entire margin. Heads in axillary or terminal panicles. 1.2 x 0.7cm in size. Rays and discs white in colour. Bracts oblong, elliptic and fimbriate. Achenes of the outer florets trigonous, thickened along the margins.

Place of collection : Bangalore, South India.
Herbarium specimen examined : SSR, 1031.

29. *Parthenium hysterophorus* L., Sp. Pl. 988. 1753.
A widespread weed in open wastelands; leaves alternate, sessile and bipinnatifid. Heads in panicles, radiate. Involucral bracts biseriate. Receptacle flat, paleaceous. Outer florets 5, female, inner few male. Corolla ligulate in female, entire and tubular in male, white in colour.

Place of collection : Bangalore, South India.
Herbarium specimen examined : SSR, 1032.

30. *Xanthium indicum* J. Koenig in Roxb. F1. Ind. 3:601. 1832 (=*Xanthium strumarium* L.)
Erect herbs up to 70 cm tall. Leaves alternate, petiolate, palmately veined. Heads unisexual, discoid with functionally male heads in the upper axils, female heads in lower axils. Bracts uniseriate; receptacle paleaceous; florets many. Female heads globose with two florets; bracts all connate forming an indurate involucre terminated by two spinescent hooks. Corolla 0. Achenes two, enclosed within the involucre.

Place of collection : Bangalore, South India.
Herbarium specimen examined : SSR, 1033.

31. *Senecio wightii* (DC) Benth.ex.Cl., Comp. Ind. 197. 1876.
Annual herbs up to 70cm tall. Leaves 5 - 10 x 0.5 - 1 cm; ovate lanceolate, acute at tip, amplexicaul at base. Heads heterogamous, rayed, ligule yellow and 4 dentate. Achene ribbed and without pappus.

Place of collection : Castle rock, Karnataka, South India.
Herbarium specimen examined : SSR, 1034.

32. *Senecio lawsoni* Gamb in Kew Bull. 1920:342. 1920
Erect herbs. Leaves scattered along the stem, linear oblong. 1.6 cm, long hispidly villous with recurved

margins. Heads in corymbs; bracts about 20, lanceolate, acuminate. Ligules 5-7 nerved. Achenes subterete. 5-10 ribbed.

Place of collection : Kemmannugundi, Karnataka, South India.
Herbarium specimen examined : SSR, 1035.

33. *Senecio belgaumensis* Cl., Comp. Ind. 200. 1876.
Annual, erect herbs 30-85 cm in height. Stems and branches usually pubescent. Leaves rhomboid ovate, coarsely toothed; strigosely hairy or glabrous. Upper heterogamous. Ligules usually eight in number, linear oblong. Involucral bracts 12-18 Pappus 0. Achenes turbinate, ribbed and hairy.

Place of collection : Shimoga, Karnataka, South India.
Herbarium specimen examined : SSR, 1036.

34. *Senecio neelgherryanus* DC. in Wt. Contrib. Bot. Ind. 23.1834.
Erect herbs. Leaves linear oblong, lower often spathulate, semiamplexicaul, margins entire to pinnatifid, upto 7 cm long; bracts 20 or more. Achenes ribbed glabrous or minutely papillose.

Place of collection : Coorg, Karnataka, South India.
Herbarium specimen examined : KRK, 4832

35. *Senecio tenuifolius* Burm, f., F1. Ind. t. 60., f.4. 1768.
A slender much branched annual with leaves cut into very narrow lobes; glabrous or nearly so, margins often reflexed. Heads small with 6-10 ray florets. The ligules 4 nerved; bracts about 12, obovate, cuspidate, glabrous, except the ciliate tips. Pappus yellowish.

Place of collection : Bangalore, South India.
Herbarium specimen examined : SSR, 1038.

36. *Senecio hohenackeri* HK. f.in HK. f. F1. Brit. India 3:345 1881.
A branching herb, up to 20 cm in height with woody root stock. Leaves deeply pinnatifid, rarely over 1 cm long. Head 0.9cm broad; ligules 8 in number, 4 nerved; bracts 10-15

Place of collection : Bangalore, South India.
Herbarium specimen examined : SSR, 1039.

37. *Senecio grahami* HK f. in HK. f., F1. Brit India 3:347. 1881.
Much branched erect herb up to 40 cm tall; stems angled. Leaves 3.5 - 5 x 1.5 - 2cm, rhomboid above, white tomentose beneath. Heads yellow 0.8 - 1cm across in sub corymbose panicles. Achenes ribbed, scabrid.

Place of collection : Bababudan giri, Chikmagalur Dist., Karnataka, South India.
Herbarium specimen examined : Simhan, 1647.

38. *Senecio edgeworthii* Hk. f., in Hk.f., F1. Brit. India. 3:346. 1881.
Erect suffruticose herbs. Stems angled. Leaves up to 2 x 1.5cm rhomboid or ovate, sub acute at the tip, hairy above cottony beneath. Flowers yellow, in terminal lax corymbose capitula. Pappus scanty.

Place of collection : Bababudan giri, Chikmagalur Dist., Karnataka, South India.
Herbarium specimen examined : Simhan, 1642.

39. *Senecio edgeworthii* (Small leaf)
This is similar to the other taxon (No.38) except for the small size of the leaf and short stature. Both the plants (normal leaf and small leaf) grow in the same locality.

Place of collection : Bababudan giri, Chikmagalur Dist., Karnataka, South India.
Herbarium specimen examined : Simhan, 1813.

40. *Senecio ludens* Cl. Comp. Ind. 199. 1876
A slender erect herb with lyrate broadly ovate leaves. Lamina hispid pubescent about 4.2 cm x 2.4. Heads heterogamous; ligules yellow 4 nerved, bracts narrow, acuminate. Achenes all similar, 5 angled; pappus sordid.

Place of collection : Kolar Gold Field., Karnataka, South India.
Herbarium specimen examined : SSR, 1043.

41. *Senecio cineraria* DC. (= *C. maritima* L., *S. acanthifolius* Hort).
White wooly herbs, specially the leaves. Stem herbaceous, branching from base. Leaves pinnatifid with oblong obtuse segments. Heads usually radiate 0.5 - 1cm across, in small compact cymes: yellow or cream coloured (cultivated plant).

Place of collection : Bangalore, South India.
Herbarium specimen examined : SSR, 1044.

42. *Senecio corymbosus* Wall. ex.DC. in Wt. Contrib.Bot. Ind. 22. 1834.
Climbers with vine like stem. Leaves pubescent and white tomentose below. Heads in corymbs, 0.6cm long with 5-15 flowers; bracts 8, oblong lanceolate cottony on the back.

Place of collection : Kemmannugundi, Karnataka, South India.
Herbarium specimen examined : SSR, 1045

43. *Senecio walkeri* Arn. in Nov. Act. Nat. cur. 18 : 349. 1836.
A slender climber with pubescent slightly cottony leaves. Leaves turn glabrous at later stages. Heads 0.5 cm long (with about 10 - 12 flowers) arranged in terminal panicles; bracts 8, linear obovate, nearly glabrous. Achenes subterete.

Place of collection : Bangalore, South India.
Herbarium specimen examined : SSR, 1046.

44. *Gynura nitida* DC ., Prodr. 6:300. 1838.
Tall, somewhat fleshy herbs with alternate, glabrous or hispid leaves. Stem angular with variable (in shape) leaves. Heads homogamous, not rayed but disciform. Involucre, cylindric with uniseriate bracts. Achenes sparsely hispid between the ribs.

Place of collection : Bangalore, South India.
Herbarium specimen examined : SSR, 1047

45. *Gynura lycopersicifolia* DC., Prodr. 6:300. 1838.

Slender erect herbs with non tuberous roots. Leaves deeply or irregularly pinnatified or lyrate, auricled at base; heads up to about 9. Achenes sparesely hispid between ribs.

Place of collection : Jog falls, Shimoga District, Karnataka, South India.
Herbarium specimen examined : SSR, 1048

46. *Gynura aurantiaca* DC. (Velvet plant)
Stout and branching herbs. 35-36 cm high; stems almost succulent, densely pubsecent with violet or purple hairs. Leaves soft, ovate. 4.5 - 7.6 cm, toothed, hairy with short of ten narrowly winged petioles. Uppermost leaves sessile, clasping. Heads 1-1.6cm across in loose terminal clusters, yellow or orange in colour. (Cultivated plant)

Place of collection : Nandi Hills, Bangalore.
Herbarium specimen examined : SSR, 1049.

47. *Notonia grandiflora* DC., in Wt. Contrib. Bot. Ind. 24. 1834.
A fleshy shrub reaching up to 1.3 meters in height; stems and leaves fleshy; leaves obovate or oblanceolate, obtuse, variable in size, entire. Heads 1-1.6 cm long, homogamous, all bisexual; corolla tube slender. Achenes long, subterete; 10 striate (cultivated plant)

Place of collection : Bangalore, South India.
Herbarium specimen examined : SSR, 1050.

48. *Notonia walkeri* Cl. Comp. Ind. 176. 1876.
A tall shrubby plant reaching more than 1.6 meters in height. Stems and leaves soft and slightly fleshy; leaves long petioled, oblong lanceolate and acuminate. Heads in large corymbs, 1-1.6cm long, the bracts long and narrow. All flowers eligulate and bisexual. Corolla tube slender 0.9cm long. Achenes sparsely hairy.

Place of collection : Nilgiri Hills, Tamil Nadu, South India.
Herbarium specimen examined : SSR, 1051.

49. *Emilia sonchifolia* (L.) DC. ex. Wt. Contrib. Bot. Ind. 24 1834.
Soft herb, leaves radical or cauline. Heads small under 0.8cm long, few flowered, homogamous, eligulate, all discoid. Corolla lobes slender short; style arms short with acute tip; achenes hispid on the angles.

Place of collection : Bangalore, South India.
Herbarium specimen examined : SSR, 1052.

50. *Emilia scabra* DC., Prodr. 6:303. 1838.
Annual or perennial herbs. Leaves chiefly radical or for a short way up the stems, lyrate, prominently crenate, crispately pubsecent 1-6-3.5cm long. Heads many flowered; bracts 3; corolla lobes 1.6cm long. Style arms truncate with a mucronate tip. Achenes hispid on the angles.

Place of collection : Bangalore, South India.
Herbarium specimen examined : SSR, 1053.

51. *Emilia ramulosa* Gamb. in Kew Bull. 1921.
Branching herb, annual or perennial. Leaves all cauline, narrow, linear oblong or spathulate, glabrous above, Hirsute beneath. Heads large, many flowered, usually solitary on 10-15 cm long peduncle: corolla lobes 1.6cm long pubescent at tip; style arms short. Achenes hispid on the angles.

Place of collection	:	Nandi Hills, Bangalore, South India.
Herbarium specimen examined	:	SSR, 10545.

LITERATAURE CITED

AHUJA, KK. and PATASKAR, R.D. 1969. *Synedrella vialis* Less. A Grey. A new record for India. *Indian Forester 95, 267.*

BATCH, A.J.G.C. 1802. *Tabula affinitatum Regni Vegetabilis.* Weimar.

BENTHAM, G. 1873. Notes on the classification, history and geographical distribution of Compositae. *J. Linn Soc. Bot, B:* Vol. XXIV.

_____________ 1915. The Phylogenetic Taxonomy of flowering plants. *Annal. Missour. Bot. Gdn.* Vol. II. Nos. 1-2.

BOULTER, D., HASTELL, B.G., PEACOCK, D., RAMSHAW, J.A.M. and SCAWEN, M.D. 1977. Chemistry, function and evolution of plastocyanin. In "Plant Bicohemistry Vol. II" (D.H. Northcote Ed.), PP 1-40. University Park Press, Baltimore, U.S.A.

CARLQUIST, S. 1976. Tribal interrelationships and phylogeny of Asteraceae, *Aliso* 8, 465-492.

CRONQUIST, A. 1955. Phylogeny and Taxonomy of the Compositae. *Ann. Midl. Nat.* 53, 478-511.

* DELPINO, F. 1871. Studi sopra an lignaggio anemofilo delle Composte. Fierenze.

ERDTMANN, G. 1952. Pollen morphology and Plant Taxonomy of Angiosperms. Hafner. N.Y., U.S.A.

GUNDERSON, A. 1943. Flower forms and groups of Dicotyledons. Waltham, Mass, U.S.A.

_____________ 1950, families of Dicotyledons, Waltham, Mass. U.S.A.

* HALLIER, H. 1903. Verwand Schaftverh der Tubifloren and Cbenalen den polyphylet. Vaturwiss. Verein Abhand Bd XVI, Hamburg.

HEGNAUER, R. 1977. The Chemistry of the Compositae. In : *The Biology and Chemistry of the Compositae.* Eds. Heywood *et al.* Academic Press. N.Y. pp. 283-335.

HEYWOOD, V.H., HARBORNE, J.B. and TURNER, B.L. 1977. An Overture to Compositae. In : *The Biology and Chemistry of the Compositae.* Eds. Heywood *et al.* Academic Press. N.Y. pp 1-20

HU.S. 1958. Statistics of Compositae in relation to the Flora of China. J. *Arnold Arbor* 39, 347-419.

HUTCHINSON, J. 1916. Notes on African Compositae, II. *Kew Bulletin.* XXXI.

JEFFREY, C. 1977. Corolla forms in Compositae, some evolutionary and taxonomic speculations. In : *The Biology and Chemistry of the Compositae.* Eds. Heywood *et al.* Academic Press. N.Y. pp 111-118.

_____________ 1978. Compositae. In : "Flowering Plants of the world by Families" (V.H. Heywood ed). Elsevier International Projects, Oxford.

LINDLEY. 1845. The Vegetable Kingdom, London.

MABRY, T.J. and BOHLMANN, F. 1977. Summary of the Chemistry of the Compositae. In : *The Biology and Chemistry of the Compositae* (Eds. Heywood *et al*) Academic Press. N.Y. pp. 1097-1104.

PHILIPSON, W.R. 1953. The relationship of the Compositae particularly as illustrated by the morphology of the inflorescence in Rubiales and Campanulatae. *Phytomorphology,* 3: 391 - 404.

RAVEN P.H. and AXELROD, D.I. 1974. Angiosperm biogeography and past continental movements. *Ann Mo. bot.* Gdn. 61, 539 - 673.

* RZEDOWSKI, J. 1972. Contributciones a la fitogeografia floristicae historica de mexico III. Algunas tendencias en la distribucion geografica ye ecologica de las Compositae mexicanas. *Eiencia Mex.* 27 123 - 132.

SKAVARLA, J.J., TURNER, B.L., PATEL, V.C. AND TOMB, A.S. 1975. Pollen morphology in the Compositae and in morphologically related families. In : *The Biology and Chemistry of the Compositae* (Eds. Heywood *et al)* Academic Press. N.Y. pp. 141 - 217.

SMALL. J. 1919. The Origin and Development of the Compositae. *New Phytologist. reprint no.* 11. William Wesley and sons. London.

STEBBINS. G.L. 1977. Development and comparative anatomy of the Compositae. In : *The Biology and Chemistry of the Compositae* (Eds. Heywood *et al*) Academic Press. N.y. pp. 91-109.

TURNER, B.L. 1977. Fossil history and geography. In : *The Biology and Chemistry of the Compositae.* (Eds. Heywood *et al*) Academic Pres. N.Y. pp. 22-39.

WERNHAM. H.F. 1912. Floral evolution, with particular reference to the sympetalous dicotyledons. *New Phytologist.* Vol. XI.

WAGENITZ. G. 1976. Systematics and Phylogeny of the Compositae. *Plant Syst. Evol.* 125-46.

WILLIS. 1960. The Dictionary of flowering Plants and ferns. Cambridge University Press, London.

2

NODAL AND FOLIAR ANATOMY

Systematic anatomy deals with the internal structure of plant organs (chiefly vegetative) in relation to their taxonomic classification. The leaf is one of the most important vegetative organs whose anatomy can be successfully exploited for systematic purposes.

Nodal and Petiolar Anatomy

Node has been defined as that part of the stem from which leaves arise. Nodal anatomy describes the pattern of vascular continuity between the stem and leaf.

Sinnott (1914), is credited with the first classification of vascular pattern at the node. He categorized the nodes into unilacunar, trilacunar and multilacunar depending on the number of gaps in the main vascular cylinder at the point of departure of the leaf trace.

Pierre (1896) classified the nodes based on the number of leaf traces, into *Monoxyless* or *monophalangoxylees* and *Trixyless* or *Triphalangoxylees,* while Hasselberg (1937) established the terms *unifaszikular insertion, Trifaszikular insertion* and *Multifaszikular insertion.* Marsden and Bailey (1955) combined the traces and gaps and proposed a classification of nodal anatomy. Pant and Mehra (1964) have recognised *alacunar* (no gap). *unilacunar* (one gap) and *multilacunar* (many gaps) nodes.

A few classical contributions to nodal anatomy include those of sinnott (1914), Bailey (1956), Bailey and Howard (1941), Swamy and Bailey (1949) etc.

Some of the recent contributions to nodal anatomy are seedling anatomy and primary vascular system in some Compositae (Deshpande and Kumar, 1968); some observations on the nodes of woody plants with special reference to split laterals (Howard, 1970); vascular anatomy of the node and flower of some Compositate (Tiyagi and Kshetrapal, 1970); node and leaf structure in the classification of some Australian shrubby Senecioneae - Compositae (Drury, 1973); node and petiole morphology of some South African plants (Coetzer *et al* 1976); nodal anatomy of Bignoniaceae (Jain, 1977); nodal anatomy in some dicotyledons (Kavathekar *et al.*, 1980); anatomy of nodal regions in some Angiosperms (Sivarama krishna and Ramappa, 1988) etc.

Sivaramakrishna and Ramappa (1988), have pointed out that so for only about a thousand species of Angiosperms have been studied for details of nodal anatomy and that, this constitutes less than 1 per cent of the total number of the species of Angiosperms.

Grew (1675) was the first botanist to recognize that the petiole can reveal various vascular patterns in cross section which can be of systematic value. DeCandolle (1879), published the first comprehensive survey of petiole anatomy, wherein he described vascular structure of the petiole in 20 families of flowering plants. Subsequently, Vesque (1885); Petit (1886, 1887, 1889); Lignier (1888) and Acqua (1887) studied petiole anatomy in several flowering plants and introduced new technical terms relating to the vascularization of the petiole. Vesque (1881, 1885) suggested that anatomy of the petiole at the middle region is most reliable for comparative purposes. Watari (1934, 1936, 1939) has published and extensive survey of petiole and leaf anatomy of Fabaceae, *Acer* and Saxifragaceae.

Hare (1943) proposed a simple classification of the vascular structure of petiole as I, U- and O- shaped. Howard (1962) has proposed a classification of the petiole anatomy relating to nodal structure. Another recent work is that of Schoefield (1968) who has studied the petiole anatomy in Guttiferales.

Foliar venation pattern

Leaf morphology has remained a virtually unexploited tool for systematic studies of dicotyledons (Hickey, 1979). This may be partly due to lack of comprehensive and unambiguous classification of leaf morphology. Some of the earlier workers who have made pioneering contributions to leaf morphology including venation pattern are Ettingshausen (1861); Lee (1948); Foster (1950); Stearn (1966); Mouton (1970); Hickey (1973); Hickey and Wolfe (1975); Melville (1976) and Hickey (1979).

Stomatal/Epidermal studies

Among the systematically useful traits of leaf surface, trichomes, stomatal types, domatia, cuticle etc., have proved to be a great help in taxonomic classification. Of the several traits on leaf surface, the stomata are perhaps the most significant from the point of view systematics and phylogeny.

The premier morphological classification of stomata was first provided by Vesque (1889) who classified mature stomata into four types - Ranunculaceous, Cruciferous, Caryophyllaceous and Rubiaceous. Metcalfe and Chalk (1950) proposed a new terminology which carried the morphological explanation of stomatal complexes to replace the terms Ranunculaceous, Cruciferous, Caryophyllaceous and Rubiaceous, with anomocytic, anisocytic, diacytic and paracytic respectively. They also added a new type actioncytic, where a ring of radiating subsidiary cells surrounds the stoma. Metcalfe (1961), identified a new stomatal complex-tetracytic which has two polar and two lateral subsidiary cells. Stace (1965), described another stomatal type cyclocytic, which differs from actinocytic type in having a narrow ring of subsidiary cells instead of radiating cells. Mickel and Lersten (1967) described a floating stoma, where a single subsidiary cell without anticlinal walls surrounds the sotma. Van Cotthem (1970) has proposed a classification of stomata into the following types.

(a) anomocytic (b) anisocytic
(c) paracytic (d) diacytic
(e) tetracytic (f) actinocytic
(g) cycloytic and (h) hexacytic

Stomata are also classified (based on their ontogeny) into mesogenous, perigenous, mesoperigenous, Helicocytic and Allelocytic (Fryns claessens and Van Cotthem 1973). Patel (1979) has proposed a new classification of stomata based on

i Presence/absence of subsidiary cells
ii. The number of whorls of the subsidiary cells around a stoma
iii. The total number of subsidiary cells in a stomatal complex and
iv. Shape and position of subsidiary cells.

Systematics and Phylogeny of stomata

There has been much speculation concerning the phylogenetic relationship of the stomatal types, but as Wilkinson (1979) points out there is no consensus on stomatal evolution. Florin (1933, 1958), believed that the perigenous stomata are more primitive than the mesogenous type.

Among the dicotyledons, the supposedly primitive Magnoliales have mesogenous and perigenous stomata (Paliwal and Bhandari 1962; Pant and Gupta 1966; Payne, 1970).

Stebbins and Khush (1961) believed that the primitive stomata had several subsidiary cells, which gave rise to stomata with less number of subsidiary cells and finally to stomata with no subsidiary cells. Cronquist (1968) who originally thought the anomocytic stomata are more primitive has revised his opinion and states 'Although anomocytic stomates are probably primitive for vascular plants as a whole, paracytic stomates are probably primitive within the angiosperms" (quoted by Wilkinson 1979). Paliwal (1969) however points to the fact that the supposedly primitive Ranales share anomocytic stomata with the supposedly advanced Asterales and Rubiales. An analysis of stomatal types in Angiosperms reveals that there is no correlation between stomatal types and phylogenetic level of families. Considering these aspects, Wilkinson (1979) observes "It would be wiser for the time being to use stomatal characters for diagnostic purposes only..........."
For systematic purposes, stomata have been proved to be of great importance. Features of stomata like the type, frequency, Index and L/B ratio of guard cells are all of diagnostic value. Stace (1965) has used stomatal characters for the sub divisions of the family Gramineae into Festucoideae and Panicoideae.

Epidermal and stomatal characters are used to separate polyploid pairs. The size of the stomatal cells very often indicates the degree of polyploidy (Sax, 1938; Carolin 1954). Timmermann (1927), opined that stomatal frequency and index are of diagnostic value. Ahmed (1964) working on solanaceous taxa has utilized the diagnostic value of stomatal characters.

A few other recent contributions to stomatal systematics included those of Chandra *et al* (1969); Mankevich (1971) etc.

In the family Asteraceae the stomata are mostly anomocytic even though anisocytic stomata have also been reported (Metcalfe and Chalk 1950). Projected stomata have been reported in *Cineraria cruenta* (Solerender, 1908). Bhatt and Inamdar (1975) have studied stomatal complexes in several members of Asteraceae.

The above account of foliar anatomy clearly reveals that there is not much work on Asteraceae and more particularly on Indian members of Asteraceae. This work proposes to fill that gap in a modest way. Three aspects of leaf anatomy viz., node and petiole anatomy, venation pattern and stomatal structure have been studied for 29 taxa (39 for stomata) spread over three tribes Eupatorieae, Senecioneae and Heliantheae. These data are made use of to analyse the tribal and subtribal characters (if any) and analysis of variation within and in between species whereever possible. Diagnostic value of these characters at the species level is highlighted. The terminology used to describe the node and petiole anatomy is that of Howard (1979); of venation pattern is that of Hickey (1979) and of stomatal structure is that of Van Cotthem (1970).

MATERIAL AND METHODS

Three aspects of foliar anatomy namely
- (a) node and petiole anatomy
- (b) venation pattern and
- (c) stomatal structure

have been studied for 29 taxa (39 for stomata) belonging to tribes Eupatorieae Heliantheae and Senecioneae (Table 1.2).

NODE AND PETIOLE ANATOMY

Free hand transverse sections have been cut at 5 different levels in order to understand the origin and course of leaf traces. A minimum of 5 samples have been studied for each taxon.

Level 1. Nodal region
2. Base of the petiole
3. Middle of the petiole
4. Base of lamina
5. Middle of lamina

Serial transverse sections and cleared petioles have been studied in several instances in order to properly understand the branching of the leaf trace bundles.

The number of gaps and number of leaf traces that originate from the stem, the details of petiole vascularization like the number of traces, their branching etc., have been studied.

Camera lucida drawings have been made from the cross sections at all the five levels. The terminology used to describe the petiole vascularization is that of Howard (1962).

VENATION PATTERN

Cleared leaves have been used in order to study the venation pattern. The technique recommended by Payne (1969) has been followed to clear the leaves. The procedure is as follows

1. Fresh leaves were boiled in alcohol for about 15 minutes to remove chlorophyll.
2. Leaf tissues were flooded with 5% NaoH and placed in an oven at about 37°C for one to many days depending on the nature of the tissue.
3. After the tissues became transparent, they were washed in water.
4. Tissues were dehydrated by treating with 30% and 50% alcohol.
5. Tissues were stained in 1% saffranin in 50% alcohol.
6. Tissues were mounted in glycerine and covered with a microcover glass and observed.
7. In some cases the tissues were dehydrated thoroughly in 70, 90 and absolute alcohol series, treated with xylene and mounted in canada balsam finally.

Camera lucida drawings have been made to record the areole pattern, veinlets etc. Other observations regarding the type of venation, branching of the primary, secondary and tertiary vein etc. have been recorded. Classification of the venation pattern and the terminology used is that of Hickey (1979). A few important parameters are described below :

(a) Determination of the size of primary vein

Determined midway between the leaf apex and base as the ratio of vein width (vw) to leaf width (lw)

$$\frac{vw}{lw} \times 100 = \text{size}$$

i. massive : > 4%
ii. stout : 2 - 4%
iii. moderate : 1.25 - 2%
iv. weak : < 1.25%

Secondary vein

Angle of divergence is measured between the primary vein and the secondary vein above the point of branching.

Loop forming branches

The behaviour of the loops, the angle of their curvature and whether they are enclosed by tertiary (3^0) arches or not is studied.

Intersecondary veins

Short branches of veins arising between the secondary veins from the primary vein are called intersecondary veins. These may be simple or have several short branches joined to one another (composite).

Angle of origin of Tertiary veins (3^0)

Branches arising from the secondaries are called tertiary veins. 3^0 veins arise from both the upper (admedial) and lower (exmedial) region of the secondaries. When the average angle of tertiary origin of the exmedial side of the secondary veins is compared with the average on the admedial side of the secondary veins, the combinations shown in the following table are possible. This trait is of diagnostic value.

Analysis of tertiary vein origin

Angle of tertiary origin on the admedial side of secondary veins	Angle of tertiary origin on the exmedial side of secondary veins		
	Acute	Right	Obtuse
Acute	AA	AR	AO
Right	RA	RR	RO
Obtuse	OA	OR	OO

Veinlets

These are the freely ending ultimate veins of the leaf which occasionally cross areoles.

Areoles

The smallest area of leaf tissue surrounded by veins, which taken together form a contiguous field over most of the area of the leaf.

Vein density

This refers to total length of the veins in a unit area of leaf. This is determined on the camera lucida diagram of venation pattern at a particular magnification of unit area for all the leaves. The total vein length is reduced to a convenient number which helps in comparison of taxa.

Stomatal/Epidermal studies

Structure of the mature stoma was studied. Epidermal peels from both upper and lower surface of the leaf were removed by mechanical stripping. They were mounted in glycerine. The number and shape of epidermal cells in unit area and length and breadth of stomata have been calculated. From these data, stomatal frequency (Timmermann 1927) and stomatal index (Salisbury 1972) have been calculated as follows

$$\text{Stomatal frequency} = \frac{\text{No. of stomata}}{\text{No. of epidermal cells}} \times 100$$

$$\text{Stomatal Index} = \frac{5}{E + S} \times 100 \text{ where}$$

S = number of stomata and E = number of epidermal cells.

Plate No. 2.1 : Node/petiole anatomy (Explanation to figures)

Note : In all the taxa included in plates 2.1 - 2.4, the figures represent; first, T.S. of node, second - T.S. at the base of the petiole, Third - T.S. at the middle of the petiole. Fourth - T.S. at the base of the lamina and fifth - T.S. at the middle of the lamina. **Figs : 1-5** *Ageratum conyzoides*, **6-10** *Ageratum houstonianum*, **11-15** *Emilia scabra*, **16-20** *Emilia sonchifolia*, **21-25** *Gynura nitida*, **26-30** *Gynura aurantiaca*, **31-35** *Notonia grandiflora*, **36-40** *Senecio ludens*, **41-41d** *Acanthospermum hispidum*

Plate No. 2.2 : Node/petiole anatomy (Explanation to Figures)

Figs : 42-45 *Senecio walkeri*, **47-51** *Bidens pilosa* (normal ray), **52-56** *Bidens pilosa* (rayless), **57-61** *Eclipta prostrata*, **62-66** *Galinsoga quadriradiata*, **67-71** *Galinsoga parviflora*, **72-76** *Lagascea mollis*, **77-81** *Moonia heterophylla*, **82-86** *Rudbeckia nitida*, **92-96** *Spilanthes calva* (calva head), **97-101** *Spilanthes calva* (large head), **102-106** *Spilanthes paniculata*, **107-111** *Synedrella nodiflora*, **112-116** *Calyptocarpus vialis*

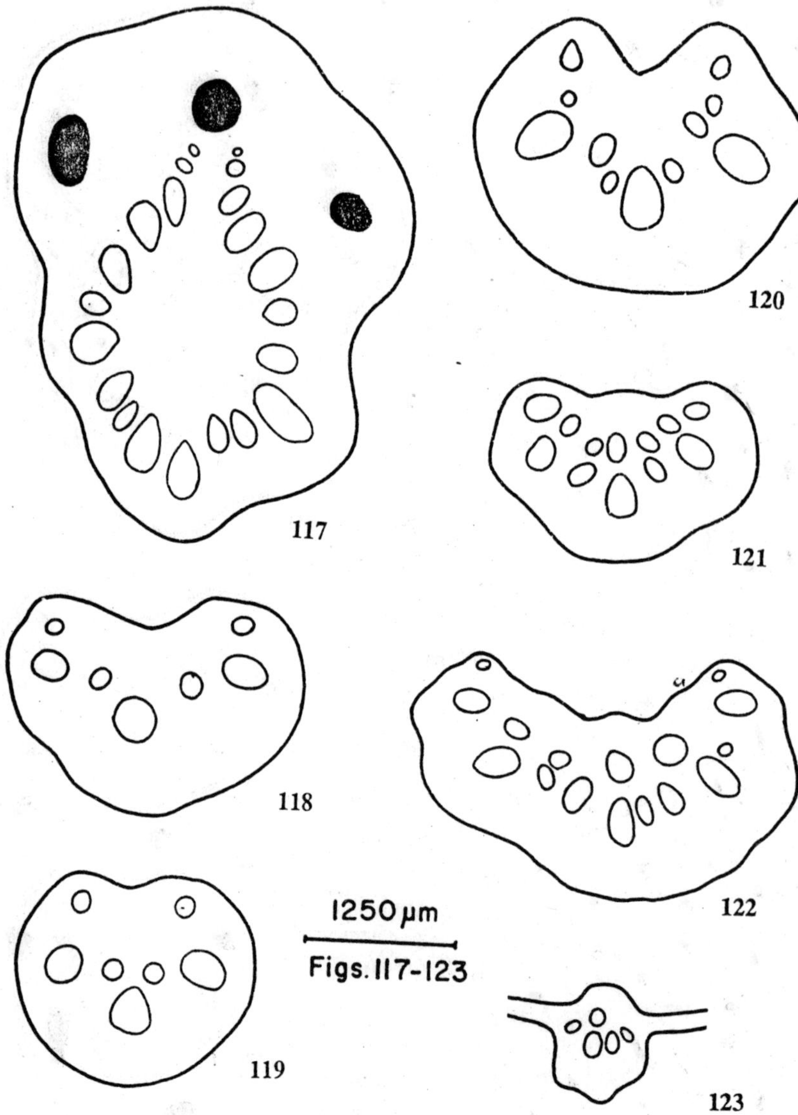

Plate No. 2.3 : Node/petiole anatomy (Explanation to Figures)

Figs 117-123. *Xanthium indicum*

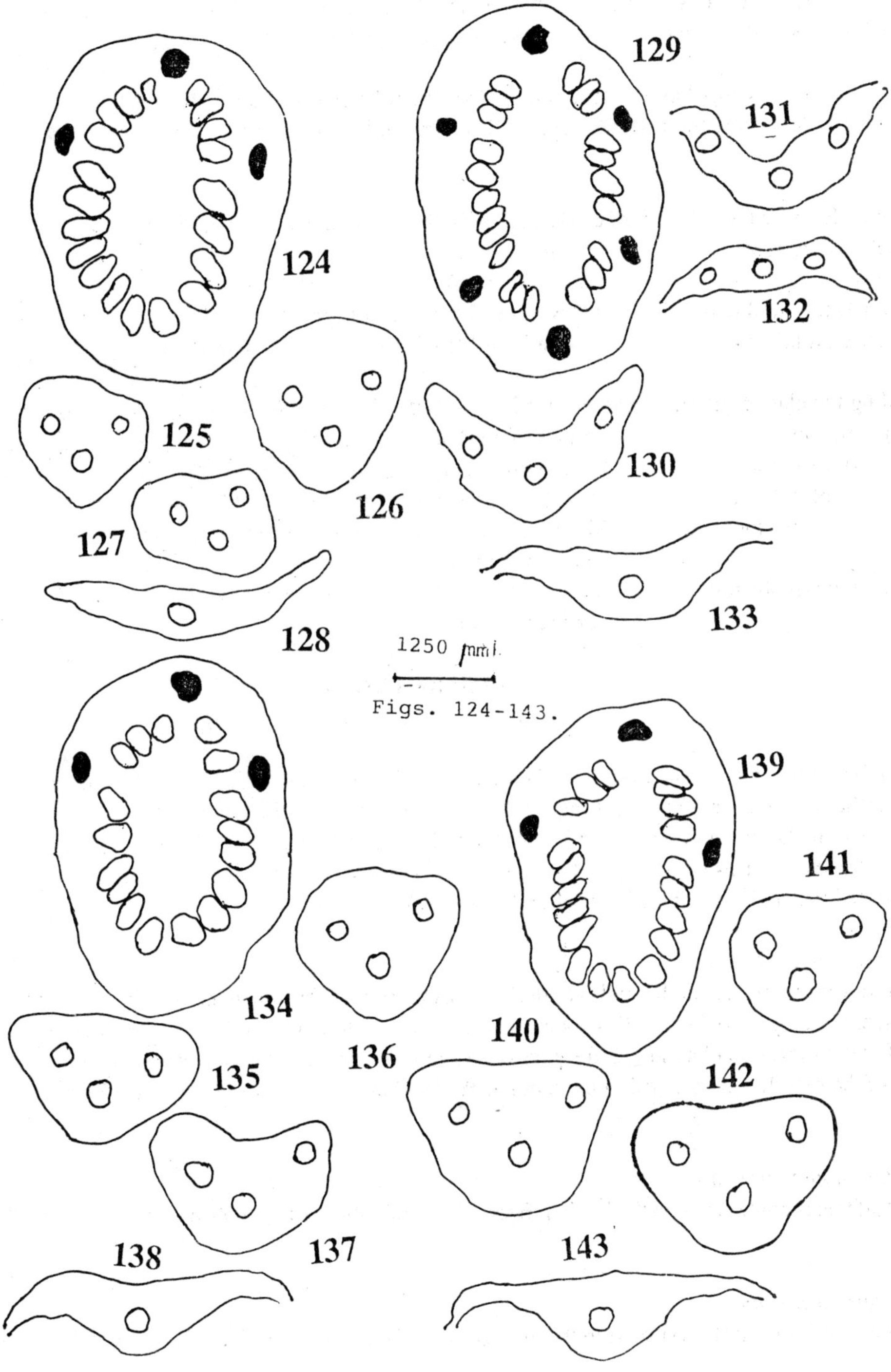

Plate No. 2.4 : Node/petiole anatomy (Explanation to Figures)

Figs : 124-128 *Siegesbeckia orientalis,* **129-133** *Galinsoga (intermediate),* **134-138** *Notina walkeri,* **139-143** *Adenostemma lavenia*

Number of epidermal cells in unit area, number of stomata, stomatal frequency, stomatal index, length and cells including standard error of mean value) are given in table 2.3.

Camera lucida drawings have been made of lower epidermal peel and upper epidermal peel (whereever necessary) to determine the type of stomata. The terminology to describe the stoma is that of Van Cotthem (1970).

In order to describe the cell wall of epidermal cells, the following terminlogy is employed.

Cell wall straight - no curvature
Wavy - cell wall curved in a shallow manner
Moderately undulated - cell wall infolded, but not very deeply
Highly undulated - cell wall infolded deeply and very frequently.

Regarding the classification of leaves based on stomatal distribution, the following terms are employed.

(a) Epistomatic - Stomata found only on the upper epidermis
(b) Hypostomatic - Stomata found only on the lower epidermis
(c) Amphistomatic - Stomata found on both the epidermal layers
(i) Epiamphistomatic - More number of stomata are found on the upper epidermis than on the lower epidermis.
(ii) Hypoamphistomatic - More number of stomata are found on the lower epidermis than on the upper epidermis.

OBSERVATIONS
(Node and Petiole anatomy)

The number of leaf traces that arise from the main vascular cylinder (node) of the stem are given in Table 2.1. Their further course, branching and multiplication if any at four different levels (base of petiole, middle of petiole, base of lamina and middle of lamina) are also given in the same table. The brief description of the vascular traces in terms of their number, multiplication and fusion if any is given below. In all members investigated the node has three lacunae and three traces.

Ageratum conyzoides
Six leaf traces are present at the nodes as the leaves are opposite in arrnagement. The median trace continues without branching up to the middle of the lamina. The two lateral traces divide soon after departure from the node, but the short branches stop at the base of the petiole. Hence only three traces are found throughout the length of the petiole. After entering the lamina, the two lateral traces do not proceed further (Plate 2.0; Figs. 1-5).

Ageratum houstonianum
Three leaf traces continue up to the lamina, from where only the median trace proceeds further (Plate 2.0; Figs. 6-10).

Adenostemma lavenia
Vasculature of the petiole is similar to that of *Ageratum houstonianum* (Plate 2.4; Figs. 139-143).

Emilia scabra
Of the three leaf traces that depart at the node the median continues into the lamina without undergoing any change. The laterals however split at the base of the petiole. The branches of the laterals continue only for some distance. The two laterals produce one more set of branches which stop at the base of the lamina. Only

the original three traces continue in the leaf (Plate 2.0; Figs. 11-15).

Emilia sonchifolia
Three leaf traces continue into the lamina without any change (Plate 2.1; Figs. 16-20).

Gynura nitida
Of the three leaf traces, the median continues into the leaf without any change. The two laterals divide, with the result there are five traces all along the petiole. The branches of the laterals do not continue up in the mid rib after entering the lamina (Plate 2.2; Figs. 21-25).

Gynura aurantiaca
Of the three traces, the two laterals branch within the stem itself. This is known as the split lateral condition. While the median continues without any alteration, the laterals divide and produce a number of traces all of which enter the lamina but do not proceed further. The median trace alone travels up to the tip of the lamina. (Plate 2.1; Figs. 26 - 30).

Notonia grandiflora
Split laterals are seen in the stem. The median trace, the two laterals and their branches travel up to the base of lamina. From there, the median and the two laterals travel up to the apex, while the two branches of laterals do not proceed further (Plate 2.1; Figs. 31-35).

Notonia walkeri
Split lateral condition is not seen. The median and the two laterals travel up to the base of lamina. From then on only the median continues up (Plate 2.5; Figs. 134 - 138).

Senecio ludens
The three leaf traces proceed up to the base of lamina. Only the median trace travels up to leaf apex (Plate 2.1; Figs. 36-40).

Senecio walkeri
The three leaf traces travel up to the middle of the petiole. The two laterals divide, but all of them (laterals and their branches) do not proceed beyond the base of lamina. Only the median continues up (plate 2.2; Figs. 42-46).

Acanthospermum hispidum
All the three traces go up to the base of the lamina without any alteration. From there, only the median proceeds up (Plate 2.1; Figs. 41, a,b,c,d).

Siegesbeckia orientalis
This is similar to what is seen in *Acanthospermum hispidum* (Plate 2.5; Figs. 124-128).

Bidens pilosa (normal ray)
The three traces enter the base of the petiole where the laterals split up. The five traces continue up to the lamina. From here only the median and the two laterals continue up (Plate 2.2; Figs. 47-51).

Bidens pilosa (rayless)
This is similar to what is seen in *B. pilosa* (normal ray) (Plate 2.2; Figs. 52-56).
Eclipta prostrata

Similar to what is seen in *Bidens pilosa* (normal ray) (Plate 2.2; Figs. 57-61)

Galinsoga quadriradiata
Similar to what is seen in *Acanthospermum hispidum*. In *Galinsoga* (intermediate) and *G.parviflora* also the course of the vascular trace is similar to that of *Acanthospermum hispidum* (Plate 2.8; Figs. 62-71).
Lagascea mollis Similar to what is seen in *Acanthospermum hispidum* (Plate 2.2; Figs. 72-76).

Monia heterophylla
Similar to what is seen in *Gynura nitida* (Plate 2.2; Figs. 77-81).

Rudbeckia nitida
Three traces depart from the node. As soon as they enter the petiole, all the three traces divide. The median trace produces an adaxial branch, while the two laterals produce 2 branches each. Of these two branches, one is thicker than the other. All the traces (2 median, 3 + 3 laterals) travel up to the middle region of the petiole. A little above the middle region of the petiole, the principal branch of the lateral splits again producing a thin trace. All these traces (10 in number) travel up to the apex of the lamina (Plate 2.2; Figs. 82-86).

Sclerocarpus africanus
Of the three traces that enter the base of the petiole, the median continues up to the apex of the leaf without any alteration. The two laterals produce a series of branches from the base of the petiole (2 branches each), which continue up to the lamina, where only the laterals proceed up to the apex of the leaf (Plate 2.2; Figs. 87-92).

Spilanthes calva (small head)
The lateral traces divide at the base of the petiole and the branches continue up to the base of the lamina. The middle region of the lamina has only the median and two laterals (Plate 2.2; Figs. 82-96).

Spilanthes calva (large head)
The three leaf traces continue right up to the apex of the leaf with no alteration (Plate 2.2; Figs. 97-101).

Spilanthes paniculata
The three traces travel up to the lamina, from where only the median continues up (Plate 2.2; Figs. 102-106).

Calyptocarpus vialis and Synedrella nodiflora
Similar to what is seen in *Spilanthes paniculata* (Plate 2.2; Figs. 107-116).

In *Xanthium indicum* (Plate 2.2; Figs. 117-123) the petiole vascularization presents a complex picture and does not appear to be uniform. There are at least two principal variations in the specimens studied. At the nodal region however, there is no variation. All the specimens examined have a trilacunar three trace node. The three traces enter the petiole and the laterals produce two branches on either side with the result there are seven traces. (Plate 2.2; Fig. 118). In some petioles however, the laterals produce only one branch each towards the adaxial side, while the median trace produces two thin traces on its adaxial side (Plate 2.3; Fig. 119).

In the middle of the petiole, in instances where the median has not divided the laterals produce another vascular trace each. At this level the median also divides and produces two branches on either side (Plate 2.3; Fig. 120).

In some petioles (where the median divides at thebase of the petiole giving rise to two branches), at the middle region there will be three branches of the median vascular trace, while the laterals have three branches each (Plate 2.3.; Fig. 121).

At the base of the lamina, further ramifications take place with the result there will be a number of branches. The median as well as laterals will have four branches each. Two thin traces produced from the laterals are present towards the extreme margin of the petiole (Plate 2.3; Fig. 122).

In the middle of the lamina (Plate 2.3; Fig. 123) there will be only five traces. The others stop a little way up the base of the lamina. Of these five traces, one is median, while the other four correspond to the two laterals and their branches.

OBSERVATIONS

(Venation pattern)

1. *Ageratum conyzoides* (Plate 2.5, Fig. 1)

Type : Pinnate, semicraspidodromous
Primary vein : Thickest vein of the leaf; occurs singly as mid vein.
(a) Size : Moderate
(a) Course : Straight, branched with a few ramifications.

Secondary vein
(a) Angle of divergence : Acute
(b) Variation in the angle of divergence : The lowest pair of secondary veins more acute than pairs above it.
(c) Relative thickness : Moderate
(d) Course : Uniformly curved
(e) Loop forming branches : Branched and curved at right angles
(f) Intersecondary veins : Simple

Tertiary vein
Angle of origin : AA and RR
Veinlets : Simple, linear and curved
Areole : Well developed, oriented, mostly quadrangular.

2. *Ageratum houstonianum* : (Plate 2.5; Fig.2)

Type : Pinnate, semicraspidodromous
Primary vein : Thickest vein occurs as a single mid vein
(a) Size : Stout
(a) Course : Straight

Secondary vein
(a) Angle of divergence : Acute
(b) Variation in the angle of divergence : Uniform
(c) Relative thickness : Moderate
(d) Course : Straight, but few curved also
(e) Loop forming branches : Curved at right and acute angles
(f) Intersecondary veins : Simple

Tertiary vein
  Angle of origin : RO, AO, and AA
Veinlets : Few areoles do not have veinlets. Otherwise, simple, linear and curved.
Areole : Well developed; mostly quadrangular.

3. *Ageratum lavenia* : (Plate 2.5; Fig. 3)
Type : Pinnate, semicraspidodromous
Primary vein : Thickest vein in the centre occurs singly as the midivein
  (a) Course : Moderate
  (a) Size : Moslty curved

Secondary vein
  (a) Angle of divergence : Acute
  (b) Variation in the angle of divergence : Upper branches less acute than the lower ones.
  (c) Relative thickness : Moderate
  (d) Course : Uniformly curved
  (e) Loop forming branches : Curved at right angles. Secondarv arches of tertiaries are also seen.
  (f) Intersecondary veins : Composite

Tertiary vein
  Angle of origin : RO, AO, and AA
Veinlets : Branched once or twice; ultimate branch often curved. When branched twice, one of the basal branches does not divide.
Areole : Developed but not compact; oriented, polygonal or pentagular.

4. *Emilia scabra* : (Plate 2.5; Fig. 4)
Type : Pinnate, simple craspidodromous
Primary vein : Thickest vein occurs as mid vein.
  (a) Size : Stout
  (a) Course : Straight

Secondary vein
  (a) Angle of divergence : Acute
  (b) Variation in the angle of divergence : Upper secondary veins more acute than lower ones.
  (c) Relative thickness : Moderate
  (d) Course : Upper ones straight, lower veins recurved
  (e) Loop forming branches : curved at right angles
  (f) Intersecondary veins : Composite
Tertiary vein
  Angle of origin : AO, and AA
Veinlets : Absent in few areoles, when present simple and bent; unbranched.
Areole : Well developed, compact, quadrangular or pentagonal.

5. *Emilia sonchifolia* : (Plate 2.5; Fig. 5)
Type : Pinnate, simple craspidodromous
Primary vein : The thickest vein occurs singly as medivein
  (a) Size · Moderate
  (a) Course : Straight and branched

Secondary vein
(a) Angle of divergence : Acute
(b) Variation in the angle of divergence : Upper veins more acute than the lower ones
(c) Relative thickness : Moderate
(d) Course : Slightly curved or arched; lowest pairs recurved.
(e) Loop forming branches : Very few, curved at right angles
(f) Intersecondary veins : Simple.

Tertiary vein
Angle of origin : RO
Veinlets : Few areoles lack veinlets. When veinlets are present they are simple or occasionally branched once.
Areole : Well developed but not compact; oriented, quadrangular or pentagular.

6. *Gynura nitida* : (Plate 2.5; Fig. 6)
Type : Pinnate, semicraspidodromous
Primary vein : The thickest vein occurs as medivein
(a) Size : Stout
(a) Course : Branched with few ramifications

Secondary vein
(a) Angle of divergence : Acute
(b) Variation in the angle of divergence : Uniform
(c) Relative thickness : Moderate
(d) Course : Uniformly curved
(e) Loop forming branches : Curved at right angles; secondary arches of 3^0 origin are seen
(f) Intersecondary veins : Simple.

Tertiary vein
Angle of origin : AA and AO
Veinlets : Simple and curved
Areole : Well developed, oriented and polygonal.

7. *Gynura aurantiaca* : (Plate 2.5; Fig. 7)
Type : Pinnate, semicraspidodromous
Primary vein : The thickest vein occurs singly as medivein
(a) Size : Massive,
(a) Course : Straight and branched

Secondary vein
(a) Angle of divergence : Acute
(b) Variation in the angle of divergence : Upper veins more acute than the lower ones
(c) Relative thickness : Moderate
(d) Course : Uniformly curved
(e) Loop forming branches : Curved at right angles, branches of secondary and tertiary form arches
(f) Intersecondary veins : Composite

Tertiary vein
 Angle of origin : AA
Veinlets : Simple, linear; absent in some areoles
Areole : Well developed but not compact; mostly pentagular.

8. *Notonia grandiflora* : (Plate 2.5; Fig. 8)
Type : Hyphodromous
Primary vein : Three in number, middle one thickest.
 (a) Size : Massive
 (a) Course : Central one straight, lateral ones curved and branched.

Secondary vein
 (a) Angle of divergence : Acute
 (b) Variation in the angle of divergence : Lower pair of veins more acute thant the upper
 (c) Relative thickness : Moderate
 (d) Course : Slightly arched or curved
 (e) Loop forming branches : Curved uniformly at right angles
 (f) Intersecondary veins : Composite

Tertiary vein
 Angle of origin : AA
Veinlets : Simple and linear. In some cases branched twice.
Areole : Well developed oriented; shape pentagonal or sometimes irregular.

9. *Notonia walkeri* : (Plate 2.5; Fig. 9)
Type : Pinnate, eucamptodromous
Primary vein : The thickest vein occurs singly as medivein
 (a) Size : Massive
 (a) Course : Straight and branched

Secondary vein
 (a) Angle of divergence : Acute
 (b) Variation in the angle of divergence : Uniform
 (c) Relative thickness : Moderate
 (d) Course : Slightly, uniformly arched or curved
 (e) Loop forming branches : Curved at right angles; secondary arches 3^0 are also seen.
 (f) Intersecondary veins : Composite

Tertiary vein
 Angle of origin : AA, rarely RR
Veinlets : Absent in most areoles, whenever present simple and linear.
Areoles : Well developed, oriented, quadrangular or pentagular.

10. *Semecop walkeri* : (Plate 2.5; Fig. 10)
Type : Pinnate, semicraspidodromous
Primary vein : Thickest vein occurs singly as the mid vein.
 (a) Size ; Massive
 (a) Course : Straight

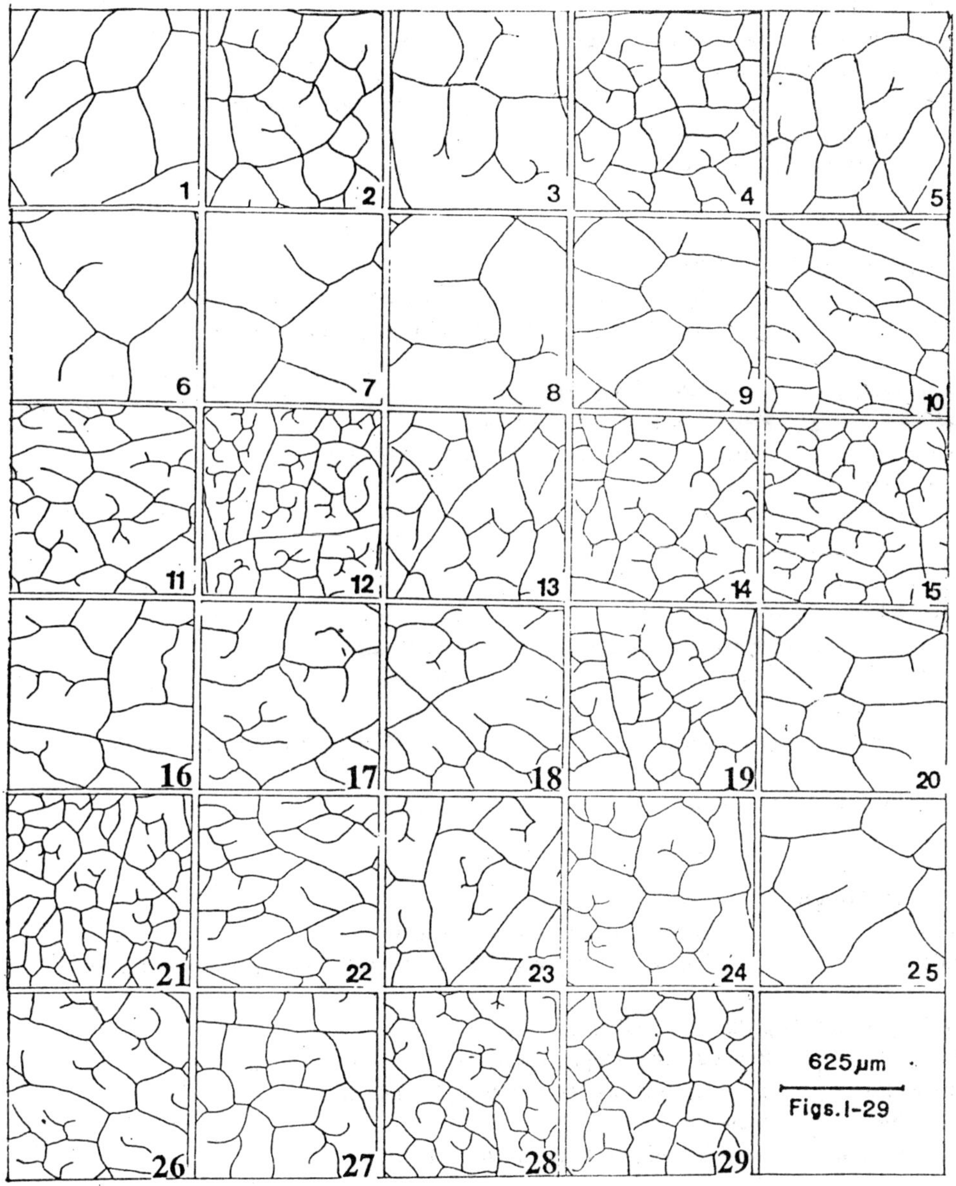

Plate No. 2.5 : Venation pattern (Explanation to figures)

Figs : 1 *Ageraaum conyzoides*, **2** *Ageratum houstonianum*, **3** *Adenostemma lavenia*, **4** *Emilia scabra*, **5** *Emilia sonchifolia*, **6** *Gynura nitida*, **7** *Gynura aurantiaca*, **8** *Notonia grandiflora*, **9** *Notonia walkeri*, **10** *Senecio cineraria*, **11** *Senecio ludens*, **12** *Acanthospermum hispidum*, **13** *Bidens pilosa* (normal ray), **14** *Bidens pilosa* (rayless), **15** *Eclipta prostrata*, **16** *Galinsoga quadriradiata*, **17** *Galinsoga* (intermediate), **18** *Galinsoga parviflora*, **19** *Lagascea mollis*, **20** *Moonia heterophylla*, **21** *Rudbeckia nitida*, **22** *Sclerocarpus africanus*, **23** *Siegesbeckia orientalis*, **24** *Spilanthes calva* (small head), **25** *Spilanthes calva* (large head), **26** *Spilanthes paniculata*, **27** *Synedrella nodiflora*, **28** *Calyptocarpus vialis* **29** *Xanthium indicum*.

Secondary vein

(a) Angle of divergence : Acute
(b) Variation in the angle of divergence : Uniform
(c) Relative thickness : Moderate
(d) Course : Straight and uniformly curved
(e) Loop forming branches : Obtuse, secondary arches of 3^0 are observed often
(f) Intersecondary veins : Composite.

Tertiary vein

Angle of origin : AA, OA and OR
Veinlets : Simple and linear. Sometimes branched once or very rarely twice.
Areole : Well developed, pentagonal to polygonal in shape.

11. *Senecio ludens* : (Plate 2.5; Fig. 11)
Type : Pinnate, semicraspidodromous
Primary vein : Thickest vein occurs singly as the mid vein.
(a) Size : Massive
(a) Course : Straight

Secondary vein

(a) Angle of divergence : Acute
(b) Variation in the angle of divergence : Nearly uniform
(c) Relative thickness : Moderate
(d) Course : Curved
(e) Loop forming branches : Curved at right angles secondary arches of 3^0 are seen.
(f) Intersecondary veins : Composite.

Tertiary vein

Angle of origin : AA and RR
Veinlets : Simple and linear, branched once.
Areole : Well developed, oriented; not compact; polygonal or pentagonal.

12. *Acanthospermum hispidum* : (Plate 2.5; Fig. 12)
Type : Pinnate, semicraspidodromous
Primary vein : Thickest vein occurs as the singly mid vein.
(a) Size : Massive
(a) Course : Straight with one or two ramifications

Secondary vein

(a) Angle of divergence : Acute
(b) Variation in the angle of divergence : The lowest pair of veins more acute than the rest.
(c) Relative thickness : Moderate
(d) Course : Uniformly curved or arched
(e) Loop forming branches : Curved at right angles.
(f) Intersecondary veins : Composite.

Tertiary vein

Angle of origin : AA

Veinlets : They are of two types, simple and branched. The simple veinlets are either linear or curved. The branched ones are branched once or twice dichotomously.

Areole : Well developed, Oriented; mostly quadrangular.

13. *Bidens pilosa* (normal ray) : (Plate 2.5; Fig. 13)

Leaves are imparipinnately compound; terminal pinna studied.

Type : Pinnate, Craspidodromous

Primary vein : Thickest vein occurs as the single mid vein.

(a) Size : Stout

(a) Course : Straight and branched

Secondary vein

(a) Angle of divergence : Acute

(b) Variation in the angle of divergence : The lowest pair more acute than the rest

(c) Relative thickness : Moderate

(d) Course : Straight

(e) Loop forming branches : Secondary arches of 3^0 are observed about 3-4 in a leaf.

(f) Intersecondary veins : Simple

Tertiary vein

Angle of origin : AA, OA and OR

Veinlets : Simple, linear or curved

Areole : Well developed, oriented; shape irregular.

14. *Bidens pilosa* (rayless) : (Plate 2.5; Fig. 14)

Venation pattern similar to *Bidens pilosa* (normal ray) except for the following

Size of Primary vein : Stout

Secondary vein : Straight as well as curved and recurved.

Loop forming branches : Curved at right angles. No secondary of 3^0, observed. Angle of origin of tertiary vein is AO. Veinlets absent in some areoles, when present simple or branched once.

15. *Eclipta prostrata* : (Plate 2.5; Fig. 15)

Type : Pinnate, semicraspidodromous

Primary vein : Thickest vein occurs singly as the mid vein.

(a) Size : Massive

(a) Course : Straight and branched with primary ramifications

Secondary vein

(a) Angle of divergence : Acute

(b) Variation in the angle of divergence : The lowest pair of secondaries more acute than the rest

(c) Relative thickness : Moderate

(d) Course : Slightly curved

(e) Loop forming branches : Secondary arches of 3^0, are seen.

(f) Intersecondary veins : Composite.

Tertiary vein
Angle of origin : AO and RR
Veinlets : Simple, linear or curved
Areole : Well developed, oriented, shape irregular.

16. *Galinsoga quadriradiata* : (Plate 2.5; Fig. 16)
Type : Pinnate, eucamptodromous
Primary vein : Thickest vein occurs singly as the mid vein.
(a) Size : Weak
(a) Course : Straight

Secondary vein
(a) Angle of divergence : Acute
(b) Variation in the angle of divergence : The lowest pair of secondaries more acute than the rest.
(c) Relative thickness : Moderate
(d) Course : Uniformly curved with smooth arches.
(e) Loop forming branches : Curved at right angles.
(f) Intersecondary veins : Simple.

Tertiary vein
Angle of origin : AA, OR and RR
Veinlets : Simple, linear. or curved.
Areole : Well developed, oriented, shape irregular.

17. *Galinsoga* (intermediate) : (Plate 2.5; Fig. 17)
Type : Pinnate, eucamptodromous
Primary vein : Thickest vein occurs singly as the mid vein.
(a) Size : Moderate
(a) Course : Straight and branched

Secondary vein
(a) Angle of divergence : Acute
(b) Variation in the angle of divergence : The lowest pair more acute than the rest.
(c) Relative thickness : Moderate
(d) Course : Uniformly curved
(e) Loop forming branches : Curved at right and obtuse angles; secondary arches of 3^0 are seen.
(f) Intersecondary veins : Simple.

Tertiary vein
Angle of origin : AA, OR and RA
Veinlets : Both simple, and branched. Simple ones are linear or curved; other veinlets are branched once.
Areole : Well developed, oriented, shape irregular.

18. *Galinsoga parviflora* : (Plate 2.5; Fig. 18)
Type : Pinnate, semicraspidodromous
Primary vein : Thickest vein occurs singly as the mid vein.
(a) Size : Moderate

(a) Course : Straight

Secondary vein

(a) Angle of divergence : Acute
(b) Variation in the angle of divergence : The lowest pair of veins more acute than the rest.
(c) Relative thickness : Stout
(d) Course : Uniformly curved; arched upwards
(e) Loop forming branches : Curved at right, obtuse and angles
(f) Intersecondary veins : Simple.

Tertiary vein

Angle of origin : AA, OR and RA
Veinlets : Simple, linear; some veinlets are branched once
Areole : Well developed, oriented, polygonal in shape

19. *Lagascea mollis* : (Plate 2.5; Fig. 19)
Type : Pinnate, simple craspidodromous
Primary vein : Thickest vein occurs singly as the mid vein.
(a) Size : Stout
(a) Course : Straight

Secondary vein

(a) Angle of divergence : Acute
(b) Variation in the angle of divergence : The lowest pair more acute than the rest.
(c) Relative thickness : Moderate
(d) Course : Basal ones straight, upper ones uniformly curved
(e) Loop forming branches : Few in number; curved at acute angles. Secondary arches of 3^0 are seen.
(f) Intersecondary veins : Simple.

Tertiary vein

Angle of origin : AA, RR and AA
Veinlets : Absent in few areoles. When present, mostly simple and linear. Very rarely branched.
Areole : Well developed, oriented, quadrangular, pentagonal or polygonal.

20. *Moonia heterophylla* : (Plate 2.5; Fig. 20)
Type : Pinnate, semicraspidodromous
Primary vein : Thickest vein occurs singly as the mid vein.
(a) Size : Massive
(a) Course : Straight

Secondary vein

(a) Angle of divergence : Acute
(b) Variation in the angle of divergence : The lowest pair more acute than the rest.
(c) Relative thickness : Moderate
(d) Course : Uniformly curved or arched
(e) Loop forming branches : Curved at right angles
(f) Intersecondary veins : Simple

Tertiary vein
Angle of origin : RR and AA
Veinlets : Simple linear or curved
Areole : Well developed, oriented, polygonal in shape.

21. *Rudbeckia nitida* : (Plate 2.5; Fig. 21)
Type : Pinnate, semicraspidodromous
Primary vein : Thickest vein occurs as the single mid vein.
(a) Size : Massive
(a) Course : Straight and branched

Secondary vein
(a) Angle of divergence : Acute
(b) Variation in the angle of divergence : The lowest pair more acute than the rest.
(c) Relative thickness : Moderate
(d) Course : Straight, lower two secondaries recurved
(e) Loop forming branches : Curved at right angles
(f) Intersecondary veins : Composite

Tertiary vein
Angle of origin : AA
Veinlets : Few areoles have no vein endings. When present veinlets are either simple, linear and curved or branched once of twice.
Areole : Well developed, oriented, compact, shape - polygonal or pentagonal

22. *Sclerocarpus africanus* : (Plate 2.5; Fig. 22)
Type : Pinnate, semicraspidodromous
Primary vein : Thickest vein occurs as the single mid vein.
(a) Size : Stout
(a) Course : Straight and branched with one or more primary ramifications

Secondary vein
(a) Angle of divergence : Acute
(b) Variation in the angle of divergence : The lowest pair more acute than the rest.
(c) Relative thickness : Moderate
(d) Course : Straight
(e) Loop forming branches : Curved at right angles
(f) Intersecondary veins : Composite

Tertiary vein
Angle of origin : OA, AA and OR
Veinlets : Few areoles have no vein endings. When present veinlets are either simple, linear and curved or linear
Areole : Well developed, oriented, quadrangular or pentangular.

23. *Siegesbeckia orientalis* : (Plate 2.5; Fig. 23)
Type : Pinnate, semicraspidodromous
Primary vein : Thickest vein occurs as the single mid vein.

(a) Size : Stout
(a) Course : Straight and branched with one or more primary ramifications

Secondary vein
(a) Angle of divergence : Acute
(b) Variation in the angle of divergence : The lowest pair more acute than the rest.
(c) Relative thickness : Moderate
(d) Course : Straight
(e) Loop forming branches : Curved at right angles
(f) Intersecondary veins : Composite

Tertiary vein
Angle of origin : AA, RR, OR and AR
Veinlets : Simple and curved; branched once or twice
Areole : Well developed, not compact; oriented, polygonal in shape

24. *Spilanthes calva* : (Plate 2.5; Fig. 24)
Type : Pinnate, semicraspidodromous
Primary vein : Thickest vein occurs as the single mid vein.
(a) Size : Massive
(a) Course : Straight and branched

Secondary vein
(a) Angle of divergence : Acute
(b) Variation in the angle of divergence : The lowest pair more acute than the rest.
(c) Relative thickness : Moderate
(d) Course : Uniformly curved
(e) Loop forming branches : Curved at right angles
(f) Intersecondary veins : Simple

Tertiary vein
Angle of origin : AA and RR
Veinlets : Rarely simple, usually branched once
Areole : Well developed, compact, oriented, quadrangular or pentangular in shape

25. *Spilanthes calva* (large head) : (Plate 2.5; Fig. 25)
Type : Pinnate, semicraspidodromous
Primary vein : Thickest vein occurs as the single mid vein.
(a) Size : Massive
(a) Course : Straight and branched with one or more primary ramifications

Secondary vein
(a) Angle of divergence : Acute
(b) Variation in the angle of divergence : The lowest pair more acute than the rest.
(c) Relative thickness : Moderate
(d) Course : Straight as well as curved; occasionally arched also
(e) Loop forming branches : Curved at right angles, secondary arches of 3^0 seen often
(f) Intersecondary veins : Composite

Tertiary vein
Angle of origin : RO, AA OR and OA
Veinlets : Simple linear or rarely curved
Areole : Well developed, oriented, not compact, polygonal in shape

26. *Spilanthes paniculata* : (Plate 2.5; Fig. 26)
Type : Pinnate, eucamptodromous
Primary vein : Thickest vein occurs as the single mid vein.
(a) Size : Massive
(a) Course : Straight

Secondary vein
(a) Angle of divergence : Acute
(b) Variation in the angle of divergence : The lowest pair more acute than the rest.
(c) Relative thickness : Moderate
(d) Course : Straight or uniformly curved or arched
(e) Loop forming branches : Curved at right angles
(f) Intersecondary veins : Composite

Tertiary vein
Angle of origin : OA, AA and RR
Veinlets : Rarely simple mostly branched once, with the branches occasionally curved back.
Areole : Well developed, oriented, not compact; polygonal, few quadrangular in shape.

27. *Synedrella nodiflora* : (Plate 2.5; Fig. 27)
Type : Pinnate, semicraspidodromous
Primary vein : Thickest vein occurs as the single mid vein.
(a) Size : Moderate
(a) Course : Straight

Secondary vein
(a) Angle of divergence : Acute
(b) Variation in the angle of divergence : The lowest pair more acute than the rest.
(c) Relative thickness : Moderate
(d) Course : Straight and uniformly curved
(e) Loop forming branches : Curved at right angles
(f) Intersecondary veins : Simple

Tertiary vein
Angle of origin : AO, OR, RR and rarely AA
Veinlets : Simple linear or occasionally curved
Areole : Well developed, oriented, penta or polygonal in shape.

28. *Calyptocarpus vialis* : (Plate 2.5; Fig. 28)
Type : Pinnate, semicraspidodromous
Primary vein : Thickest vein occurs as the single mid vein.
(a) Size : Moderate
(a) Course : Straight and branched

Secondary vein
- (a) Angle of divergence : Acute
- (b) Variation in the angle of divergence : Nearly uniform
- (c) Relative thickness : Moderate
- (d) Course : Straight as well as slightly curved
- (e) Loop forming branches : Curved at right angles
- (f) Intersecondary veins : Composite

Tertiary vein
Angle of origin : RO, AA and AO
Veinlets : Few areoles without veinlets. Rarely simple, branched once regularly
Areole : Well developed, compact, quadrangular or pentangular. in shape.

29. *Xanthium indicum* : (Plate 2.5; Fig. 29)
Type : Pinnate, semicraspidodromous
Primary vein : Thickest vein occurs singly as the mid vein.
- (a) Size : Moderate
- (a) Course : Straight and branched

Secondary vein
- (a) Angle of divergence : Acute
- (b) Variation in the angle of divergence : Nearly uniform
- (c) Relative thickness : Moderate
- (d) Course : Straight as well as slightly curved
- (e) Loop forming branches : Curved at right angles
- (f) Intersecondary veins : Composite

Tertiary vein
Angle of origin : RO, AA and AO
Veinlets : Few areoles without veinlets. When present the veinlets are simple and linear
Areole : Well developed, oriented, compact; quadrangular or pentangular shaped.

OBSERVATIONS
(stomatal/epidermal studies)

Out of the 39 taxa studied for mature epidermal structure, in all but one, the stomata are anomocytic. Only in *N.grandiflora,* the stomata are anisocytic. Brief description of the stomata of various taxa are given below. Stomatal frequency, stomatal index, length and breadth of stoma and L/B ratio are given in Table no.3.

1. *Ageratum conyzoides*
Leaves hypostomatic. Epidermal cells highly undulated (Plated 2.6; Figs. 1-2).

2. *Ageratum houstonianum*
Leaves hypoamphistomatic; cells of the lower epidermis more highly undulated than those of the upper epidermis (Plate 2.5; Figs. 3-4).

3. *Adenosteamma lavenia*
Leaves hypostomatic. Epidermal cells undulated moderately (Plate 2.6; Fig.5)

4. *Chromolaena odorata*
Leaves hypoamphistomatic. Cells of the upper epidermis show no undulation; cell walls are straight. Cell walls of lower epidemis are moderately undulated (Plate 2.6; Figs. 6-7).

5. *Emilia scabra*
Leaves hypoamphistomatic. Cells of the upper epidermis have wavy cell walls, while those of the lower epidermis have moderate undulation (Plate 2.6; Figs. 8-9)

6. *Emilia sonchifolia*
Leaves hypoamphistomatic; cell walls of upper epidermis are wavy, while those of lower epidermis are moderately undulated (Plate 2.6; Figs. 10-11).

7. *Gynura nitida*
Leaves hypoamphistomatic; cell walls of upper epidermis are moderately undulated, while those of lower epidermis highly undulated (Plate 2.6; Figs. 12-13).

8. *Gynura aurantiaca*
Leaves hypoamphistomatic; cells walls of both upper and lower epidermis are highly undulated (Plate 2.6; Figs. 14-15).

9. *Notonia grandiflora*
Leaves epiamphistomatic; stomata anisocytic. Cell walls of both lower and upper epidermis straight (Plate 2.6; Figs. 16-17).

10. *Notonia walkeri*
Leaves hypostomatic. Cell walls of both upper and lower epidermis straight (Plate 2.6; Fig. 19-20).

11. *Senecio cineraria*
Leaves hypostomatic; cell walls of both epidermal layers moderately undulated (Plate 2.6; Fig. 18).

12. *Seniecio corymbosus*
Leaves hypostomatic; cell walls of both epidermal layers are straight (Plate 2.6; Fig. 21).

13. *Senecio hohenackeri*
Leaves hypoamphistomatic. Cell walls of upper epidermis wavy, while those of lower epidermis highly undulated (Plate 2.6; Fig. 22-23).

14. *Senecio lawsoni*
Leaves epiamphistomatic or sometimes isoamphistomatic. Cell walls of both upper and lower epidermis highly undulated (Plates 2.6 and 2.7 and Figs. 24-25).

15. *Senecio ludens*
Leaves isoamphistomatic. Cell walls of both epidermal layers are straight. Cells polygonal in shape (Plate 2.7; Figs. 26-27).

16. *Senecio neelgherryanus*
Leaves hypoamphistomatic. Cell walls of upper epidermis straight, while those of lower epidermis moderately undulated (Plate 2.7; Figs. 28-29).

17. *Senecio candicans*
Leaves hypostomatic. Cell walls of upper epidermis wavy, while those of lower epidermis highly undulated (Plate 2.7; Figs. 30-31).

18. *Senecio tenuifolius*
Leaves isoamphistomatic. Cell walls of both epidermal layers moderately undulated (Plate 2.7; Figs. 32-33).

19. *Senecio walkeri*
Leaves hypostomatic. Epidermal cell walls of lower and upper layers are straight (Plate 2.7; Fig.34).

20. *Senecio wightii*
Leaves hypoamphistomatic. Cell walls of upper epidermis wavy, while those of lower epidermis moderately undulated (Plate 2.7; Figs. 35-36)

21. *Acanthospermum hispidum*
Leaves hypoamphistomatic. Cell walls of upper epidermis moderately undulated, while those of lower epidemis highly undulated. (Plate 2.7; Figs 37-38).

22. *Bidens humilis*
Leaves hypoamphistomatic. Cell walls of upper epidermis moderately undulated, while those of lower epidermis highly undulated (Plate 2.7; Figs. 39-40).

23. *Bidens pilosa* (normal ray)
Leaves hypoamphistomatic. Epidermal cell walls (both layers) moderately undulated (Plate 2.7; Figs. 41-42).

24. *Bidens pilosa* (rayless)
Same as in *B. pilosa* (normal ray) (Plate 2.7; Figs. 43-44)

25. *Eclipta prostrata*
Leaves hypoamphistomatic. Cell walls of upper epidermis wavy, while those of lower epidermis highly undulated (Plate 2.7; Figs. 45-46).

26-28. *Galinsoga quadriradiata*
Leaves hypostomatic. Cell walls of both epidermal layers highly undulated. *Galinsoga* (international) and *G.parviflora* are similar to *G.quadriradiata* (Plate 2.7 and 2.8; Figs. 47-49).

29. *Lagascea mollis*
Leaves isoamphistomatic. Cell walls of both epidermal layers highly undulated (Plate 2.7; Figs. 50-51).

30. *Moonia heterophylla*
Leaves hypostomatic. Cell walls of both epidermal layers highly undulated (Plate 2.8; Fig. 52).

31. *Rudbeckia nitida*
Leaves hypostomatic. Cell walls of both epidermal layers moderately undulated (Plate 2.8; Fig.53).

32. *Sclerocarpus africanus*
Leaves hypoamphistomatic. Cell walls of upper epidermis wavy, while those of lower epidermis highly undulated (Plate 2.8; Figs. 54-55).

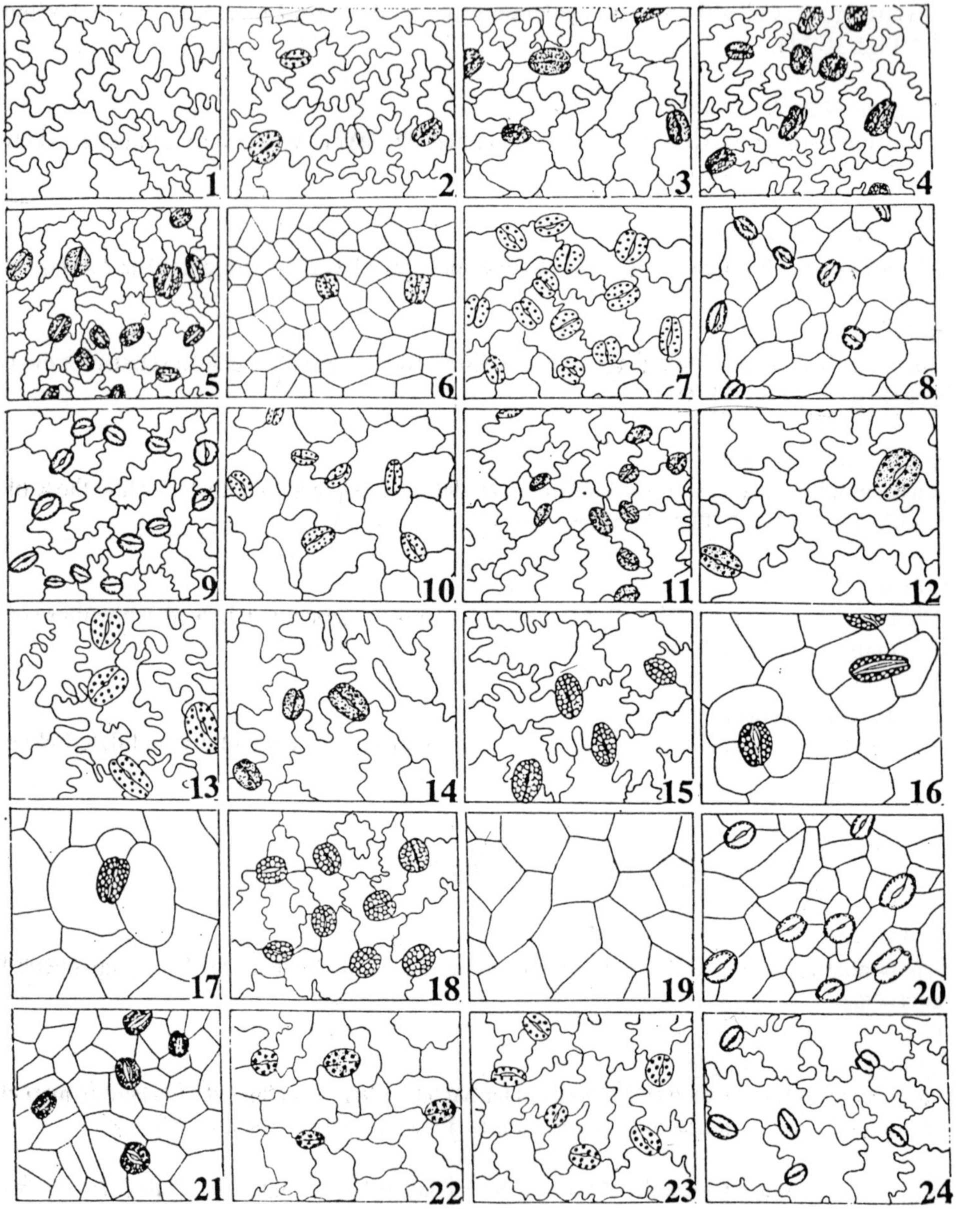

Plate 2.6 : Stomatal/epidermal studies (Explanation to Figures)

Figs	Taxon	Epidermis	Figs	Taxon	Epidermis
1-2	*Ageratum conyzoides*	(upper and Lower)	3-4	*Ageratum houstonianum*	(Upper and Lower)
5	*Adenostemma lavenia*	(Lower)	6-7	*Chromolaena odorata*	(Upper and Lower)
8-9	*Emilia scabra*	(Upper and Lower)	10-11	*Emilia sonchifolia*	(Upper and Lower)
12-13	*Gynura nitida*	(Upper and lower)	14-15	*Gynura aurantiaca*	(Upper and Lower)
16-17	*Notonia grandiflora*	(Upper and Lower)	18	*Senecio cineraria*	(Lower)
19-20	*Notonia walkeri*	(Upper and Lower)	21	*Senecio corymbosus*	(Lower)
22-23	*Senecio hohenackeri*	(Upper and Lower)	24	*Senecio lawsoni*	(Upper)

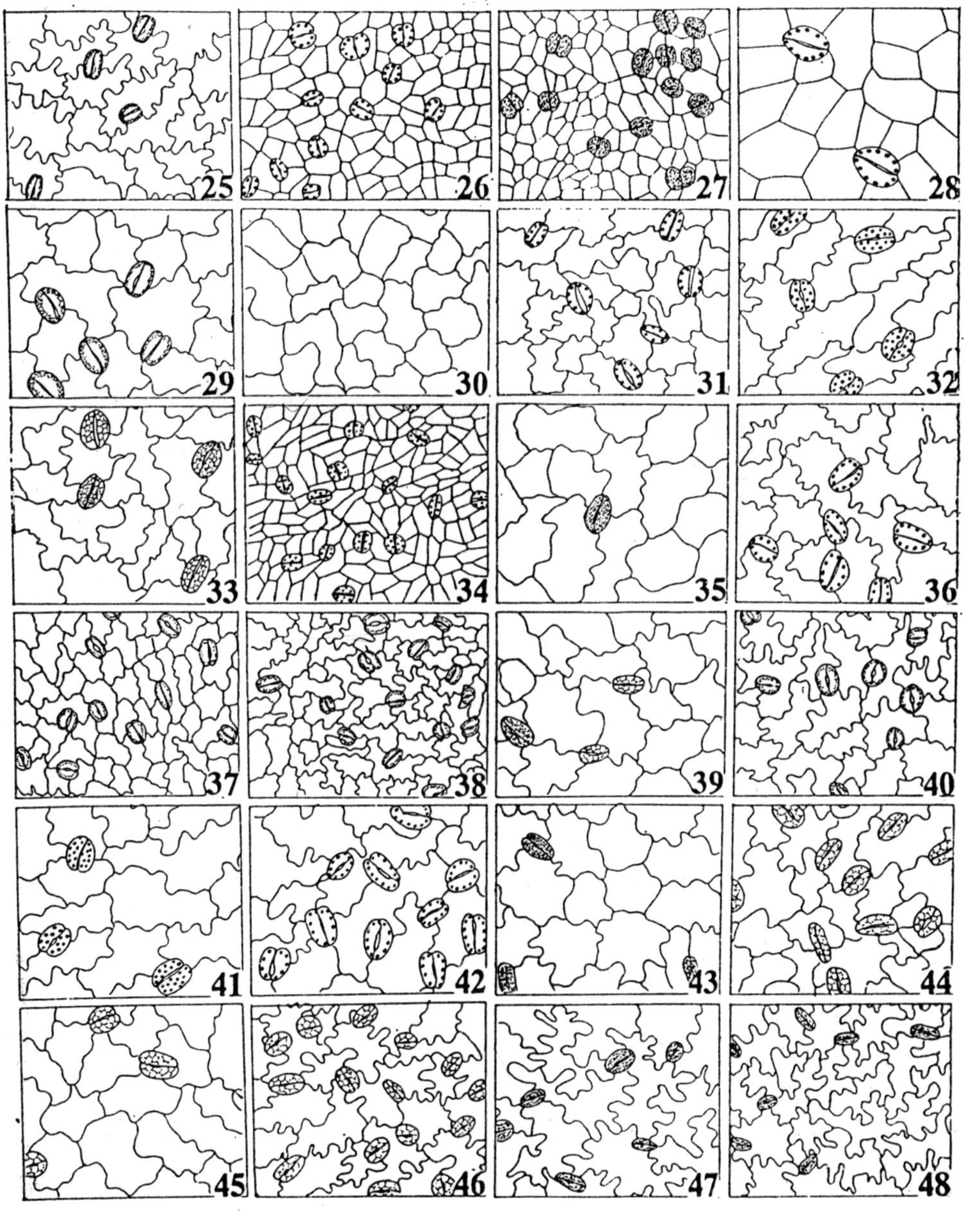

Plate No. 2.7 : Stomatal/epidermal studies (Explanation to Figures)

Figs	Taxon	Epidermis	Figs	Taxon	Epidermis
25	*Senecio lawsoni*	(Lower)	26-27	*Senecio ludens*	(Upper and Lower)
28-29	*Senecio neelgherryanus*	(Upper and Lower)	30-31	*Senecio candicans*	(Upper and Lower)
32-33	*Senecio tenuifolius*	(Upper and Lower)	34	*Senecio walkeri*	(Lower)
35-36	*Senecio wightii*	(Upper and Lower)	37-38	*Acanthospermum hispidum*	(Upper and Lower)
39-40	*Bidend humilis*	(Upper and Lower)	41-42	*Bidens pilosa* (normal ray)	(Upper and Lower)
43-44	*Bidens pilosa* (rayless)	(Upper and Lower)	45-46	*Eclipta prostrata*	(Upper and Lower)
47	*Galinsoga quadriradiata*	(Lower)	48	*Galinsoga* (intermediate)	(Lower)

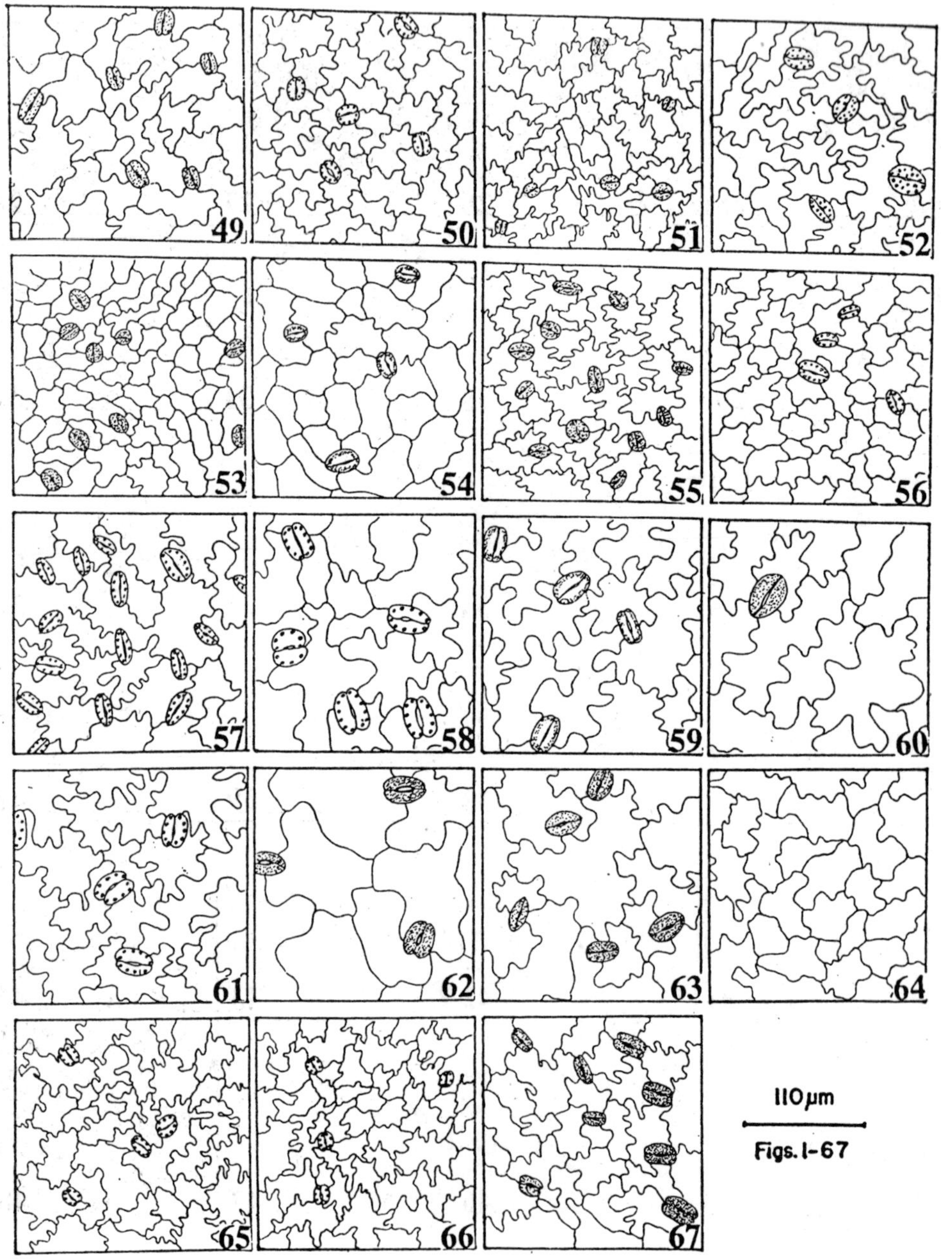

Plate No. 2.8 : Stomatal/epidermal studies (Explanation to Figures)

Figs	Taxon	Epidermis	Figs	Taxon	Epidermis
49	*Galinsoga parviflora*	(Lower)	50-51	*Lagascea mollis*	(Upper and Lower)
52	*Moonia heterophylla*	(Lower)	53	*Rudbeckia nitida*	(Lower)
54-55	*Sclerocarpus africanus*	(Upper and Lower)	56-57	*Siegesbeckia orientalis*	(Upper and Lower)
58-59	*Spilanthes calva* (small head)	(Upper and Lower)	60-61	*Spilanthes calva* (large head)	(Upper and Lower)
62-63	*Spilanthes paniculata*	(Upper and Lower)	64-65	*Synedrella nodiflora*	(Upper and Lower)
66	*Calyptocarpus vialis*	(Lower)	67	*Xanthium indicum*	(Lower)

33. *Siegesbeckia orientalis*
Leaves hypoamphistomatic. Cell walls of upper epidermis moderately undulated, while those of lower epidermis highly undulated (Plate 2.8; Figs. 56-57).

34. *Spilanthes calva* (small head)
Leaves isoamphistomatic. Cell walls of both epidermal layers highly undulated (Plate 2.8; Figs. 58-59).

35. *Spilanthes calva* (large head)
Same as in *S.calva* (small head) (Plate 2.8; Figs. 60-61).

36. *Spilanthes paniculata*
Leaves hypoamphistomatic. Cell walls of upper epidermis wavy, while those of lower epidermis moderately undulated (Plate 2.8; Figs. 62-63).

37. *Synendrella nodiflora*
Leaves hypostomatic. Cell walls of upper epidermis moderately undulated, while those of lower epidermis highly undulated (Plate 2.8; Figs. 64-65).

38. *Calyptocarpus vialis*
Same as in *Synedrella nodiflora* (Plate 2.8; Figs. 66).

39. *Xanthium indicum*
Leaves hypostomatic. Cell walls of both epidermal layers highly undulated (Plate 2.8; Figs. 67).

DISCUSSION

Foliar anatomical studies provide one of the most reliable evidences for classification and possibly for phylogenetic studies. Even though systematists are not sure of the course of evolution of foliar characters in flowering plants, their (foliar characters) diagnostic value has never been in doubt.

In the members of Asteraceae included in the present investigation, the results of anatomical studies are presented from the point of view of comparative as well as systematic anatomy. Comparative anatomical studies provide an overview of the situation while systematic anatomy analyses, the level (tribe, genus, species) at which foliar characters can be of diagnostic value).

Nodal anatomy

The literature available on the nodal anatomy of members of Asteraceae points to predominance of trilacunar 3 trace condition at foliar nodes. Exceptions are reported in the form of 7 lacunar, 7 trace type in *Elephantopus scaber* (Tiagi and Kshetrapal, 1970); 8 lacunar 8 trace node in *Brachyglottis repanda* (Drury, 1973) and a 5 lacunar 5 trace node in *Senecio myrianthos*. Most of the taxa in the present investigation conform to 3 lacunar 3 trace type except for *Gynura aurantiaca* and *Notonia grandiflora* which show split lateral condition at the foliar nodes. Howard (1970) indicated the association of split lateral condition with tree habit. Kavathekar *et. al.*, (1980), have reported split lateral condition in the foliar nodes of herbaceous species (i.e. *Carthamus* spp). Sivaramakrishna and Ramappa (1988) have reported split lateral condition in the cotyledonary nodes of several taxa of Asteraceae which however have a 3 lacunar 3 trace condition in the foliar nodes. Under the circumstances a study of the cotyledonary node structure in taxa which show split lateral condition may lead to an addition of a new dimension of nodal morphology.

Sinnott (1914) regards trilacunar nodal condition to be a primitive feature, while Marsden and Bailey (1955) regard it as advanced and derived from unilacunar double trace condition. Asteraceae being a large natural family, the amount of information available, with regard to nodal anatomy is too fragmentary to indicate which of the above views are correct.

Petiole anatomy

The members do not present a uniform picture, with regard to the number of leaf trace bundles at different levels in the petiole. 15 of the 29 taxa show 3 traces throughout the petiole with only the median extending up to the leaf tip while the laterals diverge into the basal part of lamina. The other fourteen taxa, show deviations from this common pattern with an amplification in the number of traces resulting often from a branching of the lateral (5 trace condition in 11 taxa) traces or by a branching of both the lateral and median traces in some taxa. *Gynura aurantiaca* shows branching of the lateral traces only, while in *Xanthium indicum* both the lateral and median traces branch with 5 traces extending up to the tip of the leaf. In *Rudbeckia nitida* also both the median and lateral traces branch resulting in a 10 trace condition with all these traces extending through the mid rib all the way up to the tip of the leaf. Branching of the median and lateral traces has earlier been reported for *Brachyglottis repanda* and *Senecio myrianthos* (Drury, 1973).

Venation pattern

In most of the members, the venation pattern is pinnate and semicraspidodromous. In a few taxa it is simple craspidodromous or eucamptodromous. In only *Notonia grandiflora* the venation is hyphodromous. In all the taxa the divergence of secondaries is at an acute angle, the lowest pair being more acute than the others. Areoles are well developed in all the members but compactly arranged only in some taxa. Veinlets may be simple, linear, curved or branched once or twice.

The quantification (vein density) method gives an idea of total vein length in a unit area of lamina. This value appears to be low for species of moist/shady habitats, while the taxa of xeric conditions show a higher value. Thus *Gynura nitida, G.aurantiaca, Notonia grandiflora* and *N.walkeri* (Table 2.2) show a low value and are plants growing in generally moist habitat conditions. *Rudbeckia nitida* with the highest vein density (52) occurs in open areas bordering on xeric conditions.

Stomatal/Epidermal studies

The members investigated present a uniform picture in having anomocytic type of stoma except for *Notonia grandiflora* of the tribe Senecioneae, where the stoma is of the anisocytic type. Earlier studies also indicate to a predominance of anomocytic type (Bhatt and Inamdar, 1975). In the present work, while there is near uniformity in the stomatal type, variations are met with in the shape of epidermal cells, stomatal freqency and stomatal index.

Systematic considerations

The taxa studied belong to two tribes Heliantheae and Senecioneae. None of the anatomical features can be used to distinguish these two tribes.

The following is a discussion at the sub tribal level in Heliantheae.

Melampodinae

Acanthospermum hispidum and *Siegesbeckia orientalis,* which belong to this tribe have similar vascularization of the petiole, venation pattern (pinnate, semicraspidodromous) and stomatal type. But *A.hispidum* has a higher vein density (37) than *S.orientalis* (26). Leaves are amphistomatic in both.

Ecliptinae

Eclipta prostrata, Syndrella nodiflora and *Calyptocarpus vialis* which belong to this sub tribe do not have similarity in petiole vascularization. The venation pattern however is same. Vein density is similar between

Eclipta prostrata and *Calyptocarpus vialis*. In stomatal distribution also there is variation, while leaves of Eclipta are amphistomatic, those of *Synedrella nodiflora* and *Calyptocarpus vialis* are hypostomatic.

Verbesininae

Legascea mollis and *Sclerocarpus africanus* which belong to this sub tribe have varied petiole anatomy with the latter having multiple traces. The venation pattern is also different, while the stomatal structure is same.

Coreopsidinae

Bidens pilosa and *Moonia heterophylla* belong to this sub tribe. They have identical petiole anatomy with both median and lateral traces travelling up to the leaf apex. The venation pattern however is slightly different. While it is pinnate simple craspidodromous in *B.pilosa,* it is pinnate semicraspidodromous in *M.heterophylla.* Leaves are amphistomatic in *B.pilosa,* while they are hypostomatic in *M.heterophylla.*

Galinsoginae

Galinsoga quadriradiata, G. parviflora, Spilanthes calva and *S. paniculata* belong to this sub tribe. There is similarity in petiole anatomy between *S. paniculata* and species of *Galinsoga.* Another feature common between *S. paniculata* and *G.quadriradiata* is the venation pattern; while *G. parviflora* has a different type of venation. The leaves of *Spilanthes* spp are amphistomatic, while those of *Galinsoga* are hypostomatic.

The foliar characters discussed above (at the sub tribal level) in Heliantheae reveal that the traits are not indicative of sub tribal affinity in the absolute sense. While some characters are common between the taxa included in a sub tribe, some are different. Further, some traits are distributed uniformly in all the sub tribes rendering them unsuitable as taxonomic markers at the sub tribal level. Similarities in foliar characters in taxa included in a sub tribe may at best be used as collateral evidence in conjunction with evidences from other sources.

Diagnostic foliar characters in Genera and Species

At the generic and species level, foliar anatomical characters prove to be of great significance as they can be relied upon for

(a) diagnosis of species
(b) species interrelationship and
(c) Analysis of genetic/ecological variation at the intraspecific level. The following is a diagnostic account of some of the genera and species.

Ageratum

Foliar characters can clearly demarcate the two species *A.conyzoides* and *A.houstonianum.* Petiole anatomy is similar in both except that in *A.conyzoides* two short traces are produced from the laterals in the petiole. While the venation pattern is basically similar in both the species, areoles are very compact (Plate 2.4; Fig 2). in *A. houstonianum.* Leaves are typically hypostomatic in *A. conyzoides,* while they are amphistomatic in *A.houstonianum. A. conyzoides* has a vein density of 16, while that of *A.houstonianum* is 31. The lower epidermis of *A.conyzoides* has a stomatal index of 37.84 with a L/B ratio of 1.40, while the corresponding values are 30 and 2.15 for *A.houstonianum.*

In all the members of the tribe Eupatorieae investigated, (*A. houstonianum, A. conyzoides, Adensostemma lavenia* and *Chromolaena odorata*) the epidermal cells have undulated cell walls, except for upper epidermal cells of *Chromolaena odorata* which have straight cell walls.

Emilia

Emilia

The two species, *E. scabra* and *E.sonchifolia* can be categorised as follows on foliar characters.

		E. scabra	*E. sonchifolia*
(a)	Petiole traces	lateral traces produce short branches 5 traces are seen	3
(b)	Intersecondary veins	Composite	Simple
(c)	Veinlet	Simple	Occasionally branched
(d)	Areolc	Compact	Not compact
(e)	Stomatal Index for	30.13	47.17
(f)	L/B ratio	1.40	1.38
(g)	Vein density	35	23

Gynura

The two species, *G. nitida* and *G.aurantiaca* can be distinguished on the basis of the following foliar characters.

		G. nitida	*G. aurantiaca*
1.	Node	3 traces	3 with split laterals
2.	Petiole traces	5	many
3.	Middle of lamina (traces)	3	1
4.	Vein density	13	10
5.	Stomatal Index and L?B ratio epidermis	36 and 1.30	33 and 1.31

Foliar characters, other than node and petiole are identical in *G. nitida* and *G.aurantiaca.*

Notonia

Among the members studied, the best foliar anatomical evidence for species delimitation is provided in the genus *Notinia.* The two species *N.grandiflora* and *N.walkeri* have the following diagnostic foliar characters.

		N. grandiflora	*N.walkeri*
1.	Node	3 leaf traces with split	3 traces
2.	Petiole traces	5	3
3.	Middle of lamina (traces)	3	1
4.	Venation type	Hyphodromous	Pinnate eucamptodromous
5.	Primary vein (s)	3	1
6.	Veinlets	branched once	moslty absent
7.	Leaf	epiaphistomatic	hypostomatic
8.	Stomata	anisocyticAnomocytic	
9.	Stomatal Index for upper/	15.38/9.09	-19.12
10.	L/B ratio for upper and lower epidermis	1.51/1.35	-/1.48

Senecio

In the genus *Senecio*, where 10 taxa are studied for stomatal structure, leaves are both hypostomatic as well as (for lower epidermis) amphistomatic. Stomatal structure can be correlated with their habit and also can be used to a great extent to characterize the species.

S. corymbosus, *S.walkeri* and *S.candicans* are climbers. In all these, the leaves are hypostomatic. The epidermal cells do not have undulate cell walls in *S.corymbosus* and *S.walkeri*, instead they are straight and cells are polygonal. In *S.candicans* however the cell walls are wavy. The values of stomatal index and L/B ratio for *S. corymbosus, S. candicans* and *S.walkeri* are 18.52/1.42; 30.14/1.74 and 19.71/1.37 respectively. The fact that *S.corymbosus* and *S.walkeri* with polygonal epidermal cells and straight cell walls have almost identical stomatal index and L/B ratio appears to be taxonomically significant. *S.candicans* with wavy epidermal cell walls has a different stomatal index and L/B/ ratio. Another morphological character common to *S.corymbosus* and *S.walkeri*, but absent in *S.candicans* is the white tomentose growth on the leaves. Another taxon *S.ludens* is not a climber but the epidermal cells have straight cell walls. The leaves in this taxon have white tomentose growth like in *S.corymbosus* and *S.walkeri.*

In *S.hohenackeri* and *S.tenuifolius,* the leaves are highly dissected and amphistomatic. The stomatal index on lower epidermis is 38.30 and 42.96 respectively. It has to be pointed out however that stomatal structure is not of much help here systematically. In *S. lawsoni, S. wightii* and *S.neelgheryanus* also the leaves are amphistomatic. *S.neelgherryanus* however can be characterized by the polygonal epidermal cells (Plate 2.4; Fig. 28) with straight walls on the upper epidermis, while the cells of lower epidermis have wavy walls. *S.cineraria* has hypostomatic leaves with epidermal cells having undulated cell walls.

Bidens

The normal rayed and rayless forms of *B. pilosa* have identical petiole anatomy and venation pattern. In *B.humilis* and both forms of *B.pilosa* leaves are hypoamphistomatic. Stomatal index is different in *B.pilosa* and *B.humilis.* It is 41 and 45 in *B.pilosa* normal rayed and rayless and 22 in *B.humilis* (values for lower epidermis)

Galinsoga

Nodal and petiole anatomy is identical between *G.quadriradiata, G.parviflora* and an intermediate form of *Galinsoga.* The appearance of the leaves in the intermediate is more like in *G.quadriradiata.* In the venation pattern *G.quadriradiata* and *Galinsoga* intermediate are identical (pinnate eucamptodromous venation). In all the three plants, the leaves are hypostomatic. The epidermal cells of *G.quadriradiata* and *G.parviflora* are alike in size and shape, while the cells appear narrow in *Galinsoga* intermediate.

Spilanthes

The petiole vascularization in *S.calva* (small head) and *S.calva* (large head) are different. The 3 leaf traces continue without any change right up to the leaf apex in *S.calva* (large head) while in *S.calva* (small head) the lateral traces divide at the base of the petiole and 5 traces can be seen in the petiole up to the lamina. Further up only the original 3 traces continue. In *S.paniculata* also there are 3 traces but the lateral traces stop at the base of the lamina. They contribute to the venation in the lower part of the leaf. The venation pattern is pinnate semicraspidodromous in the two forms of *S.calva* while in *S.paniculata* it is pinnate eucamptodramous. In the areole pattern however there is similarity between *S.calva* (small head) and *S.paniculata.* Veinlets in these two plants are curved and branched whereas in *S.calva* (large head) they are linear and unbranched. The leaves are amphistomatic in all the three plants. The stomatal index and L/B ratio for small and large head *S.calva* and *S.paniculata* are 40/1.44, 30.43/1.50 and 37/1.95 respectively, (values for lower epidermis).

Synedrella and Calyptocarpus

Both these taxa (*S.nodiflora C.vialis*) have identical petiole anatomy and venation pattern and Areoles are not very compact in *S.nodiflora* while they are so in *C.vialis.* The veinlets are simple in the former, but branched in the latter. Leaves are hypostomatic in both. *S.nodiflora* has stomatal index and L/B ratio of 15 and 1.38 respectively, while the corresponding values of *C.vialis* are 22.22 and 1.46 respectively.

LITERATURE CITED

ACUQA, C.D. 1881. Sulla distribuzione dei fasci fibrovascolari nel loro passiggio dal fusto alla fog lia. *Malpighia* 1 227-82.

AHMED, K.J. 1964. Cuticular studies in Solanaceae. *Can. J. Bot.* 42, 793-803.

BAILEY, I.W. 1956. Nodal anatomy in retrospect. *J.Arnold* Arbor. 25, 342-8.

BEILEY, I.W. and HOWARD, R.A. 1941. The comparative morphology of Icacinaceae. I-IV, *J. Arnold Arbor.* 2, 123-32, 171-87, 432-42, 56-68.

BHATT, D.C. and INAMDAR, J.A. 1975. Structure and development of stomata in some Compositae. *Va (J.Sau. univ)* PP. 106-120.

* CANDOLLE, C. DE. Anatomic compare des fenilles de quelques families de dicotyledones. *Mem. soc. phys. hist. nat.* Geneva 26, 427-80.

CARLOIN, R.C. 1954. Stomatal size, density and morphology in the genus *Dianthus. Kew Bull* 9 (2), 251-8.

CHANDRA, V., KAPOOR, S.L., SHARMA, P.C. and KAPOOR, L.D. 1979. Epidermal and venation studies in Apocynaceae I *Bull. Bot. Surv. India.* 11 (3 and 4) 286-9.

COETZER, L.A., ROBERTES, R.J. and REYNEKE, F.F. 1976. The node and petiole morphology of the south African representatives of the genera *Adenolobus, Piliostigma* and *Tyiosema; J.S. African* 42, 1-12.

CRONQUIST, A. 1968. The Evolution and classification of flowering plants. Houghton, Mifflin, Boston.

DESHPANDE, B.D. AND KUMAR A. 1968, Seedling anatomy and primary vascular system in some members of Compositae. *J.Indian Bot. Soc.* 47, 68-78.

DRURY, D.J. Node and leaf structure in the classification of some Australian shrubby Senecioneae, Compositae, *New. Zeal. J.Bot.* 1, 525-554.

* ETTINGSHAUSEN, C.R. Von. 1861. *Die Blattesklete der dicotyledonen* K.K. Hof. Vienna.

* FLORIN, R. 1933. Studies Uber die Cycadales des mesozokiums mebsteroterungen uber die spatoffnung sparatedes Benettitales. *K. Sensvavetenska akad, Handi,* 12 (5), 10-134.

____________ 1958. On Jurassic Taxads and Conifers from Northwestern Europe and Eastern Greenland. *Acta. Horti. Bergiani* 17, 257-402.

FOSTER. A.S. 1950. Morphology and Venation of the leaf in *Quiina acutangula.* Ducke. *Am. J. Bot.* 37, 159-71.

FRYNS Classens, E and Van Cotthem, W.R.J. 1973. A new classification of the ontogenetic types of stomata. *Bot. Rev.* 39 : 71-138.

GREW. N. 1675. The comparative anatomy of trunks, together with an account of their vegetation grounded there upon. London.

HARE, C.L. 1943. The anatomy of the Petiole and its taxonomic value. *Proc. Linn Soc. Lond.* 155, 223-9.

* HASSELBERG, G.B.F. 1937. Zur Morphologie des vegeativen sprosses der Loganiaceen. *Symb. Bot. Upsal* II, 3, 1-170.

HICKEY, L.J. 1973. Classification of the architecture of the dicotyledonous leaves. *Am. J. Bot.* 60, 17-33.

____________ 1979. A revised classification of the architecture of dicotyledonous leaves. pp. 25-39. In

Anatomy of the Dicotyledons (Eds. Metcalfe and Chalk, Vol. I), Clarendon Press, Oxford.

____________ and WOLFE, J.A. 1975. The bases of angiosperm phylogeny. Vegetative morphology. *Ann. Mo. bot. Gdn.* 68, 538-89.

HOWARD, R.A. 1962. The vascular structure of the petiole as a taxonomic character. Proc. 15th Intn. hort. Cong. Nice. 1958, pp. 7-13.

____________ 1970. Some observations on the nodes of woody plants with special reference to the problem of the split "lateral" varsus "common gap" pp. 195-214 in *New research in plant anatomy* (Eds. Robson N.K.R, Cutier D.K. and Greogy N. Longon.

____________ 1979. The petiole, pp. 88-96. In *Anatomy of the Dicotyledons.* Vol. I. Second ed., (Eds. Metcalfe and Chalk). Clarendon Press, Oxford.

JAIN, D.K. 1977. Studies in Bignoniaceae. I. Nodal anatomy; *Proc. Ind. Acad. Sci.B* 86- 375-380.

KASHYAP, M 1979. Nodal organisation in some species of the Ranunculaceae; *J. Indian Bot. Soc* 58, 148-153.

KAVATHEKAR, K.Y., GOYAL, S.C., SATIJA JAIN AND PILLAI, A. 1980. Nodal anatomy in some dicotyledons. pp. 31-39. In *Histochemistry, Developmental and structural anatomy of angiosperms.* (Ed) Periasamy, K. Trichy, India.

LEE, A.T. 1948. The genus *Swainsona. Contr. N.S.W. Herb.* 1. 131-121.

* LIGNIER, M.O. 1887. Recherches sur lanatomie Compareedes Calycanthees, des Melastomacees et des Myrtacees. *Arch. Sci Nird deka France.* 4, 455.

* MANKEVICH, O.I. 1971. Influence de la ploidie surles caracters anatomiques des plantes decoratives. *Vest. Akad. Navuk. BSSR* (V), 88-91 {Russian}.

MARSDEN M.P.F. AND BAILEY I.W. 1955. A fourth type of nodal anatomy in Dicotyledons, illustrated by *Clerodendron trichotomum* Thumb. *J.Arnold Arbor.* 36, 1-51.

MELVILLE, R. 1962. The terminology of leaf architecture, *Taxon* 25, 549-61.

METCALFE, C.R. 1961 The anatomical approach to systematics, General introduction with special referene to recent work on monocotyledons, 146-150. In : *Recent Advances in Botany.* Toronto, Canada.

__________ and Chalak L. 1950. *Anatomy of the Dicotyledons* Vols. I and II. Oxford, U.K.

MICHEL, J. I and LERSTEN, N.R. 1967. Floating stomates (Adenostomy) in ferns : Distribution and ontogeny. *Am. J. Bot.* 54 : 1181-1185.

* MOUTON, J.A. 1970. Architecture de la nervation foliaire. *C.R. 92e congr., natl. Soc. Savantes* 3, 165-76.

PAILWAL, G.S. 1969. Stomata in certain Angiosperms, their structure, ontogeny and systematic value. *In Recent advances in the anatomy of tropical seed plants* pp. 63-73. Hindusthan Publishing Corportation, Delhi, India.

BHANDARI, N.N. 1962. Stomatal development in some Magnoliaceae. *Phytomorphology.* 12,409-12.

PANT D.D., and GUPTA, K.L. 1966. Development of stomata and foliar structure of some Magnoliaceae. *J.Linn. Soc* (Bot) 59, 265-77).

PATEL, J.D. 1979. A mew morphological classification of stomatal complexes. *Phytomorphology*. pp. 218-229.

PAYNE, W.W. 1969. A quick method for clearing leaves. *Ward's Bull* 61, 4-5.

____________ 1970. Helicocytic and allelocytic stomata; unrecognised patterns in the dicotyledoneae. *Am. J.Bot.* 57, 140-7.

* PETIT. L. 1886. Sur importance taxonomique du petiole. *C.R. Acad. Sci Paris* 103, 767-9.

* PIERRE, L. 1896. Plantes du Gabon. *Bull. Mens. Soc. Linn. Paris*. 2. 1249-56.

RAMAPPA, H. 1985. Nodal anatomy of angiosperms. M. Phil Dissertation, Bangalore University.

SALISBURY, E.J. 1927. On the causes and ecological significance of stomatal frequency, with special reference to the wood land flora. *Phil. Trans. R. soc. 216b,* 1-65.

SAY. H.J. 1938. The relationship between stomatal counts and chromosome number. *J.Arn. Arbor.* 19, 437-441.

SCHOFIELD, E.K. 1968. Petiole anatomy of the Guttiferae and related families. *Mem. N.Y. Gard.* 18 (1) 1-55.

SINNOTT. E.W. 1914. The anatomy of the node as an aid in the classification of Angiosperms. *Am.J.Bot.* 1, 303-22.

SIVARAMAKRISHNA, D. AND RAMAPPA, H. 1988. Anatomy of nodal regions in Angiosperms. *J.Swamy Bot. cl.* 5 (1), 41-44.

SOLEREDER, H. 1908. Systematic anatomy of the Dicotyledons (Transl. L.A. Boodle and F.E. Fritsch, revised by D.H. Scott) 2 vols. Clarendon Press, Oxford.

STACE, C.A. 1965. The significance of the leaf epidermis in the taxonomy of the Combretaceae. I. A general view of tribal. generic and specific characters. *J.Linn. Soc. Lond. Bot.* 59, 229-52.

____________ 1966. The use of epidermal characters in Phylogenetic consideratios. *New Phytol.* 65, 304-18.

STEBBINS, G.L. and KHUSH, G.S. 1961. Variation in the organization of the stomatal complexes in the leaf epidermis of monocotyledons and its bearing on their phylogeny. *Am. J. Bot.* 48, 51-9.

STEARN, W.T. 1966. In supplement to the dictionary of gardening. pp. 318-32 (Ed. P. Synge) Oxford University press.

SWAMY, B.G.L. and BAILEY. I.W. 1949. The morphology and relationships of *Circidiphyllum*. *J. Arnold Arbor.* 34, 375-408.

TIMMERMANN, H.A. 1927. Stomatal numbers; their value for distinguishing species. *Pharm. J.* 18, 241-3.

TIYAGI, Y.D. and SASHIKALA KSHETRAPAL, 1970. A contribution to the vascular anatomy of the node and flower of certain species of the family Compositae. *Proc. Nat. Acad Sci. India.* B 40 : 113-128.

VAN COTTHEM, W.R.J. 1970. A classification of stomatal types *Bot. J. Linn. Soc.* 63 : 235-246.

* VESQUE, J. 1881. Anatomie des tissues appliquea la classification des plantes. *Nouv. Archs. Nus. Hist. nat. Paris (ser2)* 4, 1-56.

____________ 1885. Caratcers des principales families gamopetales tiers de 1 anatomie de la feuille. *Annls. Sci. Nat. Bot. (Sev* VII) 1, 183-360.

____________ 1889. De lemploi des caracters anatomiques dans la classification des vegetaux. *Bull. Soc. Bot. Fr.* 36, XLILXXXVII.

WATARI, S. 1934. Anatomical studies of some Leguminous leaves with special reference to the vasculara system in petioles and rachises. *J. Fac. Sci. Univ. Tokyo* (Sec. III) 4, 25-365.

__________ 1936. Anatomical studies on the vascular system in the petioles of some species of *Acer* with notes on the external morphological features J. Fac. Sci. Imp. Univ. Tokyo (Sec III) 5, 1-73.

__________ 1939. Anatomical studies on the leaves of some Saxifragaceous plants, with special reference to the vascular system. *J. Fac. Sci. Univ. Tokyo (See III Bot)* 5, 195-316.

WILKINSON, H.P. 1979. Plant surface (mainly leaf) ; Stomata. pp. 97-116. In : *Anatomy of the Dicotyledons*. Vol I Second ed. (eds. Metcalfe and Chalk) Clarendon Press, Oxford.

3

EMBRYOLOGY

Embryological investigations in Asteraceae are perhaps as old as embryology itself. Having occupied almost the pinnacle of evolution among sympetalous dicotyledons, the family offers a very favourable material for the inquiry into various processes that have characterized the most successful line of evolution.

The evolutionary success of the family is reflected in various facets including its embryology. Embryologically the family is most perplexing and interesting. The family does not show uniformly features that are regarded as embryologically advanced and expected of a family as highly placed as Asteraceae. There is virtually every type of embryological variation from the most primitive to the most advanced in the family. Listing of the embryological characters of the family is like enumerating the embryological diversity in the whole of flowering plants.

Some of the interesting embryological features of the family are :

(a) Occurrence of both bi and tetrasporangiate anthers

(b) Occurrence of both amoeboid and glandular tapetum (the latter is rare)

(c) Variation in embryo sac development. While the majority of the members exhibit Polygonum type, as many as six different types - Allium, Endymion, Peperomia, Adoxa, Drusa and Fritillaria occur within the family. In addition to this, aberrant and unclassified types also occur as reported in *Balsamita vulgaris* (Fagerlind 1939).

(d) Antipodals exhibit a wide range of variation in terms of their number, number of nuclei in each cell, secondary multiplication and fusion and haustorial nature.

(e) Endosperm is both nuclear and *ab initio* cellular.

(f) Embryogeny always conforms to *Senecio* variation of the Asterad type.

(g) Apomixis is of frequent occurrence in the family. Besides aneuspory, both somatic and gonial apospory have been reported in a number of general.

With such diverse embryological features, it is not surprising that Asteraceae has attracted the attention of embryologists since a long time. Embryological investigations in the family are quite extensive and intensive. Perhaps more number of taxa of Asteraceae are embryologically investigated than in any other family of flowering plants. Even then, considering the vast size of the family (950 genera and 20,000 spp.), the embryological work is still far less for any generalization. As Davis (1964) has pointed out, only about 15-20% of the genera have been investigated in the family. The situation since the appearance of the paper of Davis (1964) has changed only marginally and there is ample scope for embryological investigations in the family. There have been three reviews of the embryology of Angiosperms with the last one exclusively dealing with Asteraceae. These are :

1. *Vergliechende Embryologie der Angiospermen* (Schnarf 1931).
2. Systematic Embryology of Angiosperms (Davis 1966)
3. Embryology of the Compositae (Pullaiah 1984).

Review of earlier work

Literature of previous embryological work in Asteraceae is quite numerous. Some of the more important embryological works on the tribes of Asteraceae are briefely reviewed below. The tribes Heliantheae and Senecioneae which have a direct bearing on the present work are reviewed in detail separately in a tabular form. The earliest work on Asteraceae is that of Hofmeister (1849) who has studied the structure and development of the female gametophyte in *Helianthus annus.*

In India, Bhargava (1935) was the first to work on the embryology of Asteraceae. His work on *Eclipta erecta* reported the occurrence of multinucleate antipodal cells and *ab initio* cellular endosperm.
Earlier works on the tribes of Asteraceae other than those of Heliantheae and Senecioneae are listed below in chronological order. The tribal classification followed is that of Bentham (1873).

VERNONIEAE

Tiagi and Taimni (1963) were the first to work in detail on *Vernonia cinerascens.* They have reported a glandular tapetum in the anther, Polygonum and Allium types of embryo sac and Asterad type of embryo. The work of Palm (1925) is earliest for the tribe, but it deals with only the embryo sac development in *Vernonia chinensis.* Some of the other notable contributions for the tribe are those on *Vernonia elagnifolia* and *Elephantopus scaber* (Pullaiah 1979a), etc.

EUPATORIEAE

The earliest embryological work on Eupatorieae is of Holmgren (1919) who has worked on a number of species of the genus *Eupatorium* (*E. cannabinum, E. glandulosum, E. ageratiodes* etc). He has reported a linear tetrated of megaspores, Polygonum type embryo sac development and parthenogenetic development of embryo. Some other contributions on the tribe are those of Dutta (1939) on *Mikania cardifolia;* Misra (1947) on *Mikania scandens;* Deshpande and Kothare (1972) on *Adenostemma lavenia;* Sullivan (1976) on *Eupatorium lancifolium* and *E.cuneifolium;* Pullaiah (1979a) on *Adenostemma rugosum* and Mukherjee and Desai (1980) on *Eupatorium triplenervis.*

ASTEREAE

The earliest embryological work in the tribe is that of Guignard (1881) who reported the development of Polygonum type embryo sac with 10 antipodal cells in *Conyza ambigua.* Subsequent works include those of Holmgren (1919) on *Erigeron bonariensis;* Mcdonald (1927) on *Erigeron annus;* Bergmann (1944) and Fagerlind (1947b) on *Erigeron annus;* Mitra (1947) on *Grangea madraspatana;* Davis (1963a, b, 1964a, b, c, 1968a) on *Minuria denticulata, Brachycome ciliaris, Minuria integerrima* and *Calotis lappulacea;* Deshpande and Kothare (1976) on *Cyathocline purpurea;* Sharma and Murthy (1977a) on *Brachycome iberidifolia* and *Bellis perennis;* Pullaiah (1978a) on *Solidago canadensis* and *Conyza stricta* and Desai (1982a) on *Erigeron asteroides* and *Solidago purpurea.*

INULEAE

The work of Juel (1900) on *Antennaria dioica* is the earliest for the tribe wherein he reported the development of Polygonum type of embryo sac. The other important works on this tribe include those of Banerjee (1942) on *Blumea laciniata;* Davis (1961a, 1962b, and 1968b) on *Podolepis jaceoides, Ammobium alatum* and *Epaltes australis;* Deshpande (1962b) on *Caesulia axillaris;* Pullaiah (1979b) on *Blumea membranacea, Laggera pterodonta, Vicoa auriculata* and *Anaphalis busna;* Rajasekhar (1980) on *Blumea mollis*; Desai (1981b) on *Laggera aurita* and Lalitha Ram (1986) on *Pluchea tomentosa.*

HELENIEAE

Dahlgren's (1920) work on *Tagetes patula* and *T.signatus* is the earliest report on the tribe. He reported the development of Polygonum type embryo sac and Asterad type of embryo. Subsequent works on the tribe

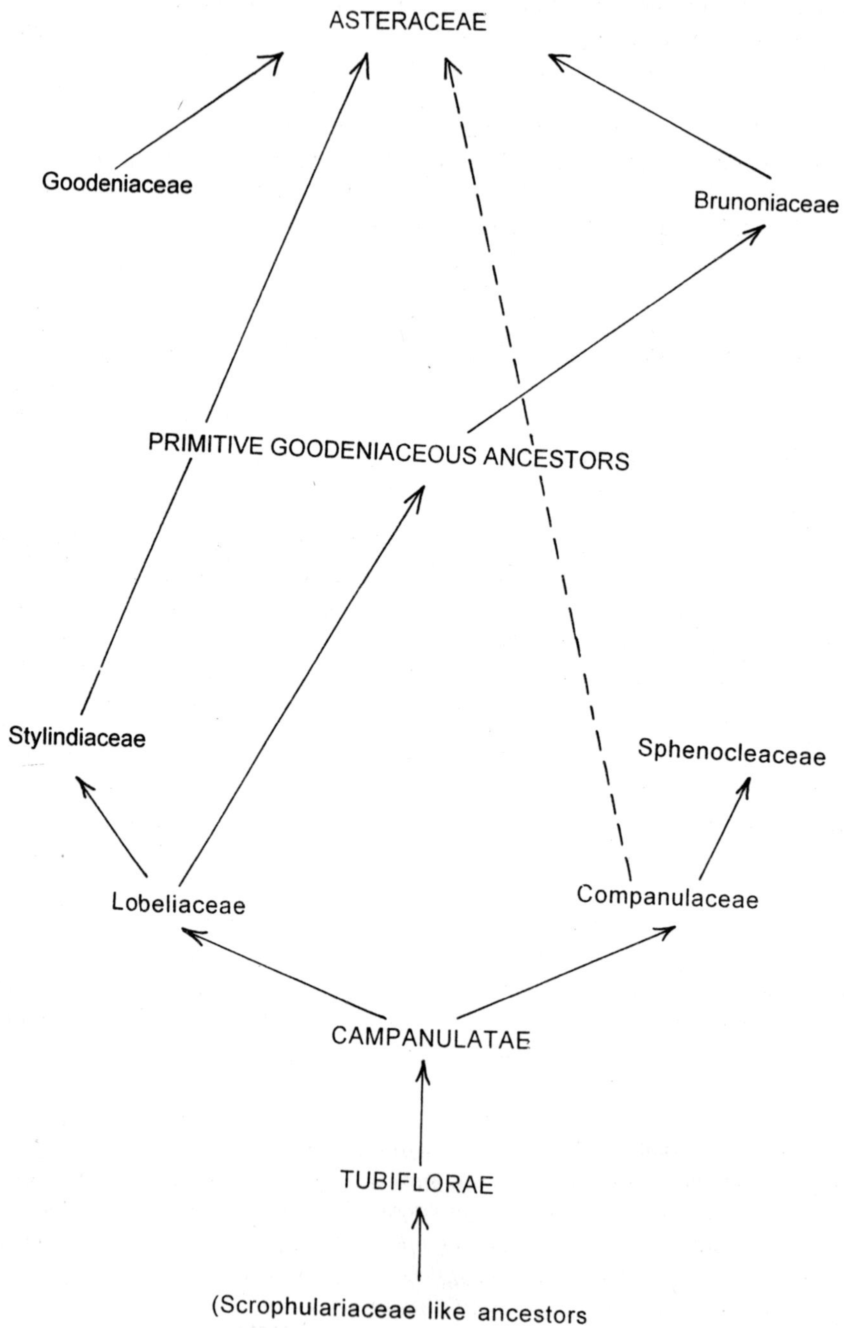

Plate 3.1 Interrelationships of Asteraceae

include those of Venkateswarulu and Maheswari Devi (1955a, b) on *Tagetes patula, Flaveria australasia* and *Gaillardia picta;* Misra (1957) on *Flaveria repanda;* Pandey and singh (1976) on *Hymenantherum tenuifolium;* Maheswari Devi and Pullaiah (1978) on *Gaillardia lorenziana* and Pandey *et al.*, (1986) on *Hymenantherum tenuifolium, Tagetes patula* and *T. erecta.*

ANTHEMIDEAE

Occurrence of Polygonum type embryo sac with two multinucleate and polyploid antipodals in *Chrysanthemum leucanthemum* is the earliest report for the tribe (Goldflus 1898). Palm (1914a) and Martin and Smith (1955) have also worked on the same taxon.

Other important embryological contributions of the tribe include those of Fagerlind (1941) on *Chrysanthemum parthenium,* Martinoli (1942a) on *Artemesia arborescens;* Harling (1950, 1951a) on *Anthemis arvensis, A.austriaca* and *A.maritima* and *Chrysanthemum* spp. Battaglia (1951c) on *Chrysanthemum viscosum;* Davis (1961d, 1962a) on *Cotula australis;* Borgen (1972) on *Chrysanthemum ferulosum, C. filifolium* and other species and Pullaiah (1983a)on *Achillea squarrosa.*

ARCTOTIDAE

The work of Gelin (1936) on *Berkheya bergicena* seems to be the earliest embryological report for the tribe wherein a Polygonum type embryo sac with numerous antipodals has been reported. Other important contributions on this tribe include those of Dahlgren (1924) on *Ursinia anthemoides* and Ahlstrand (1978, 1979a, b) on *Ursinia cakilefolia, Arctotis fastuosa, A. acaulis, Gazania maritima, G.rigens* and *G.ilurenifolia* and *Gundelia tournefortii.*

CALENDULEAE

The report of Dahlgren (1924) on *Tripteris valliantii,* wherein he reported nuclear endosperm, seems to be the earliest embryological work in the tribe. Subsequent works on this tribe include those of Dahlgren (1924) on *Dimorphotheca pluvialis;* Mestre (1963-64), on *Calendula officinalis* and Carano (1915a) on *Calendula arvensis.*

CYNAREAE

The earliest report in this tribe is perhaps on *Centaurea cirrhata* (Lavielle, 1911) wherein a Polygonum type embryo sac has been described. Some of the other works on this tribe include those of Poddubnaja Arnoldi (1931), on *Carduus acanthoides, Onopordon acanthium, Sassurea amara, Xeranthemum annum* and *Echinops sphaerocephalis;* Banerji (1940b) on *Carthamus thinctorius;* Deshpande (1964a) on *Volutarella ramosa;* Maheswari Devei and Pullaiah (1977a) on *Carthamus tinctorius* and Singh and Pandey (1985) on *Carthamus oxycantha, Centaurea moschata, Silybum marianum, Carthamus baeticus, Centaurea cyanus, Onopordon algeriense, Sassurea albescens* and *S.heteromella.*

MUTISIEAE

Very little embryological work seems to have been carried out in the tribe. The work of Maheswari Devi (1957) on *Gerbera jamesonii* is the earliest report wherein she has recorded a Polygonum type embryo sac with multinucleate antipodals. The other report on the tribe is that of Kapil and Sethi (1962a) on *Ainsliaea aptera,* wherein a Polygonum type embryo sac with three antipodal cells has been recorded. The embryogeny in this taxon is of the Asterad type.

CICHORIEAE

The work of Guignard (1882) on *Tragopogon pratensis* is the earliest report for the tribe recording the occurrence of Polygonum type embryo sac. Other important embryological contributions for the tribe are those of Rosenberg (1906, 1907), on *Taraxacum erythrospermum;* Dahlgren (1920) on *Crepis blattoroides;*

Poddubnaja Arnoldi (1935) on *Chondrilla* spp; Venkateswarulu (1939) on *Launaea nudicaulis;* Datta (1939) on *Launaea asplenifolia;* Venkateswarulu and Maheswari Devi (1955) on *Launaea pinnatifida;* Beruti (1961) on *Hypochoeris aetnensis;* Singh and Kaul (1974) on *Tragopogon gracile;* Godineau (1973) on *Crepis tectorum;* Kaul *et al.*, (1975) on *Sonchus asper* and *S.arvensis;* Gill and Iqbal (1981) on *Pterotheca falconeri;* Cichan and Palser 91982) on *Cichorium intybus* and Pullaiah (1982b) on *Sonchus oleraceus.*

Heliantheae and Senecioneae

The present work on 8 taxa of Asteraceae is on the tribes Heliantheae and Senecioneae. Tables 3.1, 3.2, 3.3, 3.4, 3.5 and 3.6 give a detailed analysis of the previous embryological work in these two tribes.

Some of the recent contributions on the embryology of these two tribes which appeared comparatively recently are on the embryology of *Senecio ambigua* and *S.gibbosus* (Gormuglio *et al.*, 1983) and on *Senecio candicans* (Rangaswamy and Pullaiah 1986).

Scope of the present work

As has been pointed out earlier, in spite of the extensive embryological work in the family, considering the vast size, there is scope for much more work before one can assess the value of embryological characters in contributing to the evolutionary success of the family. Hence there is need for further embryological work, even if it is purely descriptive and repetitive in nature. An analysis of the embryological characters of an advanced family like Asteraceae, will help in the assessment of embryological characters in terms of evolutionary advancement not only for this family but for the entire group of flowering plants.

The taxa selected for embryological investigation (see table 1.2) in the present work belong to two tribes, Heliantheae and Senecioneae. The two tribes in question have been assigned diametrically opposite ranks in the phyletic heirarchy of the family. Small (1919) whose contribution to the biology of the family may be regarded as monumental, opines after an extensive study of the evolution of styles and stamens that Senecioneae is the most primitive group. On the other hand, Bessey (1915), Hutchinson (1916) and more recently Cronquist (1955) argue that Heliantheae is the most basic tribe of the family with all other tribes being derived from it. The bases of both these diametrically opposite views appear to be morphological and chemical characters. Embryological characters do not seem to have influenced these views either because they (embryological characters) appear to be ambiguous or distributed randomly in both the tribes making them unsuitable as a character source of systematic and phyletic value. Another reason perhaps is the lack of broad based and statistically significant embryological data in both the tribes.

The present work therefore is an attempt at the phyletic appraisal of the tribes Heliantheae and Senecioneae in terms of their embryology. Attempt is made to analyse the data obtained from the present work and other published works to give an embryological view point of the advanced or primitive nature of these two tribes. Attempt is also made to compare Asteraceae with other supposed to be closely related families on embryological grounds. Finally, a rationale is provided to account for the diversity (embryo sac development) and unity (embryo development) of the embryology of the family as being symbolic of dynamism of an evolved biological system.

MATERIALS AND METHODS

Materials for embryological studies were collected from plants growing in and around Bangalore, India. The following is a list of the plants (Morphological descriptions of these plants are given in general introduction).

Heliantheae
1. *Synerella nodiflora*
2. *Eclipta alba* (= *Eclipta prostrata*)
3. *Balainvillea rhomboidea* (= *B.acmella*)

4. *Spilanthes acmella* (*Spilanthes calva*)
5. *Spilanthes paniculata*

Senecioneae
1. *Emilia sonchifolia*
2. *Senecio tenuifolius*
3. *Senecio ludens*

Materials for embryological studies consisted of young and old flower heads. These were fixed in F.A.A. (Foramlin, acetic acid, alcohol) for 24 hours and later transfered to 70% alcohol for storage.
Flower heads of different stages were selected and passed through different grades of alcohol for dehydration. After dehydration, they were embedded in paraffin blocks. Serial microtome sections were cut (longitudinally as well as transversely) at 8-12μ. Sections were deparaffinised in xylol and after passing through alcohol series, were stained in Heidehein's iron alum haemotoxylin. 4% iron alum was used for destaining. The sections were also counterstained with erythorocin and mounted in canada balsam.

Camera lucida drawings have been drawn for all the embryological stages observed.

Of the eight taxa discussed here for the purpose of embryology, *Synedrella nodiflora* (female gametophyte), *Eclipta alba* (Gametophytes and early embryogeny), *Spilanthes acmella, Blainvillea rhomboidea* and *Emilia sonchifolia* (male and female gametophytes) have been published previously by the author (Sundara Rajan 1978a, b, 1972, 1974a, b). *Spilanthes paniculata, Senecio ludens* and *S. tenuifolius* are described here for the first time.

OBSERVATIONS

Morphological description of the taxa under investigation in the present work is given in the General introduction. Embryological details of the taxa are described separately for the two tribes (Heliantheae and Senecioneae).

HELIANTHEAE

Organogeny

A young capitulum arises as a meristematic hump. Soon, floral primordia appear on it. Floral parts arise in acropetalous succession in *Synedrella nodiflora* (Plate 3.4, Figs 1-3), *Blainvillea rhomboidea* (Plate 3.11, Plate 3.8, Figs. 1-4), *Spilanthes acmella* (3.4 , Figs. 4-6). In *Spilanthes paniculata* however, corolla appears first followed by androecium, gynoecium and lastly calyx arises (Plate 3.2, Figs. 1-3).

Development of the anther

Anthers are bisporangiate in all the members investigated. A very young anther in a cross section shows an undifferentiated mass of meristematic tissue, which very soon assumes a two lobed contour. At this stage of development, the anther has a mass of cells surrounded by a layer of epidermis.

Very soon two hypodermal cells in each lobe of the anther distinguish themselves from the remainder by their dense cytoplasmic contents and prominent nuclei. These are the archesporial cells. In all the members investigated, there are always two archesporial cells (Plate 3.11, Fig 7; Plate 3.8, 5) These two cells eventually undergo a number of transverse divisions, thus forming two linear rows of archesporial cells in each lobe of the anther.

In *Spilanthes paniculata,* only one archesporial cell is seen in some of the anthers.

The archesporial cells radially enlarge and undergo periclinal division resulting in the formation of an outer primary parietal layer and an inner primary sporogenous layer. The primary parietal layer divides again periclinally to form two layers of cells. Of these,the inner (i.e. immediately external to the primary sporogenous layer) layer directly functions as the tapetal layer while theouter layer divides again to form a hypodermal endothecial layer and an inner middle layer (Plate 3.11, Figs. 5-7; Plate 3.4, Fig. 8; Figs. 6-8).

This type of anther development as described above has been classified as the Dicotyledonous type (Davis 1966). The wall development is schematically given below.

Undifferentiated mass of anther tissue

↓ Epidermis — ↓ Epidermis

↓ Archesporium
- Primary sporogenous cell → ↓ Secondary sporogenous cell
- Primary parietal cell ↓
  - ↓ Tapetum
  - ↓ Secondary parietal layer ↓
    - ↓ Middle layer
    - ↓ Endothecium

A fully developed anther in all the members consists of an epieermis, an endothecium, a middle layer and a layer of tapetum surrounding themicrospore mother cells. (Plate 3.11, Fig. 7;Plate 3.4, Fig. 8;Plate 3.8, Fig.7).

Tapetum

The tapetal cells are rectangular, generally richin cytoplasm and are uninucleate to begin with. In *Spilanthes acmella,* the tapetal cells are uninucleate throughout (Plate 3.4, Fig. 8), while in *S.paniculata* they become binucleate later. In some instances in *S. paniculata,* the nuclei of tapetum appear to be polyploid (Plate 3.2, Fig.12). In *Synedrella nodiflora* also the tapetal cells are binucleate and appear to be polyploid (Plate 3.11, Fig.8). In *Eclipta alba* and *Blainvillea rhomboidea* also the tapetal cells become binucleate (Plate 3.9, Fig. 7; Plate 3.11, Fig.1). The nuclear activity in the tapetal cells appear to coincide with the initiation of microsporogenesis.

In all the members investigated, the tapetum is of the plasmodial type. As is true of majority of Asteraceae, the tapetal cells remain intact at the tetrad stage of the microspores. When the microspores separate from the tetrad and the outer wall of thepollen grain is about to differentiate, the tapetal cells breakdown releasing the nuclei and cytoplasm into the anther loculus (Plate 3.4, Fig. 10; Plate 3.8, Fig. 7a; Plate 3.10, Fig. 7). The pollen grains are enclosed by the tapetal cytoplasm and derive nutrition from it. At the time of maturity of pollen grains, the periplasmodium disapperas having been absorbed (by the pollen grains).

The middle layer gets crushed during the microsporogenesis.

Epidermis initially keeps pace with the development of the anther, but soon the cells become streched radially and tangentially. It persists as a thin layer of cells in the mature anther. Certain granular substances are noticed in the epidermis of the mature anther in *Blainvillea rhomboidea* (Plate 3.8, Fig. 13).

In all the taxa investigated, theendothecial cells exhibit fibrillar thickenings at thetime of anther dehiscence. (Plate 3.11, Fig. 12; Plate 3.4, Fig.10; Plate 3.8, Fig.13; Plate 3.10, Fig.7).

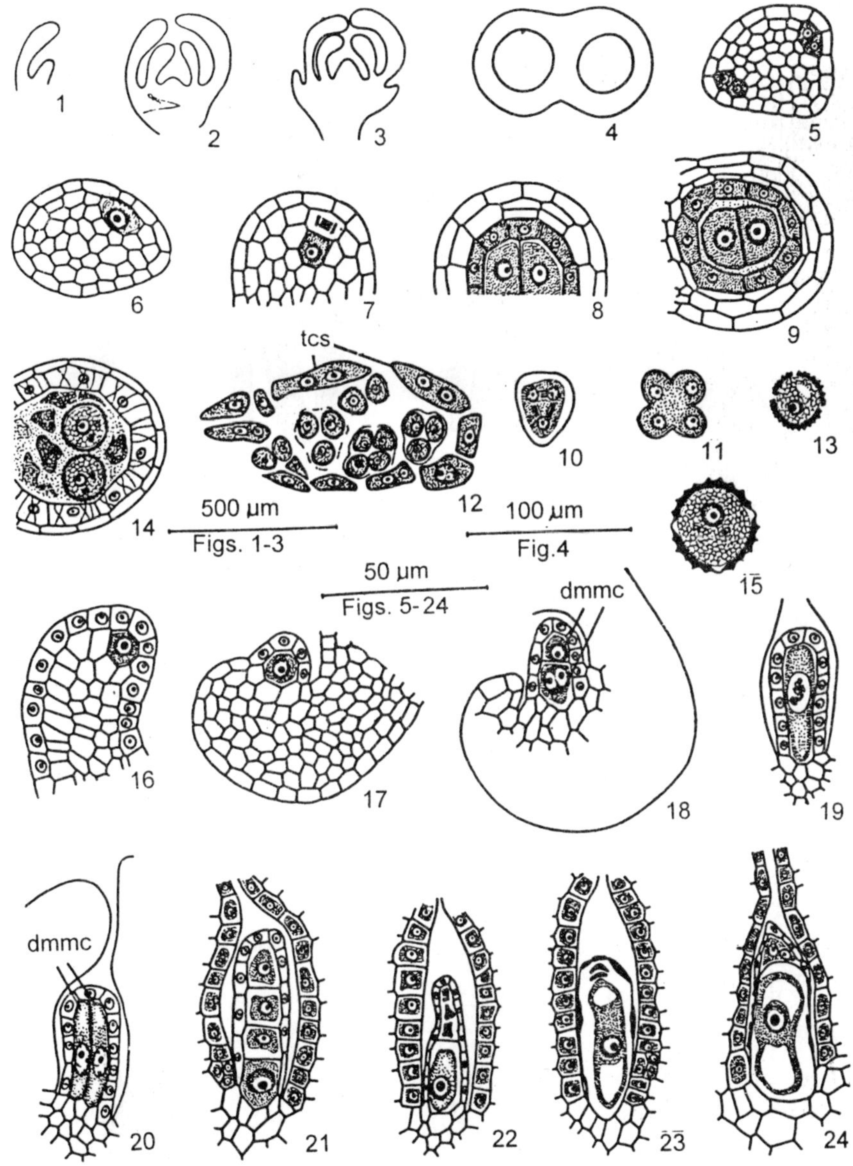

Plate No. 3.2 : *Spilanthes paniculata* (Explanation to Figures)

Figs : 1-3 Organogeny, **4** T.S. anther in outline showing bisporangiate condition, **5-8** Development of microsporangium, **9** T.S. of fully developed microsporangium, **10** Spindle formation in the dividing pollen mother cells, **11** Cytokinesis, **12** Microspore tetrads surrounded by large tapetal cells, **13** Uninucleate pollen grain, **14** T.S. of anther showing remants of periplasmodium, **15** Three celled pollen grain, **16** L.S. of ovule with a single archesporial cell, **17** Same as in Fig. 16 but with the initiation of the integument, **18** Double megaspore mother cells (superposed), **19** Single mmc with nucleus in prophase, **20** Double megaspore mother cells (juxtaposed), **21** Linear tetrad of megaspore, **22** Tetrad with functional chalazal megaspore, **23** Uninucleate embryo sac, **24** L.S. of young ovule showing lower dyad enlarging; upper dyad crushed and having two small nuclei.

(tcs = tapetal cells, dmmc = double megaspore mother cells)

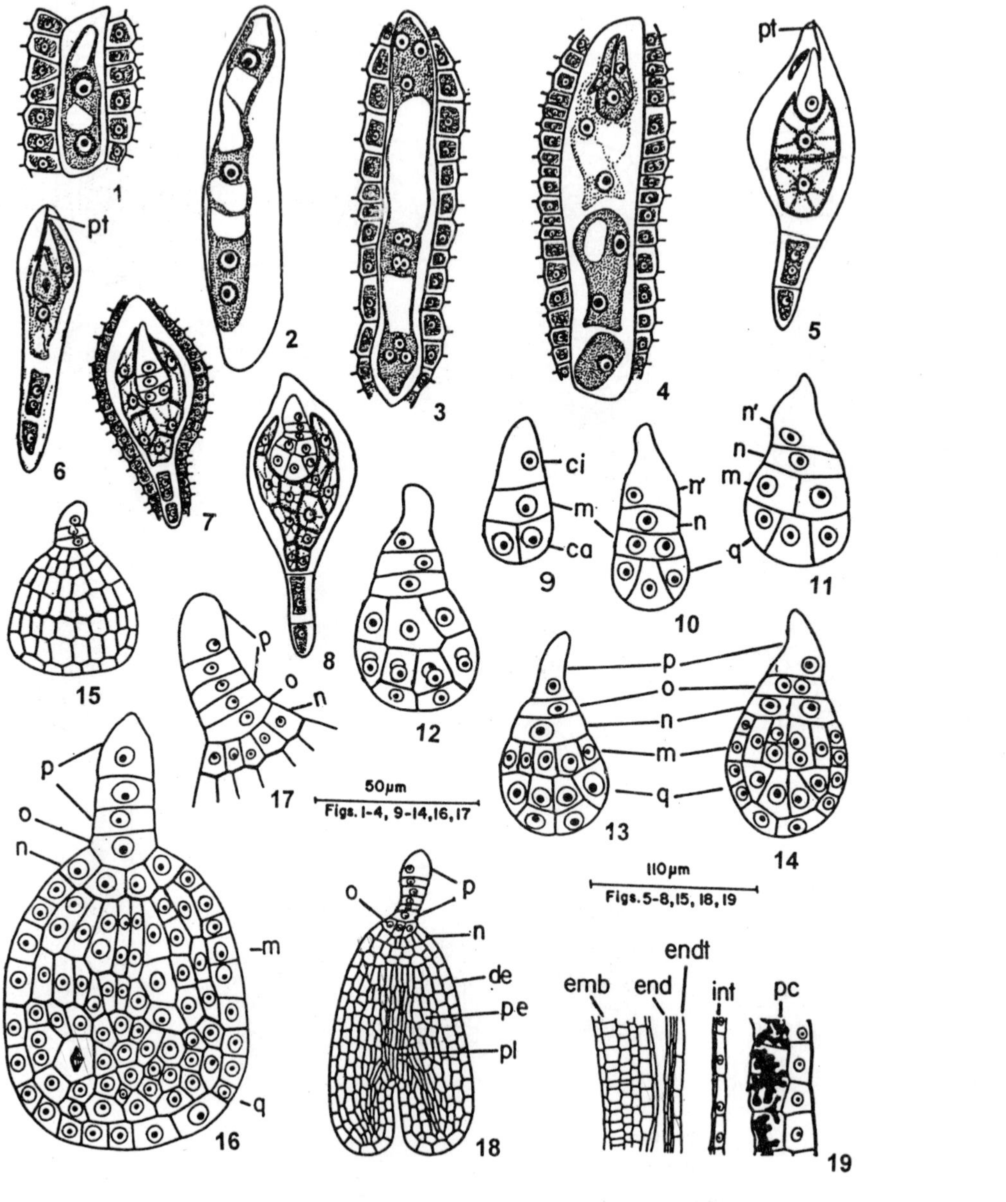

Plate No. 3.3 : *Spilanthes paniculata* (Explanation to figures)

Figs :1-2 Two and four nucleate embryo sacs, **3** Eight nucleate embryo sac, **4** Organised embryo sac with two large antipodal cells (Upper two nucleate), **5** First division of PEN with zygote and two antipodal cells, **6** First division of zygote with PEN, pollen tube and a surviving synergid, **7-8** Embryo sacs showing cellular endosperm and early embryogeny, **9-17** Embryogeny, **18** Young dicot embryo with six celled suspensor, **19** Structure of the fruit wall (in L.S.) (pt = pollen tube, pl = plerome, pe = periblem, de = dermatogen, emb = embryo, end = endosperm, endt = endothelium, int = integument, pc = pericarp)

Microsporogenesis and male gametophyte

The primary sporogenous cells divide vertically and produce the microspore mother cells. There are two rows of pollen mother cells in the anther. No cytomixis is observed in any of the members investigated.

At the initiation of meiosis, the pollen mother cells (PMC) separate from one another and appear rounded. They undergo meiosis toproduce tetrads of microspores. The tetrads are tetrahedral and decussate in *Spilanthes paniculata, Synedrella nodiflora, Eclipta alba* and *Blainvillea rhomboidea* (Plate 3.2, Fig. 12; Plate 3.11, Figs. 9-10; Fig. 9-11a; Plate 3.10, Figs. 5-6). In *Spilanthes acmella,* however the tetrads are tetrahedral and isobilateral (Plate 3.4, Fig. 9).

In all the members investigated cytokinesis is simultaneous and wall formation takes place by furrowing (Plate 3.1, Figs. 10-11; Plate 3.11, Figs. 9.10; Plate 3.4, Fig. 9; Plate 3.8, Fig. 11).

The pollen grains have a large vacuole at the uninucleate stage with the nucleus pushed towards the periphery. This is often called the signet ring stage (Davis, 1961a). The pollen grains are tricoporate and have an echinate exine. They are shed at the three celed stage as in all other members of Asteraceae. The male gametes are circular in shape in *Blainvillea rhomboidea* (Plate 3.8, Fig. 13) but may be somewhat elliptical in others (Plate 3.2, Fig.15; Plate 3.11, Fig 13; Plate 3.4, Fig. 11).

Ovule, Megasporogenesis and Female gametophyte

In all the members, the ovary is inferior, bicarpellary, syncarpous with a single, basal, anatropous ovule. The ovule arises as a small hump of erect tissue from the base of the ovary. It (ovule) later becomes anatropous with the development of the single massive integument which grows over the nucellus until it reaches the funiculus. The ovules are tenuinucellate.

Usually a single hypodermal archesporial cell is differentiated in the nucellus even before the initiation of the integument. (Plate 3.5, Fig. 16-17; Plate 3.8, Fig. 13; Plate 3.2, Fig. 14, Plate 3.10, Fig. 9; Figs. 1-2). In *Splianthes acmella* and *S.paniculata* ocassionally two superposed archesporial cells may be seen (Plate 3.2, Fig. 18, Plate 3.5, Fig. 12). The archesporial cell directly functions as the megaspore mother cell (MMC) in all the members investigated. The MMC enlarges and attains its maximum size by the time the overgrowth of the integument is completed. In *Spilanthes paniculata* double megaspore mother cells have been noticed in some ovules. These may be superposed (Plate 3.2, Fig. 18) or juxtaposed (Plate 3.2, Fig. 20).

Megasporogenesis is initiated in the ovule only after it assumes the characteristic anatropous curvature. Meiosis is normal in all the members studied and results in the formation of a linear tetrad of megaspores (Plate 3.2, Fig. 21; Plate 3.5, Fig 15; Plate 3.8, Fig. 19; Plate 3.10, Fig, 11; Plate 3.12, Fig. 4). In *Spilanthes acmella,* however occasionally the second and third megaspores may be arranged laterally (Plate 3.5, Fig. 19). This is due to the laying down of a diagnal wall in the first meiotic division followed by a wall, perpendicular to the first in the second division.

Generally the chalazal megaspore is functional and the rest degenerate due to the enlargement of the chalazal megaspore and the unyielding nucellar epidermis (Plate 3.2, Figs. 22-24, Plate 3.5, Figs. 17-18; Plate 3.9, Fig. 23; Plate 3.12, Fig. 5). A number of variations however have been noticed in the tetrad in the following members. One of the ovules in *Spilanthes paniculata* hows an enlarged lower dyad with a prominent nucleus, while the upper dyad is almost crushed. It has two small nuclei. This lone instance may perhaps point out to an occasional bisporic development of the embryo sac (Plate 3.2, Fig.24).

In *Blainvillea rhomboidea,* the following variations are seen in the megasporogenesis.

1. Two micropylar megaspores degenerate and the two chalazal megaspores show signs of enlargement (Plate 3.8, Fig. 21).

2. Chalazal and micropylar megaspores enlarge while the two middle megaspores degenerate (Plate 3.8, Fig. 20).
3. Second micropylar megaspore degenerates while the rest show signs of enlargement (Plate 3.9, Fig. 22).

In spite of all these variations, in *Blainvillea rhomboidea,* no ovule has more than one embryo sac and it may be said that eventually only one megaspore survives and contributes to the embryo sac.
In *Synedrella nodiflora* a number of variations are noticed in megasporogenesis. The following are some of them.

(a) Instead of the chalazal megaspore, the micropylar one becomes functional and consequently the lower three degenerate (Plate 3.12, Fig. 6)
(b) Chalazal and micropylar megaspores enlarge and middle two degenerate (Plate 3.12, Fig. 7).
(c) All the megaspores except the second micropylar one show signs of enlargement (Plate 3.12, Fig. 8).

In some of the ovules of *Synedrella nodiflora,* twin tetrads, both juxtaposed and superposed have been noticed (Plate 3.12, Figs. 9, 10). In both these instances, the chalazal megaspore alone is functional.
The functional megaspore in all the members undergoes 3 mitotic divisions to produce an 8 nucleate embryo sac of the Polygonum type (Plate 3.2, Fig. 23; Plate 3.3, Figs. 1-3; Plate 3.5, Figs. 20-22; Plate 3.9, Figs. 24-25; Plate 3.10, Fig. 12; Plate 3.12, Figs. 11-12).

The embryo sac shows some variations in its organisation. In *Spilanthes acmella,* the organised embryo sac is six celled with only two antipodal cells. This may be formed in two ways.

1. Instead of the formation of three ununucleate antipodals at the eight nucleate stage, the chalazal antipodal will enclose two nuclei while the micropylar, will have only one nucleus. Sometimes the chalazal antipodal may be uninucleate while the micropylar may be binucleate (Plate 3.6, Figs. 30,38).
2. After the formation of four nuclei in the embryo sac, only one nucleus at the chalazal end participates in the division, so that a seven nucleate embryo sac with four nuclei at the micropylar end and three at the chalazal end is formed. When the embryo sac is organised, only two uninucleate antipodals are fotmed (Plate 3.5, Fig.23).

In some of the ovules (about 2%), the embryo sac is seven celled with three uninucleate antipodals.
In *Spilanthes paniculata,* the organised embryo sac is always six celled with two antipodal cells. The micropylar antipodal is always binucleate while the chalazal one is uninucleate (Plate 3.3, Fig. 4).

In *Blainvillea rhomboidea* also the organised embryo sac is six celled with 2 antipodal cells in which the micropylar antipodal cell is binucleate and the chalazal one uninucleate.

In *Synedrella nodiflora*, the organised embryo sac is seven celled with three antipodal cells having varying number of nuclei while in *Eclipta alba,* the organised embryo sac is seven celled and has three uninucleate antipodal cells (Plate 3.12, Fig. 12; Plate 3.10, Fig. 12).

The egg apparatus in all the members consists of two synergids (with no hooks, beaks etc.) and an egg cell. The polar nuclei fuse a little before fertilization is effected in the embryo sac.

ANTIPODALS

In all the members investigated, the antipodals donot degenerate soon after fertilization. In fact they persist till late in embryogeny and appear healthy indicating that they might play a nutritive role. In addition to this, a great variety is noticed in the nuclear behaviour of the antipodals. There is secondary multiplication and fusion resulting in polyploidy. A few instances are discussed below.

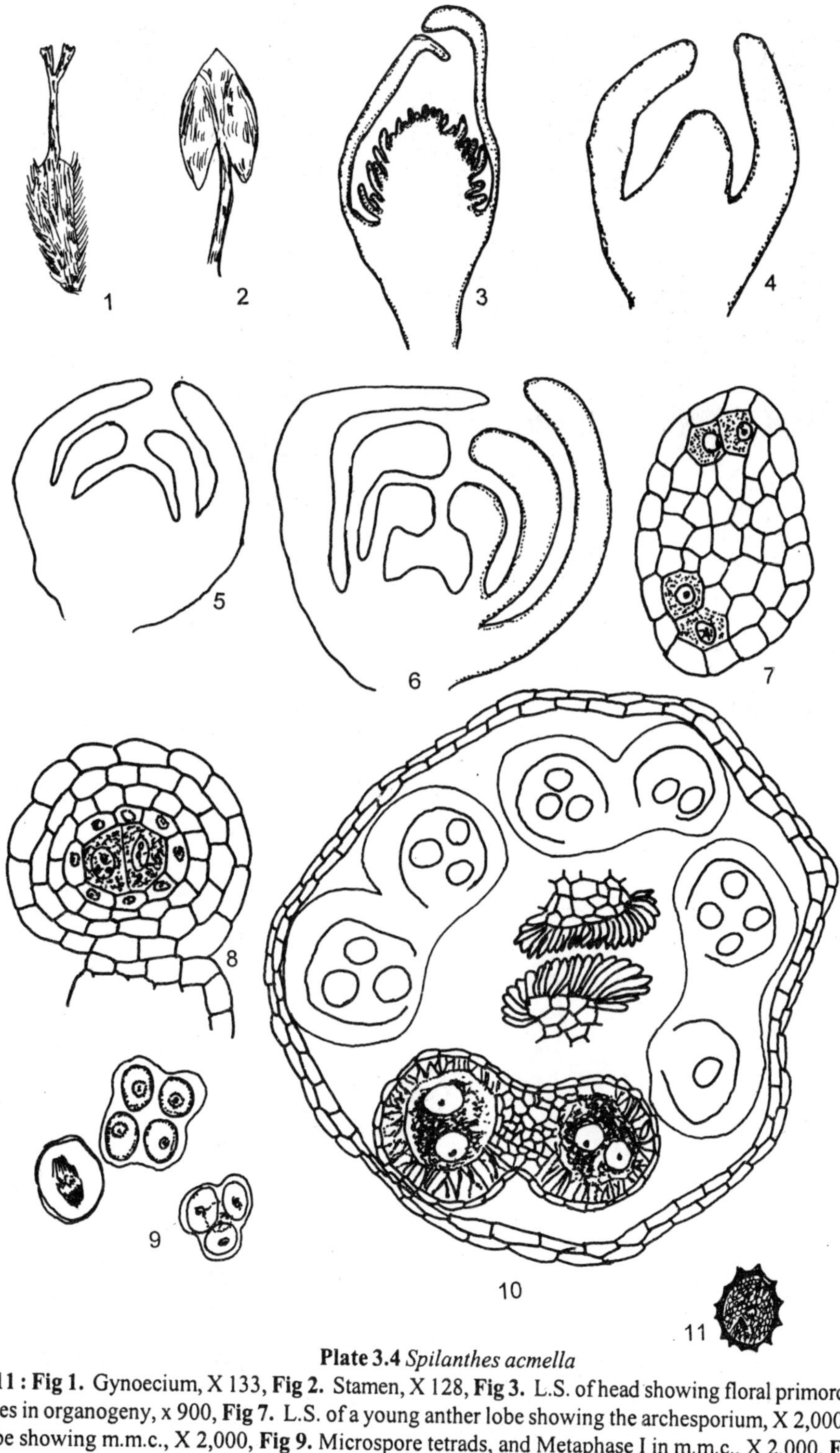

Plate 3.4 *Spilanthes acmella*

FIGS 1-11 : Fig 1. Gynoecium, X 133, **Fig 2.** Stamen, X 128, **Fig 3.** L.S. of head showing floral primordia, X 100. **Figs 4-6.** Stages in organogeny, x 900, **Fig 7.** L.S. of a young anther lobe showing the archesporium, X 2,000. **Fig 8.** T.S. of anther lobe showing m.m.c., X 2,000, **Fig 9.** Microspore tetrads, and Metaphase I in m.m.c., X 2,000. **Fig 10.** T.S. of a flower showing four bisporangiate anthers with plasmodial tapteum, X 900, **Fig 11.** A mature pollen grain, X 2,000.

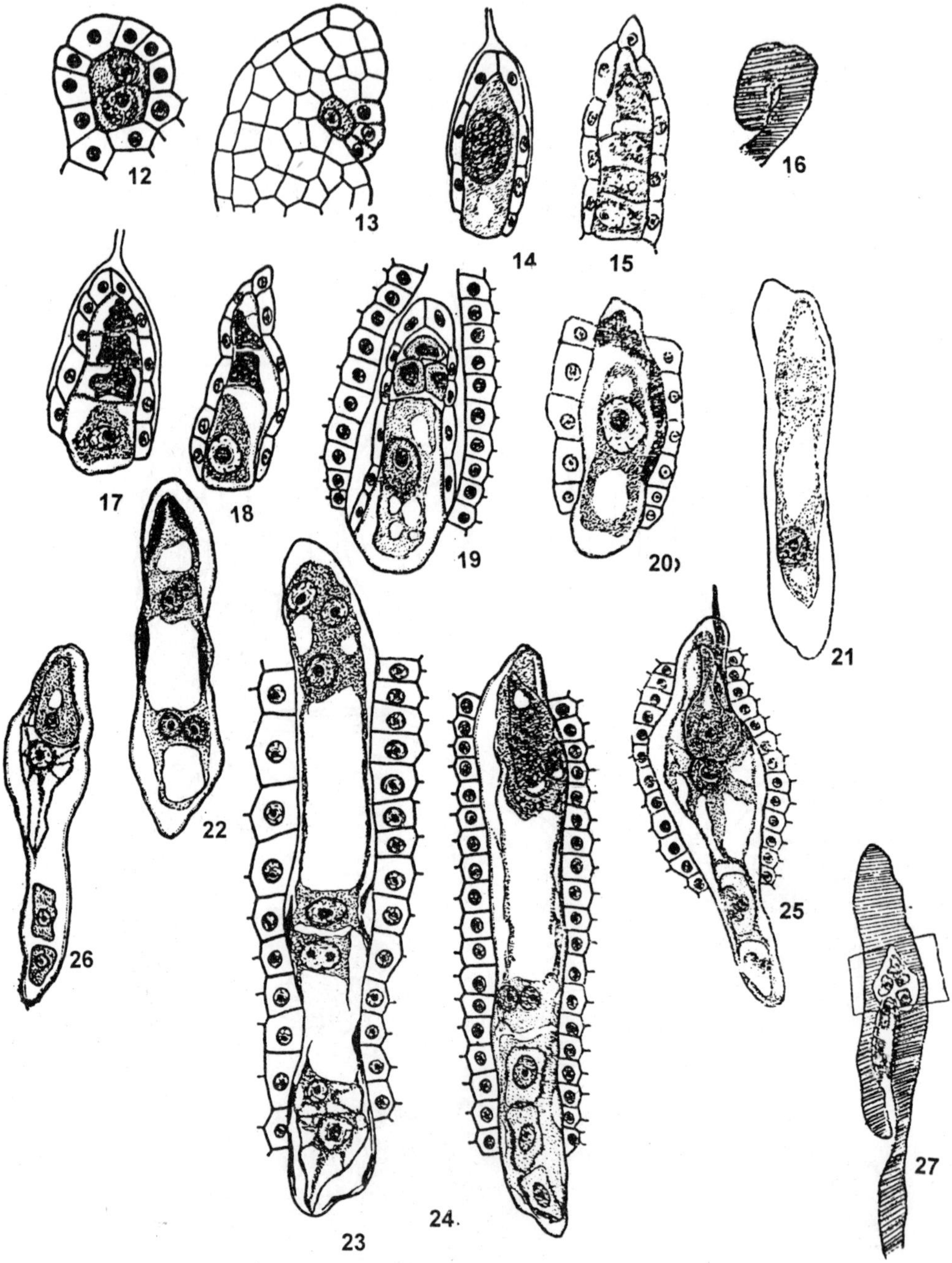

Plate 3.5 *Spilanthes acmella*

FIGS : 12-27. Fig 12. L.S. of an ovule showing two archesporial cells, X 2,000, **Fig 13.** Migaspore mother cell, X 2,000, **Fig 14.** Megaspore mother cell in prophase, X 2,000, **Fig 15.** Linear tetrad of megaspores, X 2,000, **Fig 16.** L.S. cvule showing degenerating megaspores (outline), X 200, **Fig 17.** Tetrad showing three degenerating spores, X, 2,000, **Fig 18.** Tetrad with two caps, X 2,000, **Fig 19.** Tetrad with two megaspores laterally arranged, X 2,000, **Fig 20.** Uninucleate embryosac, X 2,000, **Fig 21.** Two nucleate embryosac, X 2,000, **Fig 22.** Four nucleate embryosac, X 2,000, **Fig 23.** Seven nucleate embryosac, X 2,000 **Fig 24.** Organised embryosac with three antipodals, X 2,000, **Fig 25.** Embryosac with zygote, pollen tube and two antipodals, X 900, **Fig 26.** Emgryosac with zygote and two antipodals, X 900, **Fig 27.** L.S. of ovule showing the position of aposporic embryosacs (outline), X 200.

In *Spilanthes acmella,* the normal number of antipodals is two. This condition is metwith in about 95% of the ovules. In about 3%, the antipodals are three and in about 2%, the antipodals are four in number (Plate 3.6, Fig. 35). Regarding the number of nuclei in the antipodals, they are generally uninucleate when they are three or four in number. In one instance however, three antipodal cells with varying number of nuclei are met with (Plate 3.6, Fig. 34). Here the micropylar antipodal is longer than the other two and has a single large nucleus. Considering the (nucleus) size in comparison with the others, this appears to be polyploid produced as a result of nuclear multiplication and subsequent fusion. The middle and the chalazal antipodals have two nuclei each indicating the occurrence of division.

In instances where there are only two antipodal cells, in *Spilanthes acmella* a secondary multiplication of nuclei is noticed only in the micropylar antipodal. In the chalazal antipodal, there is neither multiplication nor fusion. It will be either uninucleate or binucleate. Even when two nuclei are present, these constitute the original component of eight nuclei. The proof fornuclear multiplication in the micropylar antipodal is obtained by the observation of anaphase stage in one of the embryo sacs (Plate 3.6, Fig. 30).

The number of nuclei in the micropylar antipodal varies fromone to six (A, Figs. 29, 32, 33 and 38). When only one nucleus is present, it could be due to two reasons; either only one nucleus is enclosed in the cell during the organisation of the embryo sac, the remaining two going to the chalazal antipodal or the nuclei in the micropylar antipodal fuse after multiplication forming a single polyploid nucleus.

In *Spilanthes paniculata,* the antipodals are always two in number and in all the instances the micropylar antipodal is binucleate and the chalazal one uninucleate. No instance of either multiplication or fusion is seen in the antipodal nuclei.

In *Blainveillea rhomboidea* also, the antipodals are always two in number (Plate 3.9, Fig. 26) with a binucleate micropylar cell and a uninucleate chalazal cell (Plate 3.9, Fig. 27). But there are variations in the number of nuclei in the antipodal cells. Both the antipodal cells may be uninucleate (Plate 3.9, Fig. 30) or both of them may be binucleate (Plate 3.9, Fig. 26). In some instances the chalazal cell is three nucleate (Plate 3.9, Fig. 29). Secondary multiplication of nuclei in the chalazal antipodal (Plate 3.9, Fig. 28) is not followed by cytokinesis which results in the multinucleate condition. Antipodals persist for a long time after fertilization.

In *Synedrella nodiflora;* normally there are three uninucleate antipodal cells (Plate 3.11, F.g 14). But there are a number of instances of variation in the number of cells and nuclei. The following are some of the variations.

(a) Three antipodal cells - with micropylar three nucleate, middle uninucleate and the chalazal four nucleate.
(b) Five uninucleate antipodal cells.
(c) Three antipodal cells; micropylar uninucleate, middle four nucleate, and chalazal three nucleate.
(d) Two nucleate antipodal cells (Plate 3.11, Figs. 18 and 19), where only nuclei can be seen.

In *Eclipta alba,* always there are only three uninucleate antipodal cells, which persist up to the quadrant stage of the embryo.

Endothelium

In all the members investigated, a layer of endothelium is distinguishable, which differentiates as early as the tetrad stage and remains till the complete formation of the endosperm. Thereafter the endothelial cells show signs of degeneration. The endothelial cells are always uninucleate.

Multiple embryo sacs

These are very rare and have been encountered only in *Synedrella nodiflora.* Here, in one of the ovules, three embryo sacs have been seen. The upper most embryo sac is organised with an egg apparatus, two polar nuclei and three antipodal nuclei, the middle embryo sac is two nucleate while the chalazal embryo sac is uninucleate. (Plate 3.12, Fig. 13). These embryo sacs are possibly the result of three megaspores of a tetrad developing simultaneously of which there are evidences (Plate 3.12, Fig. 8).

Fertilization

It is porogamous in all the members studied. In *Spilanthes paniculata,* one of the synergids gets crushed due to the entry of the pollen tube (Plate 3.3, Fig. 6). The same is true in *Synedrella nodiflora* (Plate 3.11, Fig. 14). In *Blaninvillea rhombodiea,* the remnants of the pollen tube remain up to the first division of the primary endosperm nucleus (Plate 3.9, Fig. 27). Here, one of the synergids also persists till the first division of the primary endosperm nucleus.

In *Spilanthes acmella,* the pollen tube persists even after the formation of the preombryo and six celled endosperm (Plate 3.7, Fig. 41). Here double fertilization and syngamy may be almost simultaneous (Plate 3.6, Fig. 36), or rarely the former may be delayed. In some cases the polar nuclei remain free even after the entry of the pollen tube (Plate 3.6, Fig. 37).

In *Eclipta alba,* soon after the entry of the pollen tube, both the synergids disappear (Plate 3.10, Fig. 13).

ENDOSPERM

The triploid primary endosperm nucleus (PEN) is generally close to the zygote. The first division of the PEN is followed by wall formation in *Spilanthes acmella, S.paniculata, Blanivillea rhomboidea* and *Eclipta alba.* Consequently in these members, the endosperm may be designated as the *ab initio* cellular type (Plate 3.6, Fig. 38; Plate 3.2, Fig. 5. 5; Plate 3.9, Fig. 27; Plate 3.10, Fig 14). In *Synedrella nodiflora* however the first division of the PEN is not followed by wall formation. Several free nuclear divisions take place before the cell wall formation sets in and consequently the endosperm is of the free nuclear type (Plate 3.11, Fig. 17)

There is variation in the time of cell wall formation in the endosperm in *Synedrella nodiflora.* While in a majority of the ovules cell walls are laid down in the nuclear endosperm at the globular stage of the embryo, in some ovules cell walls are laid down as early as the three celled (Plate 3.11, Fig. 18) or Asterad (Plate 3.11, Fig. 19) stage of the embryo. Eventually the endosperm becomes cellular in all the ovules.

In taxa where the endosperm is *ab initio* cellular, the first division of the PEN is at right angles to the long axis of the embryo sac resulting in the formation of two superposed cells. The micropylar cell usually divides earlier than the chalazal cell and the plane of division is parallel to thelong axis of the embryo sac as seen in *Spilanthes acmella* (Plate 3.6, Fig. 38). Subsequent divisions are not regular and ultimately a cellular endosperm completely fils the embryo sac except for the chalazal portion where the antipodals remain intact (Plate 3.1, Figs. 7-8; Plate 3.11, Figs. 18-19; Plate 3.7, Figs. 40-41; Plate 3.9, Figs. 27; Plate 3.12, Fig. 15).

The endosperm is consumed by the developing embryo and all that remains of the endosperm in the mature achene is but one or two layers of thin walled cells adhering to the margin of the embryo sac.

EMBRYOGENY

In all the species studied, embryogeny is uniform and it follows the *Senecio* variation of the Asterad type in Johansen's (1950) system of classification and grand Period I, Megarche., type II, series A, subseries A_2, in the first group according to the system of embryogeny of Soueges (1948).

The zygote normally undergoes a period of rest. It divides usually after the formation of a few endosperm cells. Sometimes the zygote may undergo division soon after the first division of the PEN (Plate 3.6, Fig. 38).

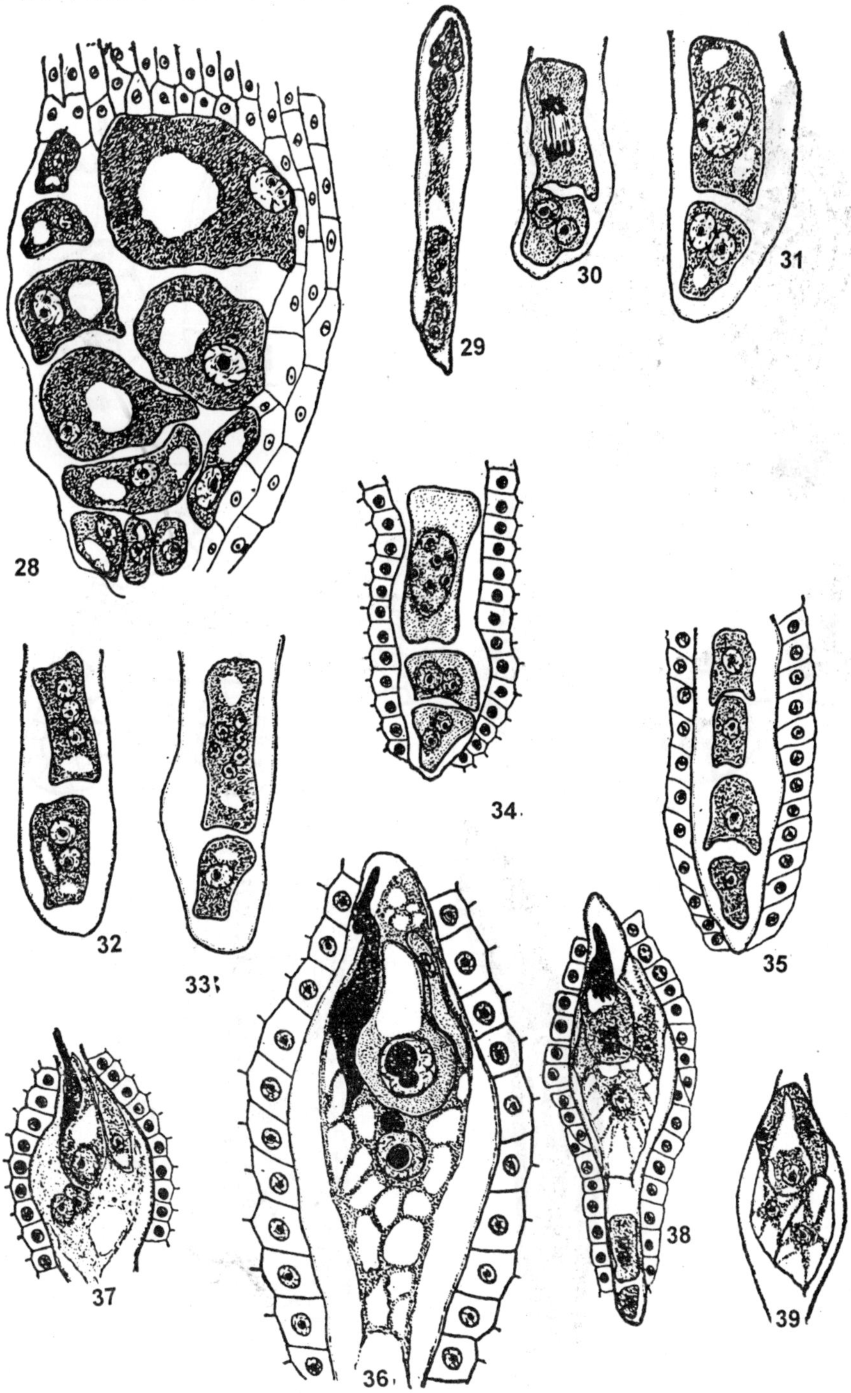

Plate 3.6 *Spilanthes acmella*

FIGS : 28-39. Fig 28. Aposporic embryosacs (enlarged), X 2,000, **Fig 29.** Organised embryosac with two antipodals, X 900, **Fig 30.** Antipodals enlarged with the upper one in division, X 2,000, **Figs 31-33.** Two antipodals with different number nuclei, X 2,000, **Fig 34.** Three antipodals, X 2,000, **Fig 35.** Four antipodals, X 2,000, **Fig 36.** Syngamy and double fertilisation, X 2,000, **Fig 37.** Zygote, pollen tube and two polar nuclei, X 900, **Fig 38.** Embryosac with 2 celled endosperm, pollen tube, persistent antipodals and zygote in metaphase, X 900, **Fig 39.** Second division of the endosperm, X 900.

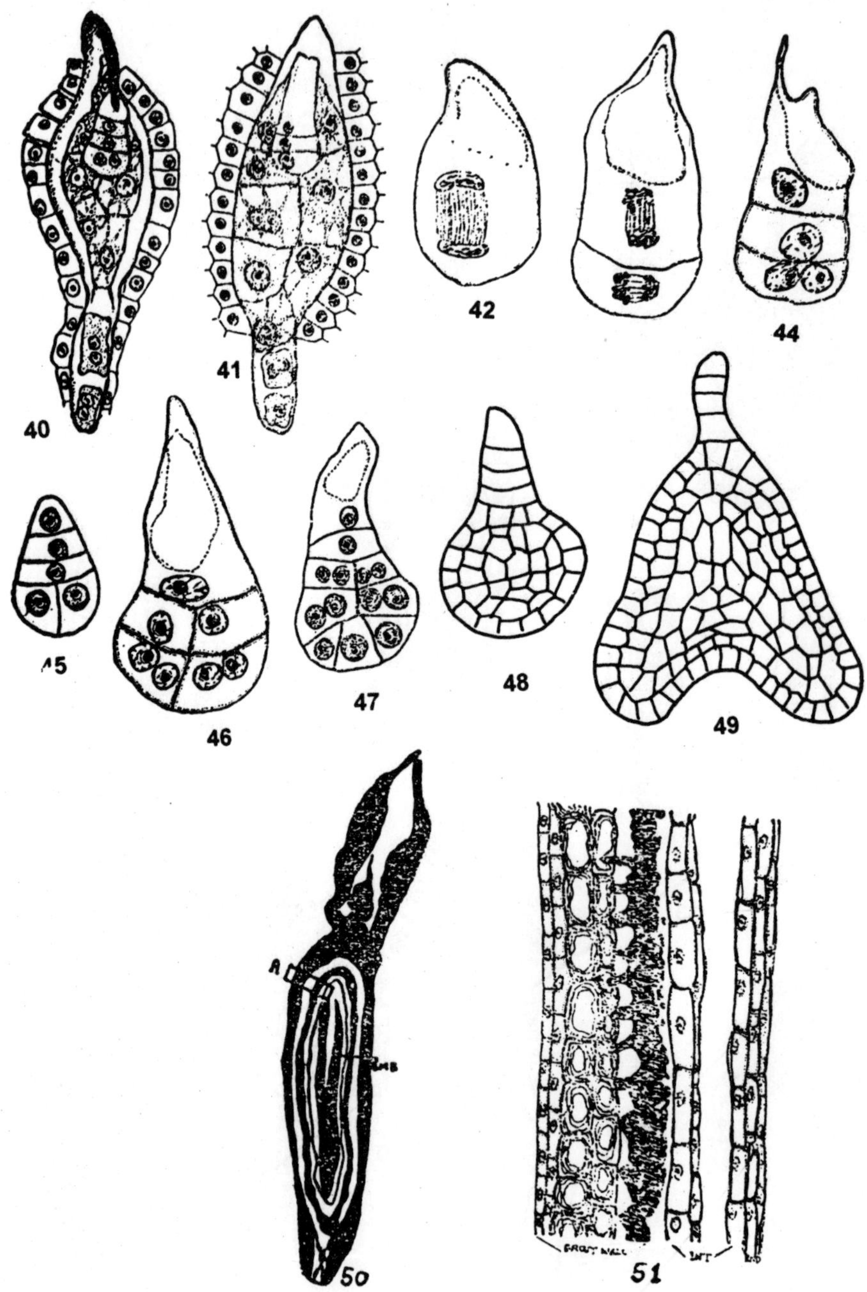

Plate 3.7- *Spilanthes acmella*

FIGS 40-51 : Fig 40. Embryosac with six-celled endosperm and proembryo, X 900, **Fig 41.** Embryosac with eight-celled endosperm, X 900, **Figs 42-49.** Embryogeny. (**Figs. 42, 43, 44, 46**, X 2,000; **Figs 45, 47-49.** X 900), **Fig 50.** LS of fruit, X 100, **Fig 51.** Portion marked A in Fig 50. magnified to show the structural details, X 900.

In one rare instance in *Synedrella nodiflora,* the division of the zygote and PEN is simultaneous (Plate 3.11, Fig. 16). The zygote divides transversely and forms two superposed cells - the terminal *ca* and basal, *cb, ca* divides vertically and *cb* divides transversely forming a 'T' shaped proembryonal tetrad consisting of two juxtaposed cells derived from *ca* and two superposed cells derived from *cb.* The derivates of *cb* are called *m* and *ci* (Plate 3.3, Fig. 9; Plate 3.11, Fig. 22; Plate 3.7, Fig. 43-44; Plate 3.9, Fig. 32). In the next stage, all the four cells divide again but in different planes. The two cells of *ca* divide by diagonal walls to form the quadrant tier *q,* while *m* divides vertically and forms two cells. The cell *ci* divides transversely to form two cells *n* and *n'*. The stage of the embryo is called the quadrant stage (Plate 3.3, Figs 10-11; Plate 3.11, Figs. 23-25; Plate 3.7, Fig. 46). The cells of the tier *q* divide again by diagnal walls to form the octant. Meanwhile the two cells of *m* divide again vertically to form four cells, which lie above the octants. By now *n* also divides by a vertical wall to form two juxtaposed cells, while *n'* divides transversely to form cells *o* and *p* (Plate 3.3, Figs. 12-13; Plate 3.11, Figs. 26-27; Plate 3.7, Fig. 47). In the next stage, tangential walls are laid down in the tier *q* and *m* to cut off an outer layer of dermatogen. The inner cells divide further in all the planes to give rise to the central plerome and surrounding periblem (Plate 3.3, Figs. 14-15; Plate 3.11, Fig. 28; Plate 3.7, Fig. 48). During further development the cell *p* divides transversely to give rise to a suspensor of variable number of cells; *o* divides vertically once or twice and gives rise to the dermatogen and cap of the root; *m* gives rise to the hypocotyledonary region and *q* to the cotyledons and stem tip. (Plate 3.3; Figs. 16-18; Plate 3.11, Figs. 29-30; Plate 3.7, Fig. 49; Plate 3.9, Fig. 31). The following is a schematic representation of the sequence of development of the embryo.

	ca — *q* —	Cotyledons and stem tip
Zygote	*cb* — *m* —	Hypocotyle and radicle
	ci — *n* —	Remaining part of root tip
	n — *o* —	Root cap
	p	Suspensor

The number of suspensor cells varies in different members. It is 6 in *Spilanthes paniculata* (Plate 3.3, Fig. 18), 4 in *S.acmella* (Plate 3.7, Fig. 48) and *Synedrella nodiflora* (Plate 3.11, Fig. 30), 5 in *Blainvillea rhomboidea* (Plate 3.9, Fig. 33) and 3 in *Eclipta alba* (Plate 3.10, Fig. 17).

There are some minor variations in some members in the sequence of development given above. In *Spilanthes paniculata, m* divides rapidly and cuts off several cells even before the divisions of cells in the tier *q* (Plate 3.3, Fig. 13). Dermatogen is cut off in the central cells of tier *q* and later it proceeds to the marginal cells. The cell *n* divides belatedly and forms two cells by vertical division only after the dermatogen is cut off in tiers *q* and *m.*

In *Spilanthes acmella,* the initial vertical divisions of *m* and *n* are almost simultaneous and further, *cb,* divides early forming *m, n* and *n'* even before the formation of the tier *q*.
In *Synedrella nodiflora,* in some ovules (Plate 3.11, Fig. 17) a linear proembryo of four cells is formed. This is due to the non division of *ca,* while *cb* has divided into *m, n* and n'.
The above account of description of embryogeny is according to the system of Johansen (195). The following is an account of the schematic representation of the origin and histogeneic functions of the proembryonal cells according to the system of Souges (1920).

1. *First cell generation* : Prombryo consists of two cells disposed in two tiers.
   - *ca* = *pco* + *pvt*
   - *cb* = *phy* + *icc* + *co* + *s*
2. *Second cell generation* : Proembryo consists of four cells disposed in three tiers.
   - *q* = *pco* + *pvt*
   - *m* = *phy* + *icc*

$ci = iec + co + s.$

3. *Third cell generation* : Proembryo consists of eight cells disposed in four tiers.

$q = pco + pvt$

$m = Phy + icc$

$n = iec$

$n = co + s$

4. *Fourth cell generation* : Peoembryo consists of 16 cells disposed in five tiers.

$q = pco + pvt$

$m = phy + icc$

$n = iec$

$o = co$

$p = s$

Mature embryo

In all the members (Plate 3.3, Fig. 18; Plate 3.11, Fig. 31; Plate 3.7, Fig. 50; Plate 3.9, Fig. 33; Plate 3.10, Fig. 18) studied, the mature embryo is straight and dicotyledonous and is usually surrounded by one or two layers of endosperm cells.

Apospory

In only one instance, in *Spilanthes acmella,* some of the nucellar cells at the chalazal region of the embryo sac (Plate 3.6, Fig. 28) are enlarged considerably with prominent nuclei and dense cytoplasm. They also show vacuolation. But in no instance, their development proceeds beyong the uninucleate stage.

STRUCTURE OF THE FRUIT WALL

In *Spilanthes acmella,* the ovary wall from the beginning is five layered with a few variations occasionally. The ovary wall up to and a little after fertilization remains uniform in having one layer of epidermis, below which is a layer of broad cells and three layers of thin walled elongated cells.

After fertilization, certain changes take place in the wall layers, though their number remains uniform throughout. The epidermis and a layer of cells below remain as thin walled elongated cells, while two layers of cells below become very broad, prominent and often prismatic. The innermost layer of cells remains as a thin walled layer until late in embryogeny. But at the stage of the globular embryo and later there is a heavy deposition of dark material which probably is tannin. This often looks as though protruding into the layer of broad cells (Plate 3.7, Figs. 50-51) with finger like projections.

In the wall of the mature fruit, thus there are 5 layers of cells. The integument in the young ovule is multilayered and at the stage of the tetrad itself, an endothelium gets differentiated all around the embryo sac. In the mature fruit, the seed coat consists of two layers of thin walled cells constituting the outer zone, while in the middle zone, there is an empty space as the cells there would have been consumed by the endothelium. In the mature fruit even the endothelium is disorganised. Internal to this, will be a layer or two of endosperm cells. In *Spilanthes paniculata,* in the mature fruit, the fruit wall is made up of only two layers of cells (Plate 3.3, Fig 19) i.e. epidermis and a hypodermal layer. The epidermis consists of rhomboidal cells with a thick outer wall. The inner layer also has rhomboidal cells and has the deposition of a dark material which assumes irregular shape. But it does not have the finger like projections seen in *Spilanthes acmella.*

The integument in the mature fruit has a layer of rectangular cells followed by a thin layer of elongated cells. Then there is an empty space representing the middle zone of the integument consumed by the endothelium.

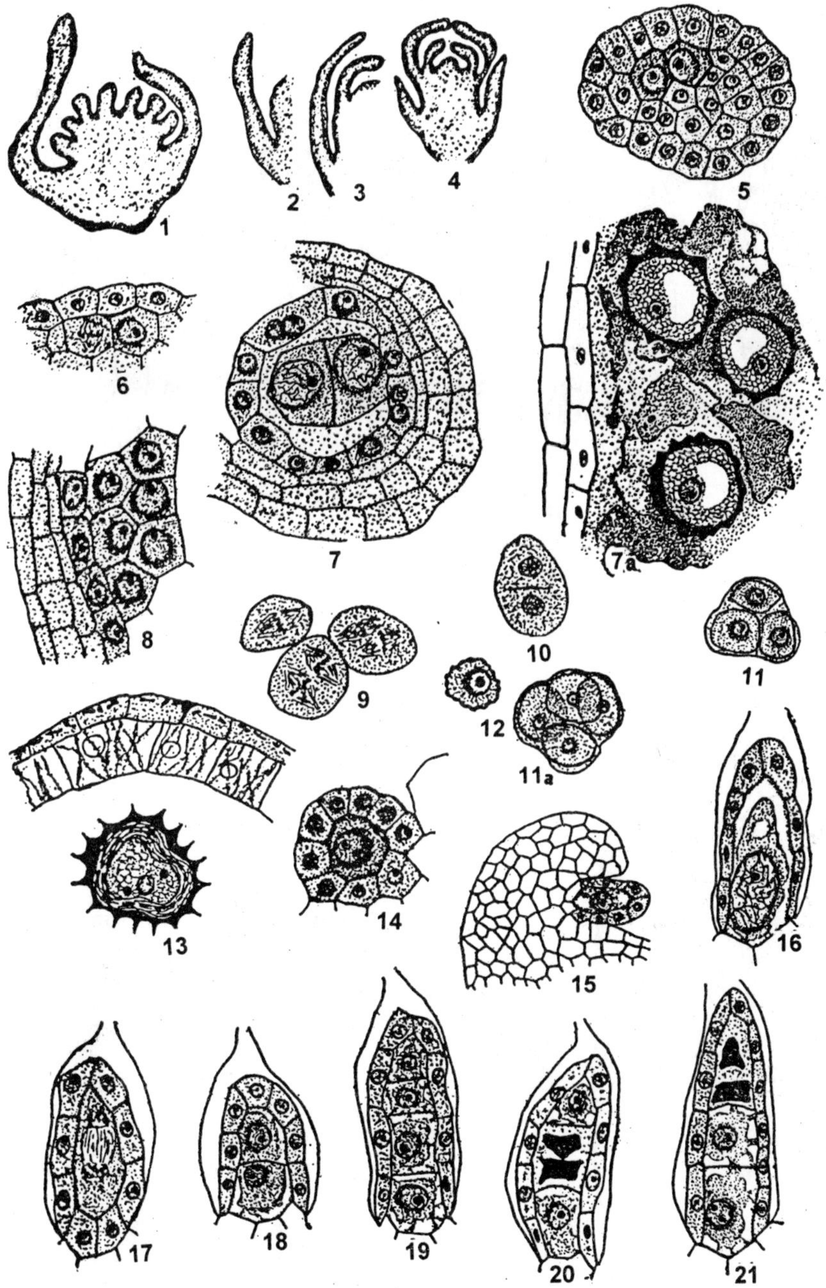

Plate 3.8 *Blain villea rhomboidea*

FIGS : 1-21 : Fig 1. L.S. of young head showing floral primordia, **Figs 2-4.** Organogeny, 5. T.S. of young anther, **Fig 6.** Periolinal division of anchesporial cell, **Figs 7-8.** T.S. and L.S. of fully formed anthers respectively, **Fig 7a.** L.S. of anther sac showing tapetal periplasmodium, **Fig 9-10.** Microsporogenesis, **Fig 11-11a.** Microspore tetrads, **Fig 12.** Microspore, **Fig 13.** Endothecium and 3 celled pollen grain, **Fig 14.** L.S. of ovule showing archesporium, **Fig 15-16.** Megaspore mother cell, **Fig 17-18.** Dyad formation and Dyad, **Fig 19-21.** Tetrads.

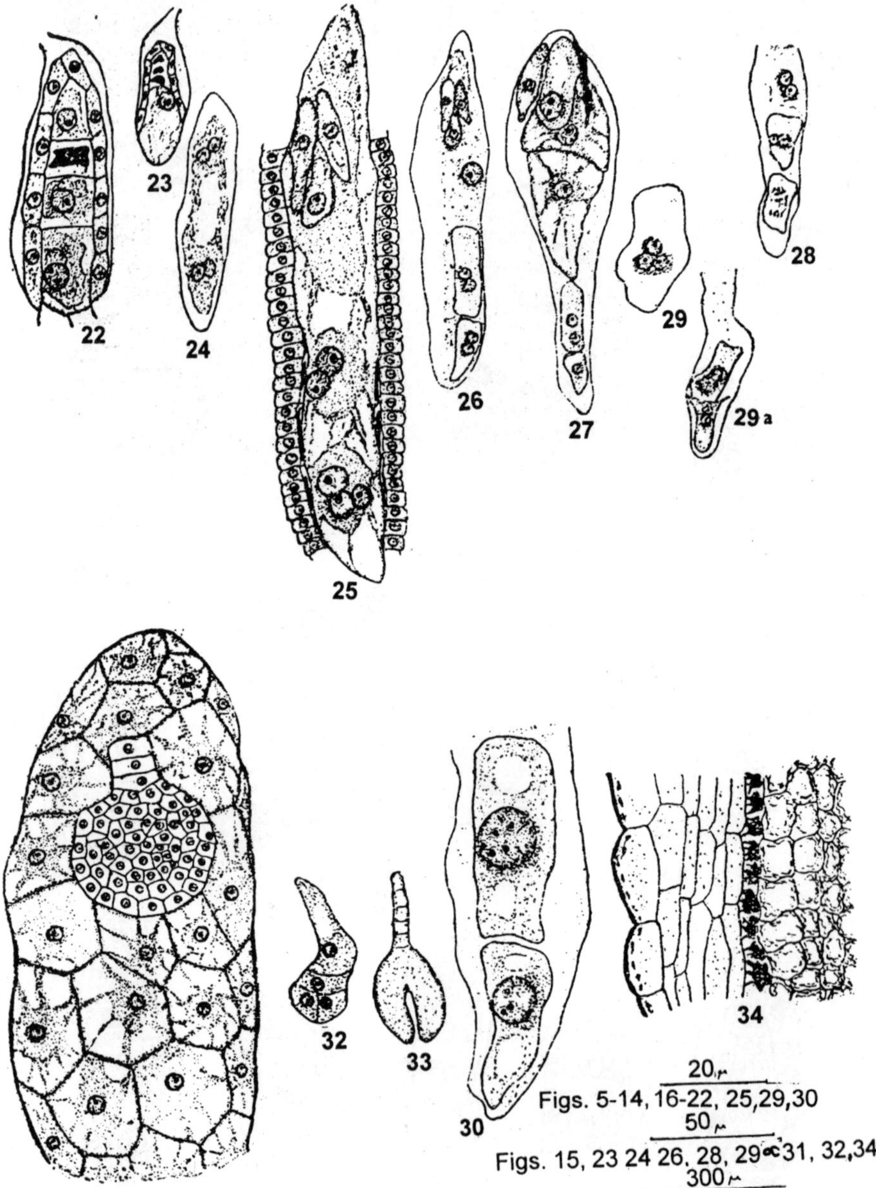

Plate 3.9 *Blain villea r\homboidea*

FIGS 22-34 : Figs 22-23. Tetrads, **Fig 24.** Four nucleate embryo sac, **Fig 25-26.** Organised embryo sacs, **Fig 27.** First division of primary endosperm nucleus, **Fig 28.** Division of antipodal nucleus, **Fig 29.** Chalazal ntipodal with 3 nuclei, **Fig 29a.** Nuclear fusion in the micropylar antipodal, **Fig 30.** Two uninucleate antipodals, **Fig 31.** Embryo sac with globular embryo and cellular endosperm, **Fig 32-33** Embryologeny, **Fig 34.** A portion of the fruit wall magnified.

Internal to the empty space is the endothelium enclosing 2 or 3 thin layers of endosperm tissue.
In *Blainvillea rhomboidea* (Plate 3.9, Fig. 34), the fruit wall is multilayered. The epidermis is made up of elongated polygonal cells with cutinized outer wall. Internal to the epidermis, the fruit wall may be distinguished into three zones. The outer zone is made up of five layers of vertically elongated thin walled cells. The middle zone is made up of a single layer of cells in which are found dark depositions with finger shaped projections much in the same way as in *Spilanthes acmella.* The inner zone is made up of 3-4 layers of isodiametric cells which are vacuolated and shiny. The integument in the mature fruit is similar to what is seen in the genus *Spilanthes.*

In *Synedrella nodiflora,* the fruit wall in the mature achene consists of many layers of cells divided into three zones. The outer zone made up of epidermis consists of a single layer of isodiametric cells. The middle zone again is made up of a single layer of hypodermis, but has some dark deposition. The inner zone is made up of 5-6 layers of thin walled elongated cells.

The integument in the mature fruit has a single layer of cells towards the interior constituting the endothelium. The endothelial cells show signs of degeneration. Abutting the endothelium internally is a layer of endosperm cells. External to the endothelium is an empty space representing the layers of integumentary cells consumed by the endothelium. The outer zone of integument consists of a layer of epidermis and 4-5 layers of elongated cells (Plate 3.11, Fig. 32).

SENECIONEAE

Organogeny

In all the members studied, the young head arises as a hump of meristematic tissue on which arise floral primordia.

In *Emilia sonchiflia,* the floral parts arise in acropetalous succession. In *Senecio tenucifolius,* corolla arises first followed by androecium, gynoecium and lastly the calyx makes its appearence (Plate 3.18, Figs. 1-4).

In *Senecio ludens,* there is variation in the development of the floral parts. In all the flowers, corolla arises first (Plate 3.15, Fig. 1). In some cases corolla is followed by androecium and then gynoecium appears (Plate 3.15, Figs. 2-3). But in some flowers gynoecium appears after corolla and then androecium follows (Plate 3.15, Figs. 4-5). In all the instances however calyx is the last to appear (Plate 3.15, Figs 6-9).

Development of the anther

Anthers are tetrasporangiate in *Senecio ludens* and *S.tenuifolius,* (Plate 3.15, Fig. 10; Plate 3.18, Fig. 5;), but bisporangiate in *Emilia sonchifolia.*

The young anther assumes a two lobed contour in *Emilia sonchifolia,* while it is four lobed in *S.ludens* and *S.tenuifolius.* A single archesporial cell arises in each lobe of the tetrasporangiate anther in *S.ludens* and *S.tenuifolius* (Plate 3.15, Fig. 11; Plate 3.18, Fig. 6). In *Emilia sonchifolia* also, each lobe of the bisporangiate anther has a single hypodermal archesporial cell.

The development of the anther conforms to the Dicotyledonous type (Plate 3.13, Figs. 6-8; Plate 3.15, Figs. 12-15; Plate 3.18, Figs. 7-9). The developmental process is similar to what has been described for the tribe Heliantheae.

In both *Senecio tenuifolius* and *S.ludens,* of the four sporangia of the anther, the peripheral two are always larger in size than the ones towards the centre of the flower.

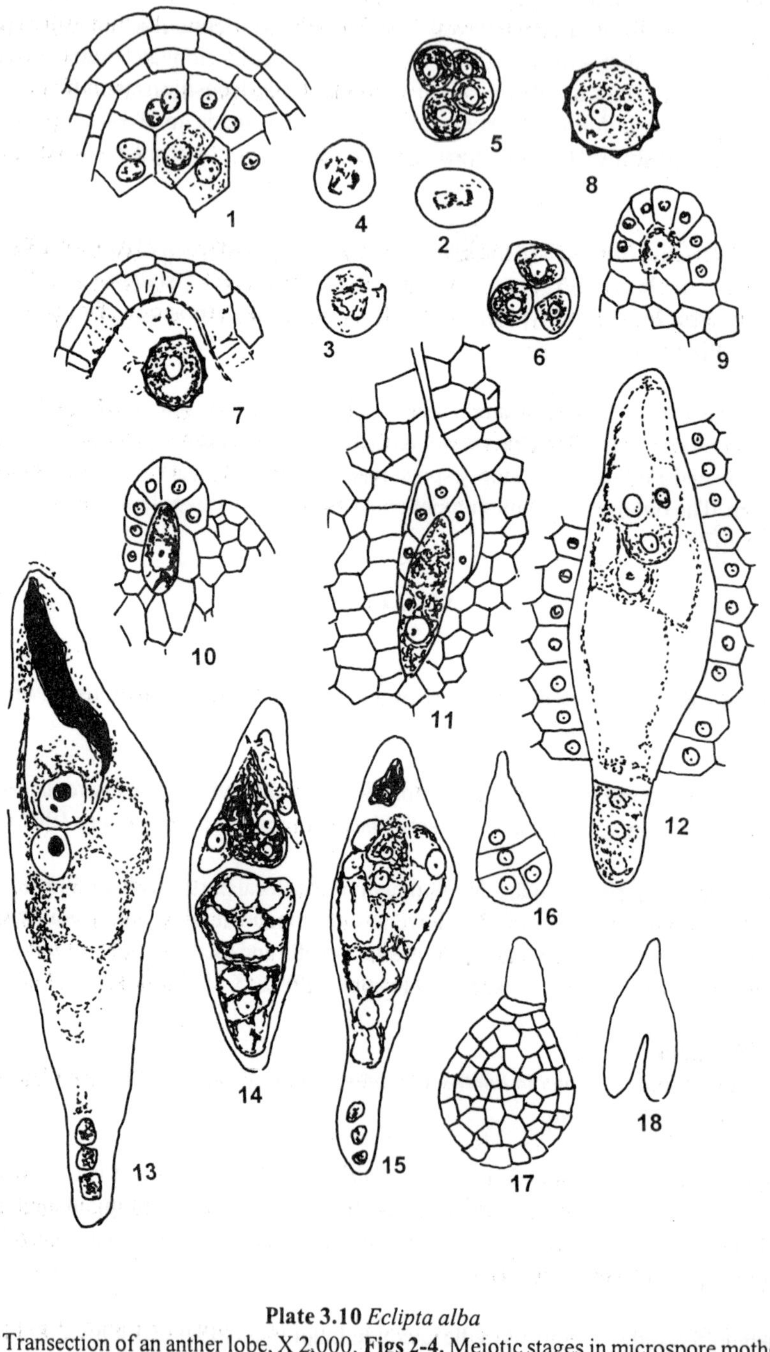

Plate 3.10 *Eclipta alba*

FIGS 1-18 : Fig 1. Transection of an anther lobe, X 2,000, **Figs 2-4.** Meiotic stages in microspore mother cells, X 2,000, **Figs 5-6.** Decussate and tetrahedral tetrads of microspores, X 2,000, **Fig 7.** T.S. of anther showing plasmodial, tapetum and fibrillar endothecium, X 900, **Figs 9-10.** Archesporium and megaspore mother cell respectively, X 2,000, **Fig 11.** Linear tetrad of megaspores, X 2,000, **Fig 12.** Organised embryo sac, megaspores, X 2,000, **Fig 13.** Embryo sac showing fertilisation, X 2,000, **Figs 14-15.** Early divisions of zygote and endosperm, X 900, **Figs 16-18.** Embryogeny, (Fig. 16, X 2,000; Fig. 17, X 900; Fig. 18, X 200).

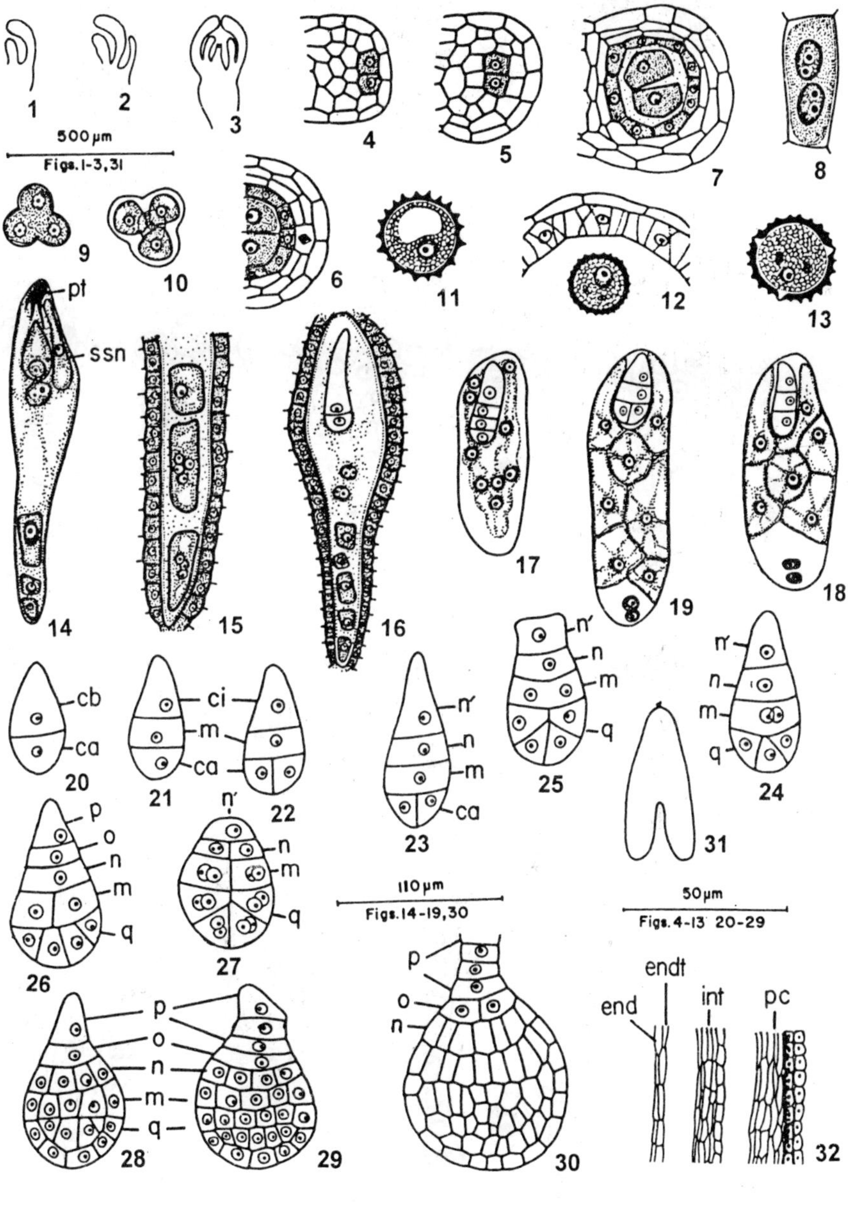

Plate No. 3.11 : *Synedrella nodiflora* (Explanation to Figures)

Figs : 1-3 Organogeny, **4-7** Transections of anthers showing development of microsporangium, **8** A binucleate tapetal cell, **9-10** Cytokinesis and microspore tetrads, **11** Uninucleate pollen grain, **12** Fibrillar endothecium, **13** 3 celled pollen grain, **14** Embryo sac with zygote, pollen tube, PEN and three uninucleate antipodal cells, **15** Lower part of the embryo sac showing variation in antipodal nuclei, **16** Two celled proembryo and first division of the PEN, **17** Linear proembryo with free nuclear endosperm, **19** Asterad embryo with cellular endosperm, **20-30** Embryogeny, **31** Dicot embryo (outline) **32** Structure of the fruit wall in long section.

(pt = pollen tube, end = endosperm, endt = endothelium, int = integument, pc = pericarp)

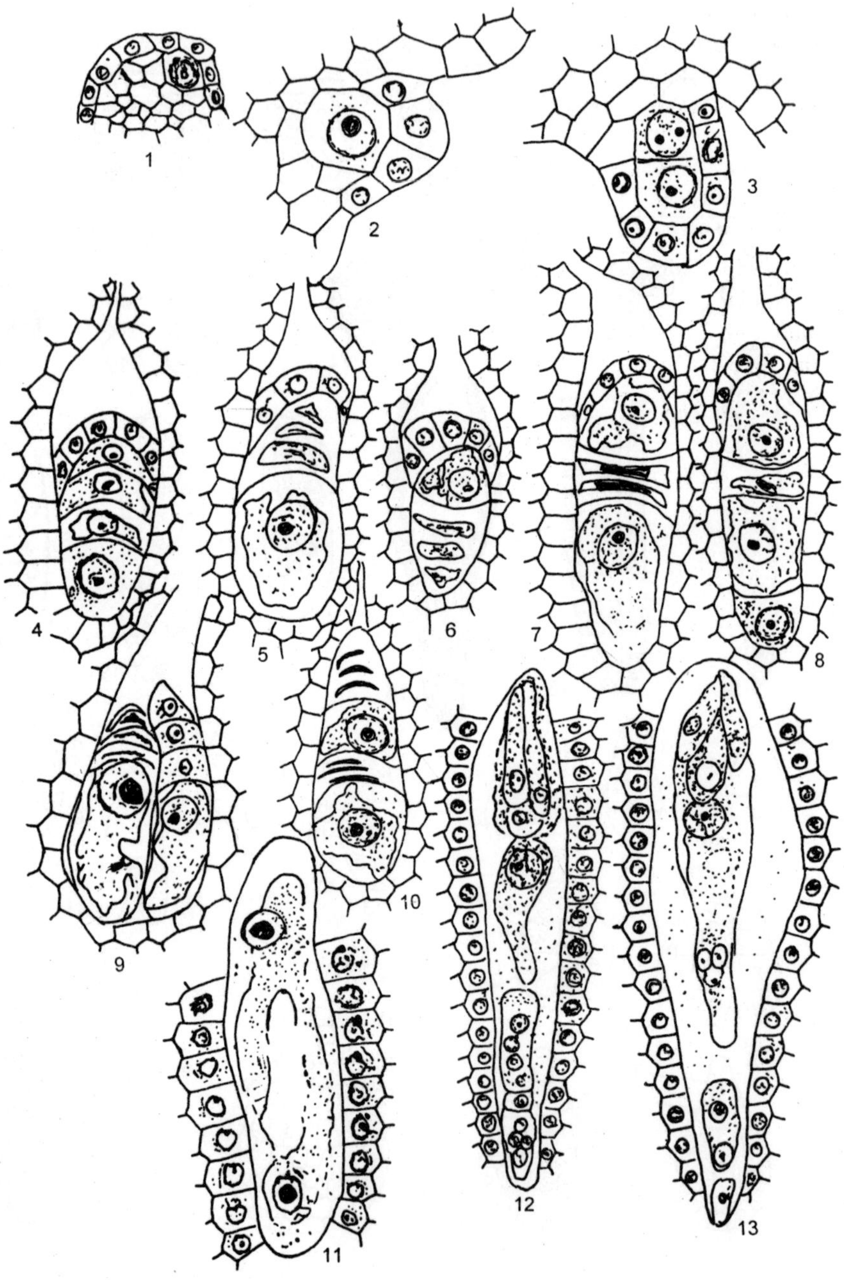

Plate 3.12 *Synedrella nodiflora*

FIGS 1-13 : Fig 1. L.S. of an ovule showing hypodermal archesporial cell, X 600. **Figs 2-3.** L.S. of ovule showing megaspore mother cell and dyad respectively, X 600, **Fig 4.** Linear tetrad of megaspores, X 600, **Fig 5.** Linear tetrad showing functional chalazal megaspore, X 600, **Fig 6.** Linear tetrad showing functional micropylar megaspore, X 600, **Fig 7.** Linear tetrad showing functional chalazal and micropylar megaspores, X 600, **Fig 8.** Linear tetrad showing the development of three megaspores, X 600, **Fig 9.** L.S. of ovule showing juxtaposed twin tetrads, X 600, **Fig 10.** L.S. of ovule showing superposed twin tetrads, X 600, **Fig 11.** Two-nucleate embryo-sac X 600, **Fig 12.** Organised embryo-sac, X 270, **Fig 13.** L.S. of ovule showing three embryo-sacs, X 270.

A transverse section of a fully developed anther in all the three taxa consists of an epidermis, an endothecium, a middle layer and a layer of tapetum surrounding two microscope mother cells representing two longitudinal rows (Plate 3.13, Fig. 8; Plate 3.15; Figs. 14-15; Plate 3.18, Fig. 9).

Tapetum

In all the members investigated, the tapetum is of the periplasmodial type. The tapetal cells are uninucleate to start with but later become binucleate (Plate 3.12, Fig. 15; Plate 3.16, Fig. 12; Plate 3.18, Fig. 16;). In *S.tenuifolius,* the tapetal nuclei may fuse resulting in a polyploid condition (Plate 3.18, Fig. 17). In *S.ludens,* three nucleate tapetal cells are seen (Plate 3.15, Fig. 18).

In *S.tenuifolius* and *S.ludens,* the tapetal breakdown occurs after the liberation of the microspores from the tetrad and the appearence of the exine. In *Emilia sonchifolia* however, the plasmodium is short lived. The tapetal cells remain intact even after the formation of uninucleate pollen grains. The tapetal breakdown begins only after the exine is laid over the intine. By the time the pollen grains reach the two nucleate stage, the plasmodium disappears. In this taxon, the plasmodium does not appear to contribute to the formation of the exine.

The middle layer gets crushed very soon in *Senecio tenuifolius* even before the initiation of microsporogenesis (Plate 3.18, Fig. 9). In the other two taxa, the middle layer disappears during microsporogenesis.
The epidermal cells get radially extended in the mature anther. While the endothecium exhibits fibrillar thickenings during anther dehiscence in *Senecio tenuifolius* and *S.ludens,* in *Emilia sonchifolia,* no such fibrillar thickenings are noticed.

Microsporogenesis and male gametophyte

In all the three taxa investigated, there are only two rows of microspore mother cells in the anther. These are produced as a result of one division in the row of sporogenous cells. No cytomixis is observed in any taxon of the present work.

At the beginning of meiosis, the microspore mother cells separate from one another and round off. They undergo reduction division to produce tetrahedral and decussate tetrads (Plate 3.13, Fig. 10-11; Plate 3.15, Fig. 20-22; Plate 3.18, Figs. 14-15). In all the members, cytokinesis is simultaneous and wall formation takes place by furrowing (Plate 3.13, Fig. 9; Plate 3.15, Fig. 19; Plate 3.18, Fig. 13).

The pollen grains are tricolporate and three celled at the time of shedding in all the three taxa (Plate 3.14. Fig. 4, Plate 3.15, Fig. 24; Plate 3.18, Fig. 20). They have a smooth intine and an echinate exine. The male gametes are circular in shape in *Emilia sonchifolia* but slightly elliptical in *Senecio ludens* and *S.tenuifolius.* In *S.ludens,* in some of the anthers pollen grains are dimorphic. In one or two instances double pollen grains (Plate 3.15, Fig. 25), have been observed. This is due to the failure of second meiotic division with the consequent non separation of the two cells. These two cells remain attached and develop the characteristic exine.

Megasporogenesis and Female Gametophyte

The structure of the ovary and ovule is similar to what is described for the tribe Heliantheae.
Usually a single hypodermal archesporial cell is differentiated when the ovule is still erect and the integument, yet to originate (Plate 3.15, Fig. 27; Plate 3.28, Fig. 21;). But in some ovules of *Senecio ludens,* multiple archesporium has been seen (Plate 3.15, Figs. 2,8). The presence of multiple archesporium may be inferred by the presence of double megaspore mother cell in *Senecio tenuifolius* (Plate 3.24, Figs. 23-24;).

In all the three taxa, the archesporial cell directly functions as the megaspore mother cell (MMC) and consequently the ovule is tenuinucellate. Multiple MMCs are seen both in *Senecio ludens* (Plate 3.16, Figs. 1-2) and *S.tenuifolius* (Plate 3.18, Figs. 23-2). In *Emilia sonchifolia,* always a single MMC is encountered. After the ovule becomes fully antropous, megasporogenesis is initiated. The MMC undeergoes reduction division to produce a linear tetrad of megaspores (Plate 3.16, Fig. 7; Plate 3.18, Fig. 25; E, Fig. 5).

Generally the chalazal megaspore is functional and the rest degenerate (Plate 3.16, Fig. 9; Plate 3.18, Fig. 26; Plate 3.14, Fig.7). But variations are noticed in several instances.

In *Senecio ludens,* in a few ovules two megaspores show signs of development (Plate 3.16, Fig. 8), or the micropylar megaspore may become functional and the rest degenerate (Plate 3.16, Fig. 12). In one ovule, it has been observed that the second micropylar megaspore has developed into a two nucleate embryo sac; chalazal megaspore is healthy, while the micropylar megaspore and the second chalazal megaspore have degenerated (Plate 3.16, Fig. 11).

In *Senecio tenuifolius,* whenever there is a single tetrad in the ovule, only the chalazal megaspore is functional, except for a lone instance, where the two micropylar megaspores are still healthy, when the chalazal megaspore has already enlarged and undergone vacuolation.

Twin tetrads are encountered in both *Senecio ludens* and *S.tenuifolius.* In *S.ludens,* in a few ovules superposed twin tetrads are noticed. Here the megaspores have different arrangements in the two tetrads (Plate 3.16, Fig. 10). In the lower tetrad the spores are arranged in 'T' shape, while in the upper tetrad the two middle megaspores are arranged laterally.

Superposed twin tetrads are also noticed in *S.tenuifolius.* Here in one instance, all the spores of the lower tetrad show signs of degeneration (due to the presence of an aposporic cell), while all the megaspores in the upper tetrad are healthy (Plate 3.18, Fig. 27). In another instance, in the lower tetrad, the chalazal megaspore has undergone vacuolation, with other megaspores having degenerated, while in the upper tetrad, the chalazal megaspore is enlarging with the rest of the spores showing indications of degeneration.
The functional megaspore in all the three taxa divides three times mitotically to produce an eight nucleate embryo sac of the Polygonum type.

Structure of the organised embryo sac

In *Emilia sonchifolia,* the organised embryo sac is seven celled. The egg apparatus consists of a median egg and two synergids. The polar nuclei fuse at the time of the embryo sac organisation, and the secondary nucleus is close to the egg apparatus. The antipodals are three in number, two of the antipodals are laterally arranged, while the lower one is elongated and top shaped (Plate 3.14, Figs. 8-10).

In *Senecio ludens,* the organised embryo sac is seven celled and is similar in structure to that of *Emilia sonchifoliam* except that all the three antipodals are of the same size and placed one above the other (Plate 3.16, Fig. 18). In one of the ovules of *S.ludens,* the egg apparatus has a peculiar disposition. The egg has descended halfway down the length of the embryo sac. One of the synergids is near the micropyle in its normal position, while the other one is above the egg (Plate 3.16, Fig. 19). In this embryo sac there are three antipodals of which the lowermost is binucleate, while the other two are uninucleate.

In *Senecio tenuifolius,* the organised embryo sac is similar to what is described for the other two taxa except that the antipodals may be two or three in number (more details about the antipodals is described later).

The description of megasporogenesis and the development of gametophyte given above is found in a large majority of ovules in *Senecio ludens, S. tenuifolius* and *Emilia sonchifolia.* But in some of the ovules in all the three taxa, some unusual features are noticed. These are described below.

1. MMC haustorium

In an isolated instance in *Senecio ludens,* one of the ovules exhibits, what may be called a megaspore mother cell (MMC) haustorium. (Plate 3.16, Figs 5-6;)

In this ovule, the nucellar epidermis is torn with the cell (MMC) coming out of micropyle in the form of a sac and digesting a portion of the funiculus, confirming the haustorial nature of the MMC. The nucleus of MMC however is retained within the nucellar epidermis. In this ovule, there are three nucellar cells at the chalazal region showing signs of enlargement. This is an indication of somatic apospory.

APOSPORY

Senecio ludens and *S.tenuifolius* appear to be facultative apomicts. Two types of apospory viz., gonial and somatic is found in both the species.

In *Senecio ludens,* In a number of instances, 2-3 nucellar cells show signs of enlargement in the chalazal region. These may be differentiated as early as the MMC stage. In one ovule, two such aposporic cells have been seen at the chalazal region when the MMC is in the metaphase stage of meiosis I (Plate 3.16, Figs. 3-4;). Eventually the aposporic cells enlarge and show vacuolation and remain at the chalazal end of the normal embryo sac (Plate 3.16, Figs. 14 and 17). In no instance however, the aposporic embryo sacs proceed beyong the uninucleate stage.

Occurrence of gonial apospory in *S.ludens* (is noticed in one or two instances, where the embryo sacs have developed up to two and four nucleate (Plate 3.16, Figs. 14 and 16) stages. In the four nucleate aposporic embryo sac, the nuclei are distributed in 1 + 1 + 1 + 1 arrangement, while in the normal embryo sac the nuclei show 2 + 2 arrangement.

Senecio tenuifolius shows a number of instances of both gonial and somatic apospory, the latter occurring more frequenlty than the former.

In several ovules, the MMC enlarges (Plate 3.19, Fig.10;) and without undergoing meiotic divisions directly develops into a uninucleate embryo sac. The degenerated nucellar epidermis, size of the nucleus and vacuolation provide proof of the direct development of the MMC. A four nucleate embryo sac has been noticed in some ovules which is derived from an unreduced MMC (Plate 3.19, Fig. 13).
Somatic apospory is encountered in *Senecio tenuifolius,* in about 2 to 3% of the ovules. Somatic apospory here may be categorized into two :

(a) Non invasive and
(b) Invasive.

In non invasive aposory, the aposporic cells (one or more) enlarge and develop into unreduced embryo sacs without disturbing the sexual embryo sac located in the nucellar lobe.

Both can be seen developing in the ovule. Aposporic embryo sacs can be easily distinguished from the normal ones as the former always arise in the chalazal region of the ovule.

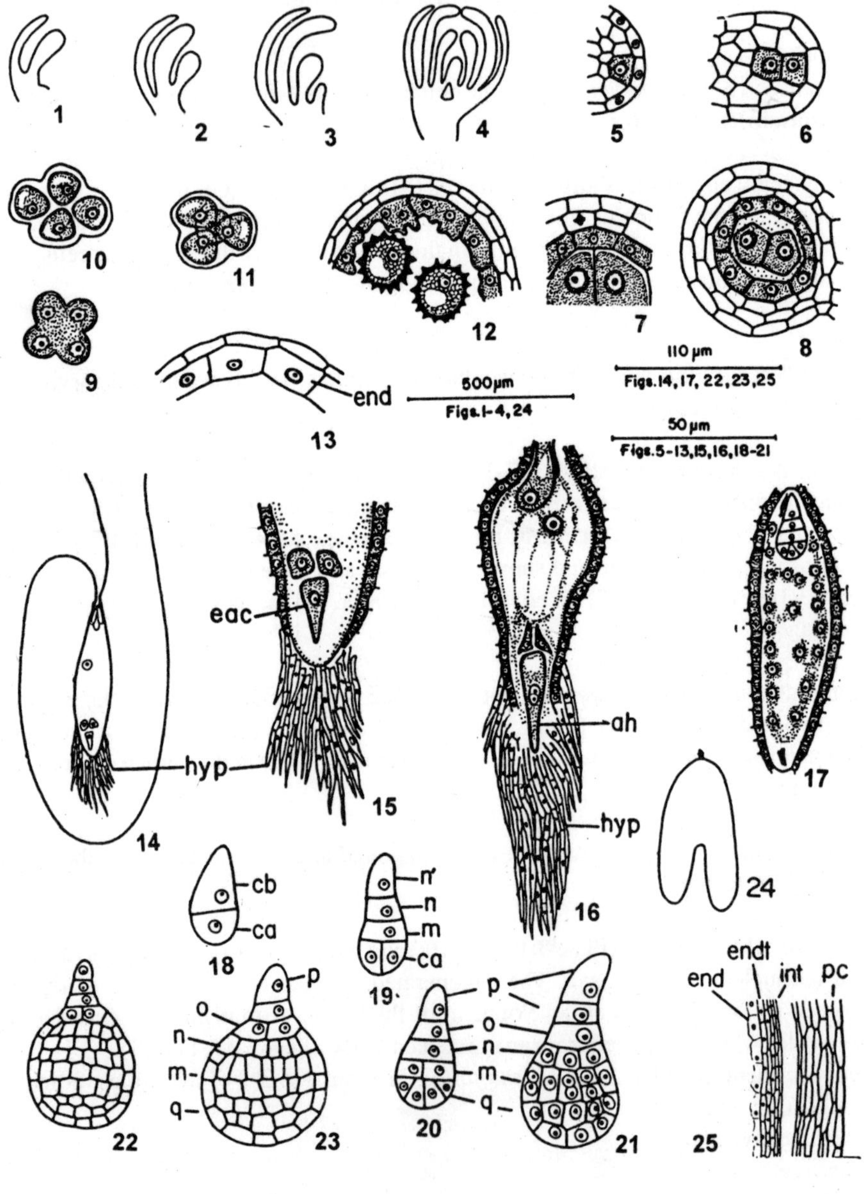

Plate No. 3.13 : *Emilia sonchifolia* (Explanation to Figures)

Figs : 1-4 Organogeny, **5-8** Development of the microsporangium, **9** Cytokinesis, **10-11** Microspore tetrads, **12** T.S. of anther showing tapteum (periplasmodial formation has just begun), **13** Endothecium showing absence of fibrillar thickenings, **14** L.S. of ovule showing organised embryo sac and hypostase, **15** Same as in Fig. 14 ; showing an enlarged view, **16** Embryo sac showing antipodal haustorium and hypostase, **17** Young embryo (quadrant) and free nuclear endosperm, **18-23** Embryogeny, **24** Dicotyledonous embryo (outline), **25** Structure of the fruit wall (in L.S.)

(hyp – hypostase, eac = enlarged antipodal cell, ah = antipodal haustorium, end = endosperm, endt = endothelium, int = integument, pc = pericarp)

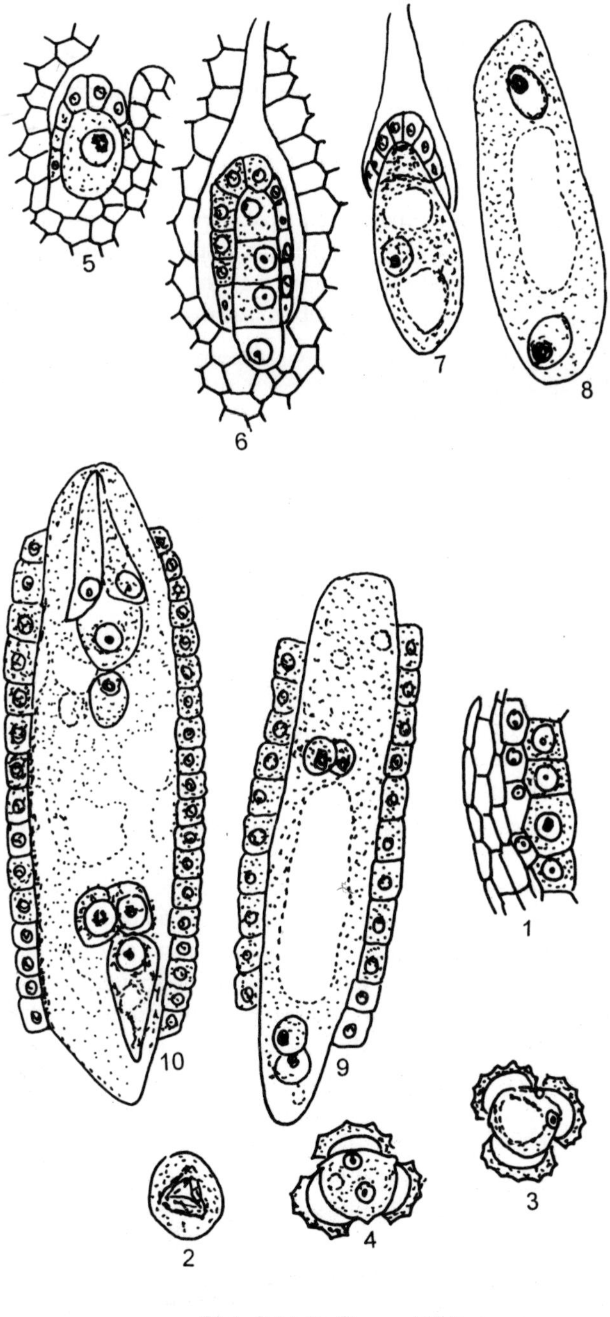

Plate 3.14 *Emilia sonchifolia*

FIGS 1-10 : Fig 1. L.S. of a young anther lobe, X 666, **Fig 2.** Arrangement of the spindles during the formation of microspores to whow tetrahedral arrangement, X 666, **Fig 3.** Uninucleate pollen grain, X 666, **Fig 4.** Two-celled pollen grain, X 666, **Fig 5.** L.S. of ovule showing megaspore mother cell, X 666, **Fig 6.** Linear tetrad of megaspores, X 666, **Fig 7.** Linear tetrad of megaspores showing functional megaspore and degenerating megaspores, X 666, **Fig 8-9.** Two and four-nucleate embryo sacs, X 666, **Fig 10.** Organised embryo-sac X 666.

In invasive apospory, one or two chalazal cells immediately below the MMC enlarge and one of them (or rarely both) protrude(s) into the MMC or tetrad and undergoes enlargement. In the process, the sexual cells get destroyed and the aposporic cell occupies the normal position of the embryo sac. In such instances, unless the initial stages of invasion are traced, nothing seems to be irregular for the aposporic cell divides in the normal course and develops into an embryo sac. However it will be unreduced and all the components are diploid.

In the non invasive type of apospory in *S.tenuifolius,* the aposporic embryo sacs develop up to two nucleate stage only, though they enlarge occupying considerable space and destroying the nucellar tissue in the ovule (Plate 3.19, Fig. 12; Plate 3.20, Figs. 3, 4, 12 and 51;). In one instance, while the normal embryo sac is four nucleate, the aposporic embryo sac is two nucleate, with the wall in between the two having been dissolved completely. At this stage, the aposporic two nucleate embryo sac may easily be mistaken for the haustorial outgrowth of the normal embryo sac (Plate 3.20, Fig. 4).

In another instance, the aposporic cell at the base of a four nucleate embryo sac appears like a caecum as its (aposporic cell) nucleus has been disorganised (Plate 3.20, Fig. 2).

The invasive type of apospory has been traced in detail in *S. tenufolius.* Instances of aposporic cell invasion have been observed at several stages from MMC up to tetrad. The various instances of invasion are described below.

1. A single aposporic cell protrudes into the cavity of the nucellar epidermis and the MMC shows signs of disintegration (Plate 3.29, Figs. 1-2;).
2. One or two cells invade the normal tetrad (Plate 3.19, Fig. 7;). The tetrad shows degeneration of megaspores (Plate 3.19, Figs. 3 and 5), due to such invasion. In one instance, the aposporic cell has developed into a two nucleate embryo sac within the cavity of the nucellar epidermis destroying two chalazal megaspores, while the micropylar megaspores are still surviving (Plate 3.19, Fig. 8). In another instance, two aposporic cells have protruded into the cavity and remnants of megaspores are visible (Plate 3.19, Fig.6).

In a very interesting case of invasion, a two nucleate aposporic embryo sac has deeply invaded the embryo sac cavity over the tetrad in which perhaps the chalazal megaspore had become functional with the upper three spores undergoing degeneration. The invasion of the aposporic cell has destroyed the functional chalazal megaspore also. The position of the two nuclei of the aposporic embryo sac clearly indicates the invasion. The aposporic cell having invaded has enlarged up to the micropyle and the two nuclei are clearly away from the chalazal region of the embryo sac which houses the tetrad (Plate 3.20, Fig. 1;). The original position of the chalazal cell is indicated by what looks like a caecum at the chalazal region of the embryo sac.

Hypostase

This is the first report for the occurrence of hypostase in Asteraceae. In *Emilia sonchifolia,* a group of cells at the chalazal region of the embryo sac can be distinguished by their deep staining character from the remainder of the nucellar cells (Plate 3.13, Figs. 15-16;). They have dense cytoplasm and appear fibrous in being narrow with tapering end walls. The hypostase gets organised in the ovule, a little before the organisation of the embryo sac and remains till late in embrygogeny.

Endothelium

In all the three species, the innermost layer of the integument (after the disorganisation of the nucellar epidermis) comes into direct contact with the embryo sac. It can be distinguished from the other cell layers by their dense cytoplasmic content and prominent nuclei.

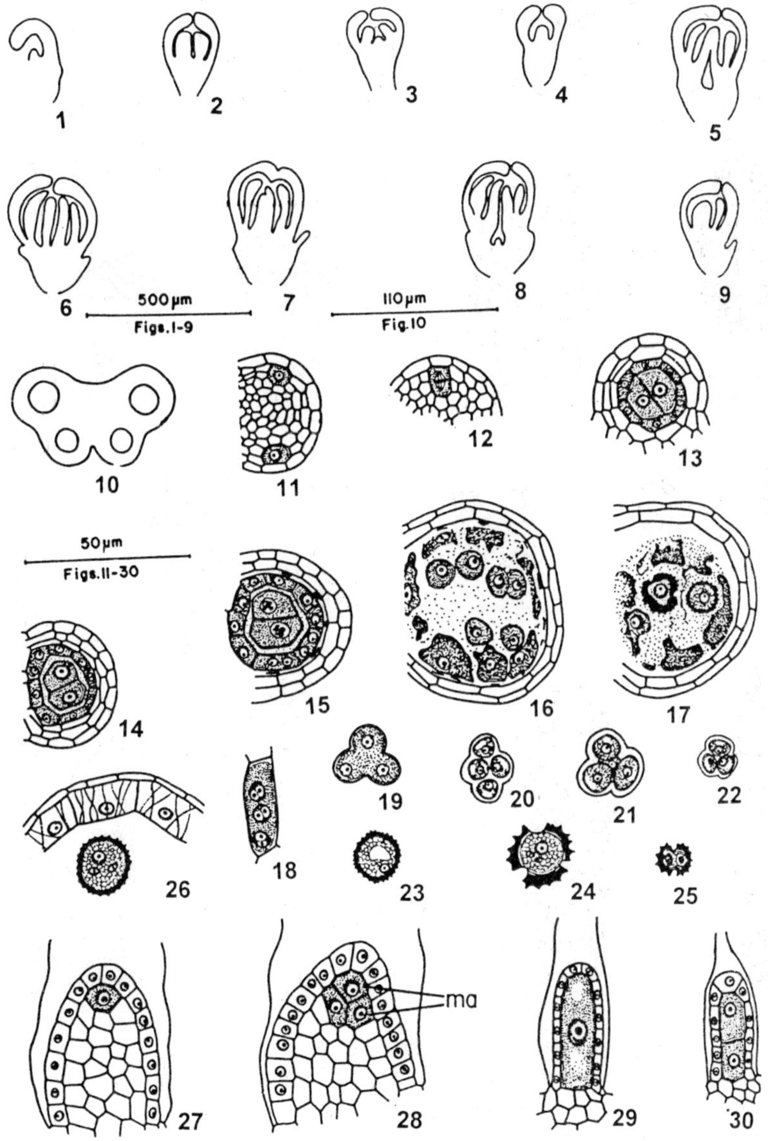

Plate No. 3.15 : *Senecio ludens* (Explanation to Figures)

Figs : 1-9 Organogeny (for details see text), **10** T.S. of young anther showing tetrasporangiate condition, **11-15** Development of microsporangium (T.S. of anthers), **19** T.S. of anthers showing tapetal periplasmodium, **20-22** Microspore tetrads, **23** Uninucleate pollen grain, **24** 3 celled pollen grain, **25** Abnormal pollen grain, **26** T.S. of mature anther showing fibrillar endothecium, **27** L.S. of ovule with a single archesporial cell, **28** L.S. of ovule showing multiple archesporial cells, **29** Megaspore Mother cell, **30** Dayed,

(ma = multiple archespoiral cells)

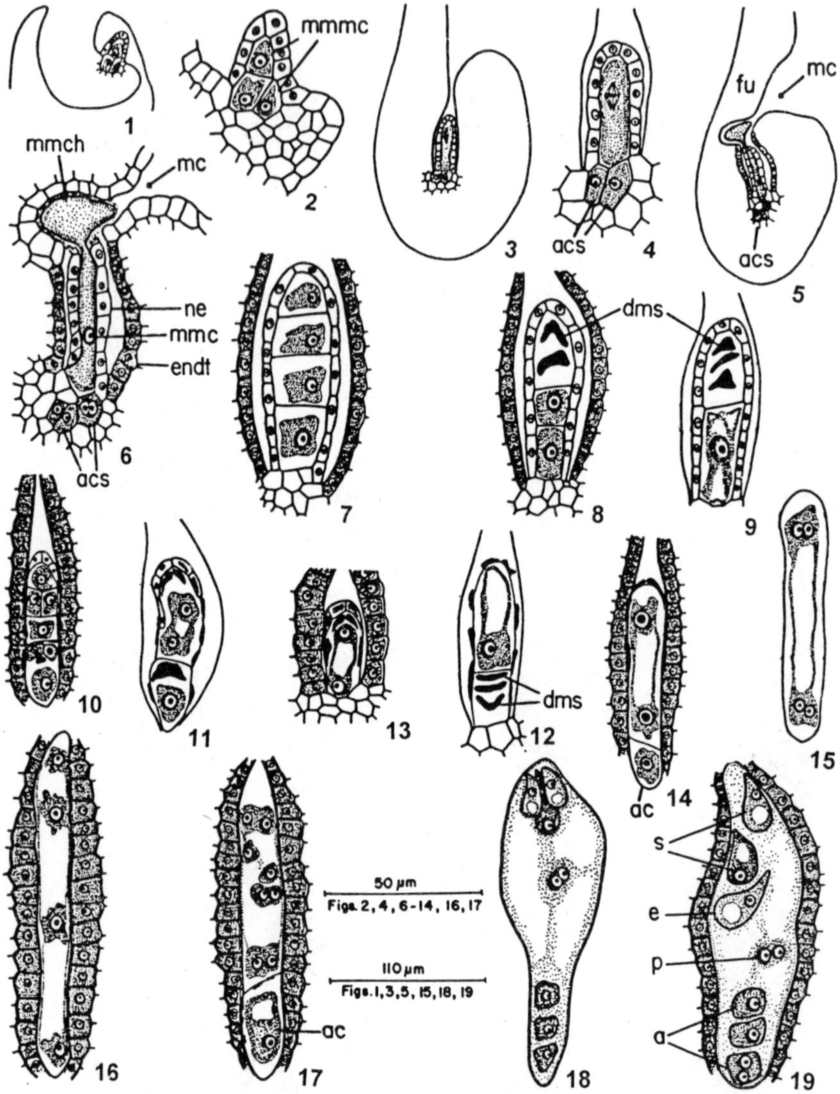

Plate No. 3.16 : *Senecio ludens* (Explanation of Figures)

Figs : 1 Ovule showing multiple megaspore mother cells, **2** L.S. of ovule (outline) showing MMC in division and two aposporic cells, **3** L.S. of ovule (outline) showing MMC in division and two aposporic cells, **4** Same as in Fig. 3; enlarged, showing details, **5** L.S. of ovule showing MMC haustorium and two aposporic cells, **6** Same as in Fig. 5; enlarged view, **7** Linear tetrad of megaspores, **8-9** Tetrads showing degenerating megaspores, **10** Twin superposed tetrads; lower tetrad "T" shaped, **11** Tetrad with functional micropylar megaspore (developed into a 2 nucleate embryo sac), **12** Tetrad showing functional micropylar megaspore, **13** Two nucleate embryo sac, **14** Two nucleate embryo sac with an aposporic cell, **15** Four nucleate embryo sac (normal), **16** Four nucleate embryo sac (apospoic), **17** Eight nucleate embryo sac with an aposporic cell, **18** Organised embryo sac, **19** Organised embryo sac with peculiar disposition of the egg apparatus.

(mc = micropyle, mmme = multiple megaspore mother cells, mmch = megaspore mother cell haustorium, ne = nucellar epidermis, mmc = megaspore mother cell, endt = endothelium, acs = aposporic cells, ac = aposporic cell, dms = degenerating megaspores, fu = funiculus)

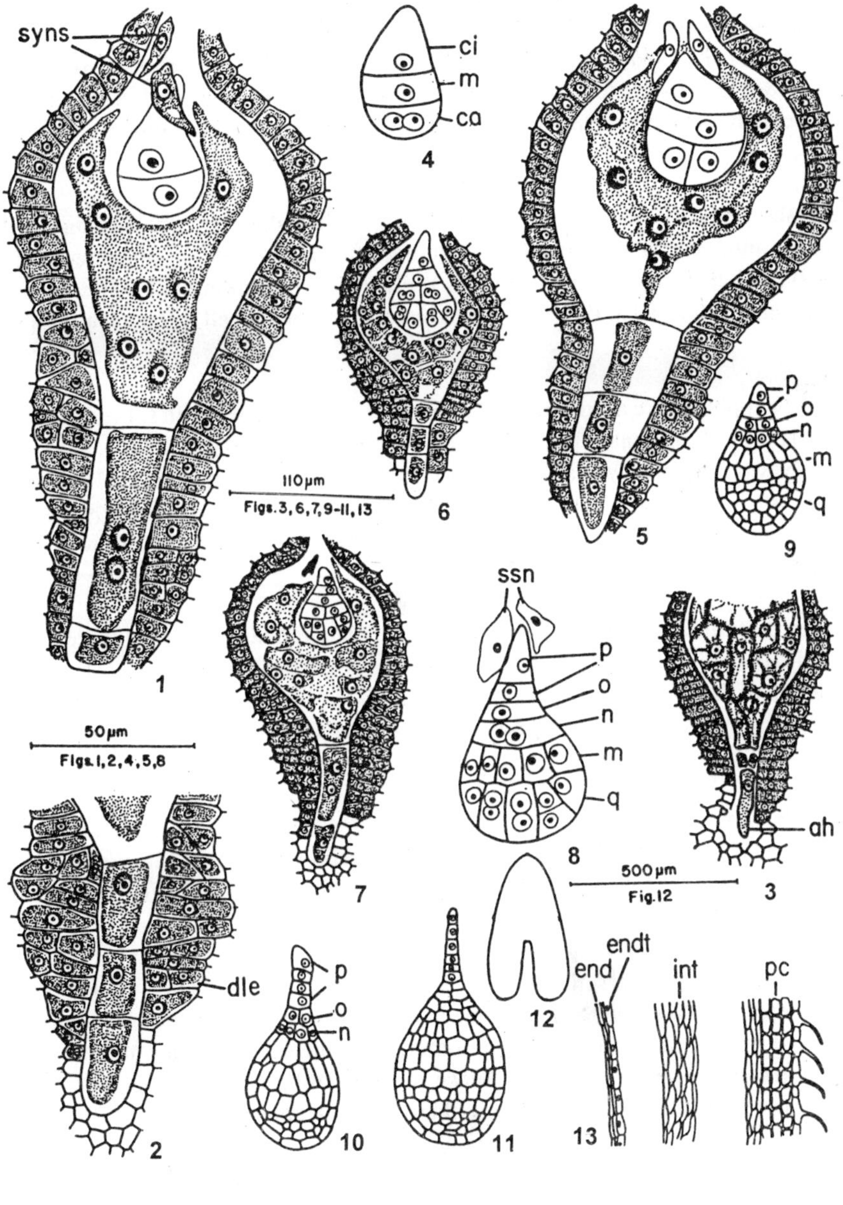

Plate No. 3.17 : *Senecio ludens* (Explanation to Figures)

Figs : 1 Embryo sac showing free nuclear endosperm, surviving synergids, two celled proembryo and two antipodal cells, **2** Lower part of the embryo sac showing three uninucleate antipodal cells, **3** Lower part of the embryo sac showing cellular endosperm and one antipodal cell exhibiting haustorial nature, **4** Three celled proembryo, **5** Embryo sac showing Asterad embryo, synergids, and free nuclear endosperm, **6-11** Different stages in embryogony, **12** Dicot embryo (outline) **13** Structure of the fruit wall in long section.

(Syns = synergids, ssn = surviving synergids. ah = antipodal haustorium, die = double layered endothelium, end = endosperm, endt = endothelium, int = integument, pc = pericarp).

In all the three members, the endothelium gets organised very soon i.e. at the tetrad stage. In some ovules of *Senecio ludens*, endothelium may be seen even at the MMC stage (Plate 3.16, Fig. 6). In *Emilia sonchifolia*, there is always a single layer of endothelium, while in both the species of *Senecio*, it may be even two layered in some ovules. In *S.ludens*, the endothelium at the middle region of the embryo sac may become two or three celled thick (Plate 3.18, Fig. 7) after fertilization, while it is always single layered in the young stages. In *S.tenuifolius*, (in several ovules) the endothelium is two layered right from the beginning.

Antipodals

The antipodal cells show several unusual features in terms of their number, structure and behaviour. In *Emilia sonchifolia*, the antipodals are always three in number and uninucleate. In some ovules, the chalazal antipodal shows a slight haustorial behaviour in that it elongates, protruding out of the embryo sac and penetrating into hypostase (Plate 3.13, Fig. 16;). The haustorial antipodal cell is binucleate, while the other two are uninucleate.

In *Senecio ludens*, the antipodals are normally three in number and uninucleate. In some instances however, the chalazal antipodal may be binucleate (Plate 3.16, Figs. 18-19; Plate 3.17, Figs. 2 and 5). In some ovules, the antipodals may be only two in number with a very large binucleate micropylar antipodal and a small uninucleate chalazal antipodal (Plate 3.17, Figs. 1 and 7). In one of the ovules with three antipodals, the chalazal one shows elongation coming out of the embryo sac and disintegrating the nucellar cells around. This indicates its (antipodal) haustorial natural (Plate 3.16, Fig. 3;).

In *Senecio tenuifolius*, the number of antipodals may be 2 or 3. There is both nuclear multiplication and fusion resulting in polyploidy (Plate 3.20, Figs. 8-11). In one of the ovules, the two antipodals are very large occupying about 60% of the length of the embryo sac (Plate 3.19 Fig. 15).
In *Emilia sonchifolia*, the antipodals generally do not persist long after fertilization, while they do so in the species of *Senecio*.

Synergids

There is nothing unusual about the synergids, except that in several instances in *Senecio ludens*, either one or both of them (synergids) persist for a long time during embryogeny (Plate 3.17, Figs. 1, 5 and 8).

Fertilization

It is porogamous in all the members (Plate 3.13, Fig. 16; Plate 3.17, Fig. 7; Plate 3.10, Fig. 7). In *Senecio tenuifolius*, where gonial and invasive types of somatic apospory are of frequent occurrence, it has been seen that the unreduced embryo sacs do not develop further, nor is there any instance of fertilization in them. Consequently, there is some percentage of disturbance in fruit setting in *Senecio tenuifolius*. In *S.ludens* where apospory is not as frequent as in *S.tenuifolius*, there does not appear to be much disturbance in fruit setting.

Endosperm

In all the three members studied, endosperm formation is free nuclear (Plate 3.13, Fig, 17; Plate 3.17, Fig 1; Plate 3.20 Fig 7). The endosperm becomes cellular at later stages.

Embryogeny

It is similar to what is described for the tribe Heliantheae and conforms to *Senecio* variation of the Asterad type (Plate 3.13, Figs. 18-24; Plate 3.17, Figs. 4-12; Plate 3.21, Figs. 1-14).

In *Emilia sonchifolia* the suspensor is short and is made up of only two cells. In *Senecio ludens*, the

suspensor derived from *p* is long and is made up of six cells. Here, the cell *o* divides vertically and transversely even before the suspensor is fully formed. In some ovules of *S.ludens,* it has been observed (Plate 3.17, Fig.8), that dermatogen is cut off in the tier *m* earlier than it does so in the tier *q*.
In *S.tenuifolius* also the suspensor is short being made up of only three cells. In several instances, division of *n* is delayed. It has undergone only vertical divisions when dermatogen has been fully differentiated in tiers *q* and *m.* The mature embryo is straight and dicotyledonous in all the members (Plate 3.13, Fig. 24; Plate 3.17, Fig 12; Plate 3.21, Fig. 14).

STRUCTURE OF THE FRUIT WALL

In *Emilia sonchifolia* (Plate 3.13, Fig. 25), the fruit wall is made up of 6-8 layers of cells with a clearly demarcated epidermis. The hypodermal cells are elongated and narrow. There is no chemical deposition in any of the cells. In the integument (seed coat), there is no empty space in the middle region, which is noticeable in a good majority of Asteraceae. The endothelium constituting the innermost layer of the integument does not disorganise fully in the mature achene. A layer of endosperm persists adhering to the inner side of the endothelium.

In *Senecio ludens,* the fruit wall (Plate 3.17, Fig. 13) consists of three zones i.e. epidermis, hypodermis and inner zone. Epidermis is made up of a single layer of barrel shaped cells, which give out unicellular hairs. Hypodermis consists of regular rectangular cells which are somewhat loosely arranged. They appear to be collenchymatous in nature. The inner zone of the fruit wall (pericarp) has 2-3 layers of elongated and narrow parenchyma cells. The integument has three regions. The inner region constitutes the endothelium which remains intact though in the form of a single layer of narrow cells. The middle region of the integument is empty, the cells there having been consumed by the endothelium. The outer region of the integument is made up of 5-6 layers of elongated thin walled cells. One or two layers of endosperm persist in the mature fruit.

In *Senecio tenuifolius* also the fruit wall is made up of three zones (Plate 3.21, Fig. 15). The epidermis is single layered and made up of very narrow cells. There are no hairs on the epidermis. The hypodermis constituting the middle region is made up of two layers of collenchymatous cells. The inner region has 3-4 layers of narrow, thin walled parenchyma. The integument has an outer zone made up of 2-3 layers of narrow, thin walled parenchyma. The middle region is empty and the inner zone has the single layered endothelium. A thin layer of cellular endosperm persists in the mature fruit.

DISCUSSION

There is uniformity in anther development in the members of Asteraceae in that all of them show Dicotyledonous type of development (Davis, 1965). In the present work also members of both the tribes Heliantheae and Senecioneae exhibit Dicotyledonous type of anther development.

Both bisporangiate and tetrasporangiate anthers have been reported for the family, the latter being more frequent. Bisporangiate anthers are reported in *Parthenium hysterophorus* (Deshpande, 1960b; Pullaiah, 1982a); *Cotula australis, C.coronopifolia, Calotis hispidula* and *C.squammigera* (Davis, 1962a, 1966); *Spilanthes acmella* (Sundara Rajan, 1974a); *Blumea membranacea* (Pullaiah, 1979b), *Ageratum conyzoides* (Pullaiah 1982c); *Gynura nitida* (Pullaiah 1976) etc. In *Blumea lacera* Lalitha Ram (1981), has reported both bi and tetrasporangiate anthers.

Even though the tribe Heliantheae has been the subject of numerous embryological investigations, unfortunately in many reports, the number of sporangia are not recorded. Hence it is difficult to generalize the number of sporangia in the tribe. The available reports indicate the occurrence of both the types of anthers.

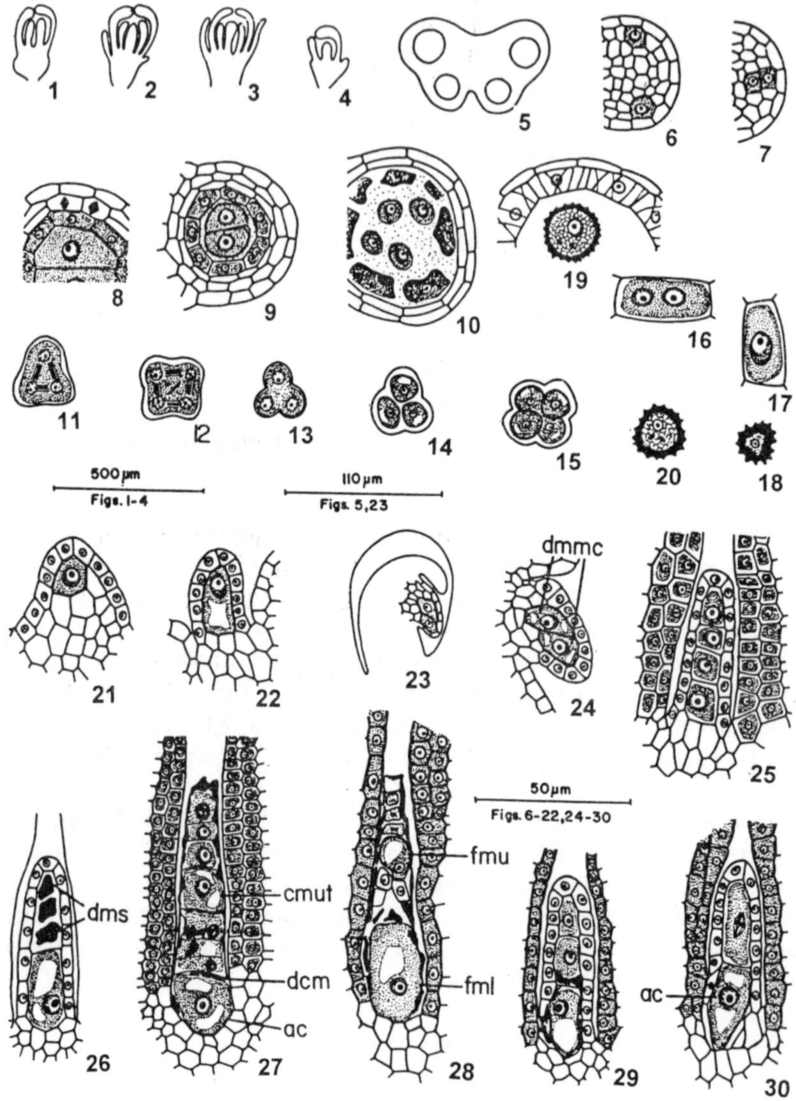

Plate No. 3.18 : *Senecio tenuifolius* (Explanation to Figures)

Figs : 1-4 Organogeny, **5** T.S. of young anther showing tetrasporanigate condition, **6-9** Development of microsporangium, **10** T.S. of anther with microspores and large tapetal cells (periplasmodium yet to be formed), **11-12** Meiotic divisions in pollen mother cells showing spindle orientation, **13** Cytokinesis in pollen mother cells, **14-15** Microspore tetrads, **16-17** Tapetal cells (enlarged, **28-20** Pollen grains (Fig. 19, shows fibrillar endothecium), **21** L.S. of ovule showing a single archesporial cell, **22** Megaspore mother cell, **23** Double MMC (outline), **24** Same as in Fig.23 (enlarged view), **25** Linear tetrad of megaspore, **26** L.S. of ovule showing a tetrad with a functional chalazal megaspore and three degenerating megaspores, **27** Twin tetrads. Lower tetrad shows degenerating megaspores. Upper tetrad shows all four healthy megaspores. Note the enlarging chalazal megaspore, **28** Twin tetrads with chalazal functional megaspores, **29** Tetrad with functional chalazal megaspore. Two other megaspores healthy. Second chalazal megaspore degenerated, **30** MMC in division with an aposporic cell protruding into the cavity of the nucellar epidermis.

(dmmc = double megaspore mother cells, dms – degenerating megaspores, ac = aposporic cell, dec = degenerating chalazal megaspore, cmut = chalazal megaspore of the upper tetrad, fmu = functional megaspore of the upper tetrad, fml = functional megaspore of the lower tetrad.

In the present work however, the anthers are bisporangiate in all the investigated members of Heliantheae. In the tribe Senecioneae *Emilia sonchiofolia* has bisporangiate anthers while the taxa *Senecio tenuifolius* and *S.ludens* have tetrasporangiate anthers. In *Emilia flammea* (Prakasha Rao *et al* 1979) a tetrasporangiate anther has been reported. In all the reports on the genus *Senecio,* the available information indicates a tetrasporangiate anther. One of the recent works on *Senecio candicans* (Rangaswamy and Pullaiah, 1986) also reports a tetrasporangiate anther. However, tetrasporangiate anther does not seem to be uniform for the tribe, as *Gynura nitida* another member of the tribe has a bisporangiate anther (Pullaiah 1976).

In almost all members of Asteraceae, the endothecium exhibits fibrillar nature at the time of anther dehiscence. In a few instance like *Conyza leucantha, C. stricta, Anaphalis subdecurrens. A. wightiana, Blumea, belangeriana, B. mollis, B.virens, Laggera aurita, Pluchea tomentosa* (Lalitha Ram 1979), *Emilia sonchifolia* (Sundara Rajan, 1968a), *E.flammea* (Prakasha Rao *et al* 1979) and in *Erigeron bonariensis* (Sehgal, 1966) no fibrillar thickenings are reported in the endothecium. Maheswari Devi (1957), who reported the absence of fibrous thickenings in endothecium has subsequently (Maheswari Devi, 1963) revised her opinion and noted the presence of fibrous thickenings in *Tagetes patula, Gaillardia picta, Flaveria australasica, Launaea Pinnatifida* and *Gerbera jamesonii.* In the present work, fibrous endothecium is present in all the members except in *Emilia sonchifolia.*

The tapteum in Asteraceae is overwhelmingly plasmodial except for a few instances like *Ainsliea aptera* (Kapil and Sethi, 1962); *Chrysothamnus* spp. (Snow 1945; Anderson, 1970); *Vernonia cinerea, V.cinerascens* (Tiagi and Taimni, 1963); *Sonchus oleraceus* (Walter and Kuta, 1971); *S.asper* (Walter and Kuta, 1971; Kaul *et al.,* 1975); *Hypochoeris radicata* (Kaul, 1972); *Epaltes divaricata* and *Pulicaria angustifolia* (Kothare, 1972); *Youngia japonica* (Kaul, 1973); *Tragopogon gracile* (Singh and Kaul, 1974); *Sonchus arvensis* (Kaul *et al.,* 1975) and *Emilia flammea* (Prakasha Rao *et al.,* 1979), where it is reported to be glandular. The observations of Davis, Ellison, Horner, Pullaiah, Maheswari Devi and Venkateswarulu and Maheswari Devi (quoted by Pullaiah, 1984) however show that the tapetal breakdown in Asteraceae occurs only after the formation of uninucleate pollen grains. In addition to this, the life of the plasmodium is short as it is quickly consumed by the pollen grains. In some instances the tapetal cells remain intact even at thestage of uninucleate pollen grain with exine already laid out. This may give an impression that the tapetal break down does not occur. Pullaiah (1984), opines "Probably the (above) authors might have missed the stages of formation of periplasmodium and have mistaken the tapetum for glandular type". In view of this observation, it seems desirable that the instances of glandular tapetum in Asteraceae need critical re-examination.

In the present work, all the members show a clearly plasmodial type of tapetum. In *Emilia sonchifolia,* however the plasmodial formation is delayed and even at the uninucleate pollen stage with fully formed exine, the tapetal cells appear to be intact with one or two cells showing slight signs of breakdown. In this instance, the tapetum may be easily mistaken for a glandular type. Another feature of tapetum in *Emilia sonchifolia* is that the plasmodium disappears very quickly, sometimes before the formation of male gametes.

Regarding the function of tapetal periplasmodium, Bambacioni (1941) observes that it contributes to the exine formation. Davis (1961a), and Deshpande (1960b, 1962b) however opine that in Asteraceae the tapetum is purely nutritive in function as exine would have already been formed. According to Pullaiah (1979a), as the major portion of the exine is formed only after the tapetal breakdown, the periplasmodium plays a dual role i.e., nutritive and contributor to exine formation.

In the present work, the opinion of Pullaiah (1979a) seems to be true in all instances except for *Emilia sonchifolia,* where the exine formation would have been completed when the tapetal cells are still intact. Hence tapetal function is purely nutritive.

The tapetal cells in Asteraceae are uninucleate to start with and remain so in some members like *Spilanthes acmella* (Sundara Rajan 1974a) and *Solidago canadensis* (Pullaiah, 1978a). Frequently, the tapetal cells may become 2-4 nucleate or the nuclei fuse to become polyploid. 2-4 nucleate tapetal cells are found in *Lactuca sativa* (Gates, 1920), *Chrysothamnus naseosus* (Snow, 1945) *Synedrella nodiflora* (Banerji and Pal, 1959), *Caesulia axillaris* (Deshpande, 1962), *Brachycome ciliaris* (Davis, 1964) and *Conyza stricta* (Pullaiah 1978a, Laitha Ram, 1979). Multinucleate tapetal cells have been reported in *Podolepis jaceoides, Ammobium alatum, Minuria integerrima* (Davis, 1961a, 1962b, 1964b); In *Eupatorium odoratum,* Maheswari Devi and Pullaiah (1977b) have reported 8 nuclei arranged in a linear fashion. Bhargava (1935), has reported nuclear division and fusion in *Eclipta erecta.* In *Glossocardia bosvallea,* Deshpande (1962d) has reported the occurrence of two large barrel shaped nuclei in the tapetal cells. The most abnormal tapetal behavious has been reported for *Calotis lappulacea* (Davis, 1968a), where the tapetal nuclei multiply and organise themselves like an embryo sac.

In the present work, nuclear multiplication and fusion is seen in all the members, but the nuclei rarely exceed two in number.

In the tribe Heliantheae, a single row of archesporial cells is reported in members like *Rudbeckia bicolor* (Maheswari and Srinivasan, 1944); *Synedrella nodiflora* (Banerji and Pal, 1959); *Acanthospermum hispidum* (Vaidya, 1961); *Bidens pilosa, Xanthium strumarium, Zinnia pauciflora, Eclipta erecta, Cosmos sulphureus,* and *Helianthus annus* (Maheswari Devi, 1963).

In the present work however 2 rows of archesporial cells are seen in *Synedrella nodiflora, Spilanthes acmella* and *Blainvillea rhomboidea,* while in *Spilanthes paniculata* there may be one or two rows of archesporial cells. In *Senecio ludens, S. tenuifolius* and *Emilia sonchifolia* of the tribe Senecioneae which are included in the present work there is a single row of archesporial cells in each lobe of the anther. This is in conformity with the earlier reports for the tribe Senecioneae (Pullaiah, 1984; Rangaswamy and Pullaiah, 1986).

In all the members studied, microsporogenesis is regular and results in the formation of tetrahedral or decussate tetrads. Cytokinesis is simultaneous and wall formation is by furrowing. This is in agreement with the rest of the reports on Asteraceae in the tribes Heliantheae and Senecioneae. However in *Bellis perennis, Conyza stricta* and *Erigeron bonariensis* (Sharma, 1975), and *Blumea lacera* (Lalitha Ram, 1979) both simultaneous and successive cytokinesis have been reported.

The pollen grains are 3 celled at the time of shedding in all the members studied and this conforms to what is reported for the rest of Asteraceae.

The shape of the male gametes in Asteraceae is variable. It ranges in shape from spherical to elliptical or elongated (Pandey and Singh, 1973). Davis (1961a) has reported variable shapes of male gametes in *Podolepis jaceoides,* a feature also observed in *Guizotia abyssinica* (Chopra 1972). In the present work both circular and elliptical male gametes have been noticed but not in the same taxon. Bicarpellary, syncarpous, inferior ovary with a single anatropous ovule is the characteristic feature of the family. Abnormalities in the form of multiovulate condition, ovarian fusion etc., have been reported in a number of instances (*Tagetes patula,* Venkateswarulu and Maheswari Devi, 1955; *Helichrysum bracteatum,* Pandey 1977 and *Blumea malabarica;* Pullaiah 1979). In *Carthamus tinctorius* (Maheswari Devi and Pullaiah, 1977a) about 4% of the ovaries in a capitulum show ovarian fusion, while 6% of the florets exhibit multiovulate ovaries.

In all the taxa of present investigation no such abnormalitics have been noticed.

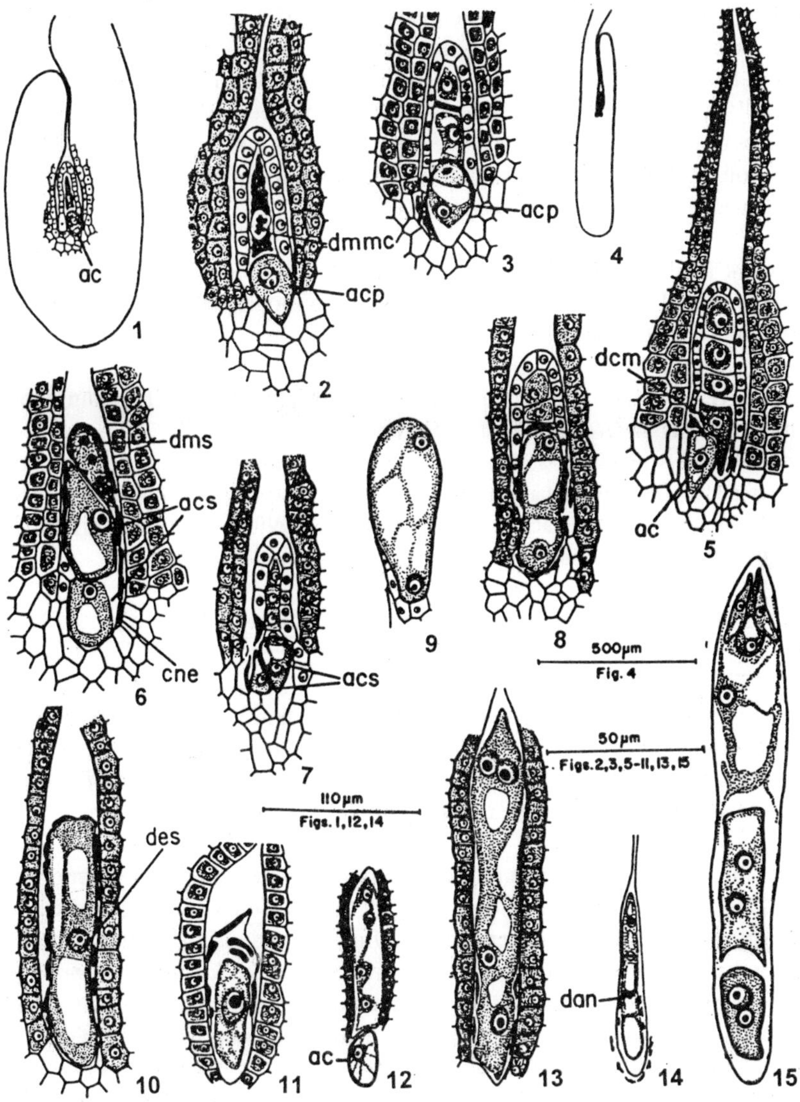

Plate No. 3.19 : *Senecio tenuifolius* (Explanation of Figures)

Figs : 1 L.S. of ovule (outline) showing an aposporic cell protruding into the MMC. The nucleus of MMC appears to be generating, **2** Same as in Fig.1 enlarged to show the details, **3** Normal tetrad with an aposporic cell overlapping the chalazal megaspore, **3** Normal tetrad with an aposporic cell overlapping the chalazal megaspore, **4** Tetrad with protruding aposporic cell and degenerating chalazal megaspore (ovule in outline), **5** Same as in Fig. 4; enlarged to show the details, **6** Ovule showing degenerating megaspores and two large aposporic cells protruding into the cavity of the embryo sac, **7** Tetrad with two aposporic cells, **8** Two nucleate aposporic embryo sac with one degenerated and two healthy megaspores, **9** Two nucleate embryo sac. Nucellar epidermis is shown below. The embryo sac is shown in an inverted position, **10** Diplospory (Gonial apospory), **11** Uninucleate embryo sac, **12** Four nucleate embryo sac with an aposporic cell, **13** Four nucleate aposporic (diplosporic) embryo sac, **14** Two nucleate embryo sac pushed towards the micropyle of the ovule. The nucleus of the protruded aposporic cell has degenerated, **15** Organised embryo sac, with two large binucleate antipodal cells.

(dmmc = degenerating megaspore mother cell, ac = aposporic cell, acp = aposporic cell protruding, dcm = degenerating chalazal megaspore, acs = aposporic cells, dms = degenerating megaspores, cne = crushed nucellar epidermis, dan = degenerating aposporic nucleus, des = diplosporic (gonial apospory) embryo sac

The female archesporium in Heliantheae is generally unicellular, but multicellular archesporium has been reportged in *Spilanthes acmella* (Mukherjee and Kothare, 1973); *Melampodium divaricatum* (Maheswari Devi and Pullaiah 1976b); *Melampodium cinereum* (Maheswari Devi and Pullaiah 1980) and *Bidens pilosa* (Maheswari Devi, 1963). In the tribe *Senecioneae,* there does not seem to be any earlier report of a multicellular archesporium. Elsewhere in the family multicellular archesporium is reported in *Sanvitalia procumbens* (Hjelmqvist and Holberg 1961), *Achillea millefolium* (Dahlgren, 1927), *Chrysanthemum* spp. (Harling, 1951a), *Erigeron bonariensis* (Sehgal, 1966, 1969) etc.

In the present work multicellular archesporium is reported in *Spilanthes acmella, S.paniculata, Senecio ludens* and *S.tenuifolius.* The presence of multiple archesporium (2 celled) can be visualized in *Synedrella nodiflora,* by the presence of twin tetrads.

The embryo sac development in the tribe Heliantheae is mostly of the Polygonum type while there is an occasional Allium type as in *Tridax trilobata* (Hjelmqvist, 1951). Fritillaria type is reported in *Rudbeckia bicolor* (Maheswari and Srinivasan, 1944). In *Sanvitalia procumbens,* both Allium and Polygonum types occur in the same species (Hjelmqvist and Holmberg, 1961). In the tribe *Senecioneae,* all the reports indicate only a Polygonum type of embryo sac. In *Arnica alpina* (Afzelius, 1936), where MMC does not undergo reduction division, *Antennaria* type of gonial apospory has been noticed.

Elsewhere in the family Allium type of embryo sac development has been reported in *Ammobium alatum* (Davis, 1962b).

The other types of embryo sac development reported in the family are Fritillaria type in *Gaillardia picta* (Venkateswarulu and Maheswari Devi, 1955b) and *Ratibida tagetes* (Howe 1964); Drusa type in *Chrysanthemum* spp and *Erigeron* spp (Harling, 1951a, b) and Adoxa type in *Rudbeckia hirta* (Palm, 1934). In the present work only Polygonum type is seen in all the members.

Irregularities in the megaspore functioning and tetrad formation have been reported for a number of Heliantheae. In *Helianthus annus* (New Comb, 1973a), the division of the megasporocyte is unequal so that the lower dyad cell is larger. This also undergoes an asymmetrical division resulting in the formation of a large chalazal cell, which engenders the embryo sac. In *Bidens pilosa* and *Volutarella ramosa,* Deshpande (1964a, b) has reported the presence of linear and **T** shaped tetrads. In *Rudbeckia bicolor,* a coenomegaspore has been reported (Maheswari and Srinivasan, 1944). In *Tridax trilobata,* Hjelmqvist (1951) a tetrad has been reported in which the middle megaspore is binucleate, while the two spores towards the micropyle and chalaza are uninucleate. In *Melampodium divaricatum* (Maheswari Devi and Pullaiah, 1976b) and *Tithonia rotundifolia* (Pullaiah, 1978b), linear as well as isobilateral tetrads have been reported with more than one megaspore becoming functional.

In the tribe Senecioneae also, mostly the tetrad is linear with the chalazal megaspore becoming functional. In *Senecio abrotanifolius* (Afzelius, 1924), the two middle megaspores may be functional. In *Emilia flammea* occasional **T** shaped tetrads have been reported (Prakasha Rao, *et al.*, 1979).
In the present work, the members of tribe Heliantheа like *Spilanthes acmella* and *Blainvillea rhomboidea* show indications of more than one megaspore functioning. In *Synedrella nodiflora*, there are a number of variations in the functioning of the megaspores. In some ovules, twin tetrads (both superposed and juxtaposed) also have been noticed.

Among the members of Senecioneae investigated in the present work, no abnormalities are noticed in *Emilia sonchifolia,* while the two species of *Senecio* show a number of unusual features like, twin tetrads, many functioning megaspores, micropylar megaspore functioning etc.

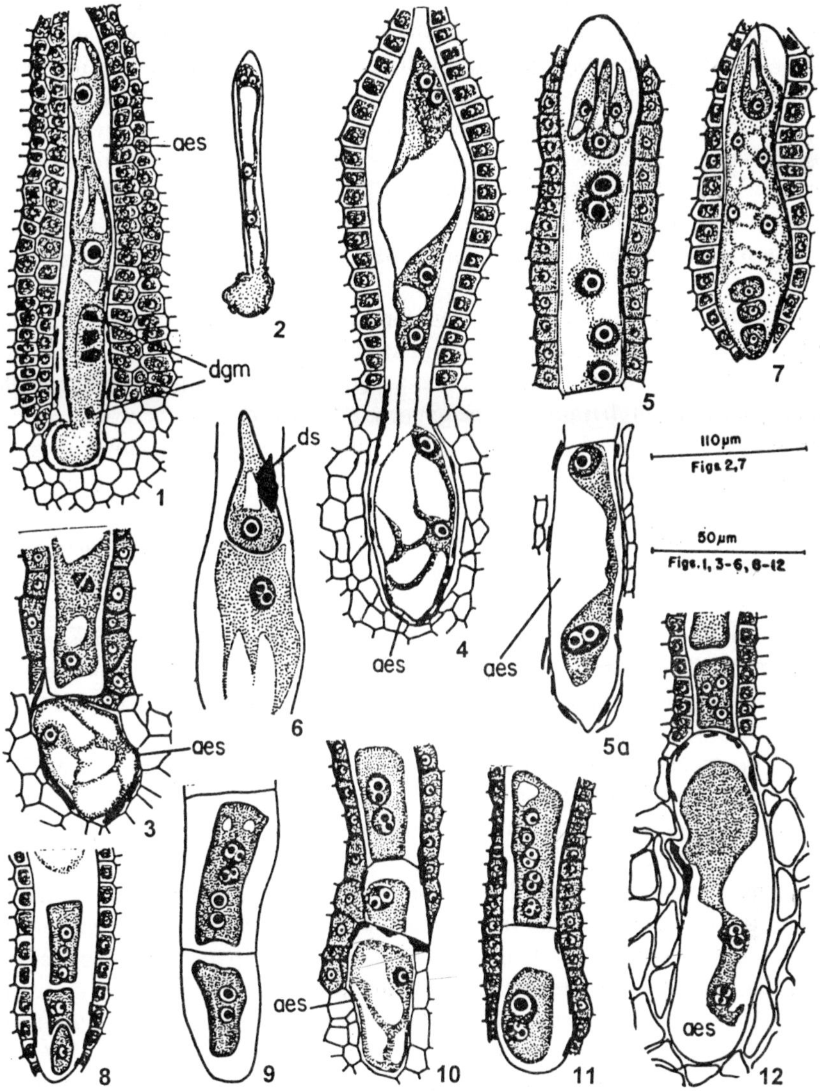

Plate No. 3.20 : *Senecio tenuifolius* (Explanation of Figures)

Figs : 1 2 nucleate aposporic embryo sac protruding into the embryo sac cavity. The two nuclei are towards the micropylar end of the ovule. The normal tetrad of the ovule is degenerating, **2** Four nucleate embryo sac with an aposporic cell at the base. The walls between the two are dissolved. The nucleus of the aposporic cell has degenerated, **3** Lower part of a four nucleate embryo sac. (upper nucleus is in division) with a uninucleate aposporic embryo sac at its chalazal region, **4** A four nucleate embryo sac. A two nucleate aposporic ambryo sac is attached to the normal embryo sac at the chalazal region. With the dissolution of the wall, the aposporic embryo sac appears to be integrated with the normal embryo sac, **5** Mature embryo sac; antipodals are still in the nuclear condition, **6** Upper part of the embryo sac showing zygote. PEN and degenerated synergid, **7** Embryo sac showing zygote, pollen tube, free nuclear endosperm and three antipodal cells, **8** Lower part of the embryo sac showing three antipodal cells with varying number of nuclei, **9** Two large antipodal cells showing variation in nuclear number, **10** Lower part of the embryo sac showing two antipodal cells and a uninucleate aposporic embryo sac, **11** Two large antipodal cells; upper 5 nucleate and lower two nucleate, **12** Lower part of the embryo sac showing antipodal cells and a two nucleate aposporic embryo sac.

(aes = aposporic embryo sac, dgm = degenerating megaspore, acs = aposporic embryo sac, ds = degenerating synergid)

The most unusual feature that is reported for the first time not only for Asteraceae, but perhaps for the entire group of flowering plants is the occurrence of what may be regarded as a **megaspore mother cell haustorium** in *Senecio ludens*. As far as the author is aware such a structure is unheard of in Angiosperm embryology. The record of a haustorium in the earliest stage of the development of the female gametophyte is in the megaspore. In *Sedum chrysanthum* and *S. sempervivoides*, (Subramanyam, 1967), each megaspore produces a tube, which grows towards the apical portion of the nucellus with only one tube reaching the micropyle. In the present instance, the MMC grows out of the nucellar lobe and digests a part of the funiculus. This is noticed however only in few isolated instances. Nevertheless, the dissolution of a part of the funiculus is sufficient indication of the haustorial nature of the MMC.

Another unusual feature recorded for the first time for Asteraceae is the occurrence of hypostase in *Emilia sonchifolia*. Elsewhere hypostase is recorded in families like Crossosomataceae, Eleganaceae, Umbelliferae, Loranthaceae, Aristolochiaceae (Bhojwani and Bhatnagar, 1985) etc. In the Loranthaceae the hypostase is a collenchymatous pad, while in *Aristolochia* the hypostase persists even in the mature seed. In Eleagnaceae a prominent hypostase is organised after fertilization, while in *Bupleurum tenue* (Gupta and Gupta 1964), hypostase is organised at the tetrad stage itself.

In *Emilia sonchifolia*, the hypostase gets differentiated a little before the organization of the embryo sac. The cells of the hypostase are narrow and elongated with tapering and walls. They stain deeply.

Van Tieghem (1901), who first coined the term hypostase, suggested that it acts as a barrier preventing the further chalazal growth of the embryo sac. Other opinions regarding the role of hypostase are water balancing (Johansen, 1928); helping in the rapid transport of the food material (Venkata Rao, 1952) etc.

In the opinion of the author, the sporadic occurrence of hypostase, may at best be an unusual feature. If hypostase were to perform such important roles like checking the growth of the embryo sac, food transportation etc., it should have been of universal occurrence in Angiosperms.

Endothelium is a characteristic feature in all sympetalous docotyledons including the Asteraceae. The nucellar layer degenerates at an early stage, and the cells of the innermost layer of the integument develop into endothelium. The cells have dense cytoplasmic contents and envelop the developing female gametophyte. Usually the endothelium starts differentiating at the tetrad stage and is generally uniseriate. But in some cases like *Bidens biternata* (Deshpande, 1964b); *Siegesbeckia orientalis* (Misra, 1965) an endothelium is formed at the megaspore mother cell stage. A uniseriate endothelium is reported in *Coreopsis tinctoria, Bidens pilosa, Xanthium strumarium, Zinnia pauciflora, Eclipta erecta, Cosmos sulphureus, Helianthus annua* and *H. debilis* (Maheswari Devi, 1963); *Eclipta prostrata, Flaveria repanda, Glossocordia bosvallea, Volutarella ramosa* and *Bidens biternata* (Deshpande, 1961a, b, 1963, 1964, b); *Sonchus oleraceus* (Hiremath, 1980) and a number of taxa.

Biseriate or multiseriate endothelium with uninucleate or binucleate cells has been reported in *Tridax trilobata* (Hjelmqist, 1951); *Tridax procumbens* (Maheswari and Roy, 1952); *Parthenium hysterophorus* and *Caesulia axillaris* (Deshpande, 1960, 1962b); *Cyathocline purpurea* (Deshpande and Kothare, 1976) and *Guizotia abyssinica* (Chopra and Singh, 1976).

In *Helianthus debilis, H. annuus* and *Dahlia pinnata*, the endothelium becomes more than one layer thick at places owing to periclinal or slightly oblique divisions (Maheswari Devi 1963; Pandey 1980). In *Ainsliea aptera* (Kapil and Sethi, 1962a), the endothelial cells become binucleate, while in *Bidens pilosa* (Maheswari Devi, 1963), the cells become multinucleate and polyploid. Endothelium becomes multiseriate only in the chalazal side in *Coreopsis stillmani, Cosmos bipinnatus, Dimorphotheca sinuata,* and *Eupatorium adenophorum* (Pandey, 1980). In *Lagascea mollis* (Pullaiah 1981a), the multiseriate endothelium is noticed in the middle region of the embryo sac. Multilayered endothelium has also been reported for *Elephantopus scaber* and *Tithonia rotundifolia* (Pullaiah, 1978b, 1979a).

In *Cyathocline purpurea* the endothelium persists as a cuticular layer in the mature seed (Deshpande and Kothare, 1976). The endothelial cells are radially elongated either at the micropylar end as in *Bidens biternata* (Deshpande, 1964a) or at the chalazal end as in *Tridax procumbens* (Padmanabhan, 1962).
Maheswari Devi (1979) who has made a detailed study of endothelium in Asteraceae recognises four types, based on their (endothelium) structure and function.

I (a) Biseriate with Uninucleate cells (Eg. *Zinnia linearis*)
  (b) Biseriate with multinucleate cells (Eg. *Bidens pilosa*)
II (a) Multiseriate with uninucleate cells (Eg. *Lagascea mollis*)
  (b) Multiseriate with multinucleate cells (Eg. *Helianthus annus*)
III Cells metamorphosed into embryos (Eg. *Melampodium divaricatum, Carthamus tinctorius* etc.)
IV Cells become rich in food materials divide repeatedly and form a massive tissue (Eg. *Melampodium divaricatum*

In the present work, the endothelium is uniseriate with uninucleate cells in the members of the tribe Heliantheae, while in Senecioneae only in *Emilia sonchifolia* the endothelium is uniseriate with uninucleate cells. In these members, endothelium gets differentiated at the tetrad stage. In *Senecio ludens,* an endothelium is recognisable even at the MMC stage. In *Senecio tenufolius* the endothelium is biseriate with uninucleate cells, while in *S.ludens,* the uniseriate layer becomes biseriate at the middle region of the embryo sac after fertilization.
Antipodals in the family Asteraceae show a great variety in terms of their number, nuclei in them and their behaviour. Initially the antipodals are either two or three in number.

Pullaiah (1984), has classified the variation in antipodal cells into four categories. These are :

1. The number of cells remains constant, but ploidy varies due to nuclear divisions and fusions.
2. Secondary multiplication of antipodal cells occurs due to cytokinesis following nuclear division.
3. Combination of one and two above (multiple antipodals with multiple nuclei).
4. The numberof antipodal cells remains constant but the cells become haustorial and crush the cells of the integument.

In the tribe Heliantheae three of the above types (except-4) occur as seen in *Melampodium divaricatum* (Maheswari Devi and Pullaiah, 1976b), *Tithonia rotundifolia* (Pullaiah 1978b), *Acanthospermum hispidum* (Maheswari Devi and Pullaiah, 1980) etc.

In the present work on the tribe Heliantheae, the number of antipodals are either two or three. When only two antipodals are present, one of them is binucleate. Occasionally as has been seen in *Spilanthes acmella,* there may be only two uninucleate antipodals. This is due to the phenomenon of **strike** in one of the chalazal nuclei (at the four nucleate embryo sac stage) which fails to undergo division. Nuclear multiplication and fusion is also seen in the members investigated.

In the tribe Senecioneae, the antipodals are normally three in number, but a number of variations have been reported. In *Senecio aureus* (Motteir, 1893) the antipodals may undergo both karyokinesis and cytokinesis, resulting in six (often polyploid) antipodal cells. In *S.nebrodensis,* two binucleate antipodals are seen (Afzelius, 1924). Other variations are 20 to 40 antipodals in *S.jacobea* (Afzelius, 1924) and *S.cineraria* (Afzelius, 1924); three polyploid antipodals in *S.leucanthemum* var. *pinnatifidum* (Bachetti, 1961); three binucleate antipodals in *Erechtites heiracifolium* (Cooper, 1936); six multinucleate and polyploid antipodals in *Emilia sonchifolia*

(Pullaiah, 1976); three multinucleate and polyploid antipodals in *Gynura nitida* and *Notonia grandiflora* (Pullaiah 1976); three uninucleate antipodals in *Emilia flammea* (Prakasha Rao *et al* 1979). A recent report on *Senecio candicans* (Rangaswamy and Pullaiah, 1986) indicates multiple antipodals with polyploid nuclei.
In the present work involving *Emilia sonchifolia, Senecio ludens* and *S.tenuifolius,* antipodal haustoria are observed in the first two. In *Emilia sonchifolia,* the haustorium penetrates into the hypostase. Antipodal haustoria have been reported earlier in *Grindelia squarrosa* (Norris, 1892); *Aster novae angliae* (Chamberlian, 1895); *Senecio vulgaris* (Small, 1919); *Artemesia tridentata* (Diettert, 1938); *Bidens pilosa Coreopsis tinctoria* and *Cosmos sulphureus* (Maheswari Devi, 1963); *Grindelia stricta, G. arizonica, G.lanceolata* and *Prionapsis ciliata* (Howe, 1975); *Chrysothamnus viscidiflorus* and *C.lnifolius* (Anderson, 1970); *Coreopsis lanceolaria* (Pullaiah, 1976); *Elephantopus scaber* (Pullaiah, 1979a) and *Grangea madraspatana* (Desai, 1981a).
In *Coreopsis tinctoria* (Maheswari Devi, 1963) and *C.lanceolaria* (Pullaiah, 1976), the two antipodals grow enormously in size and absorb the integument but for a thin membrane.

In *Senecio ludens* and *S. tenuifolius* of the present work, the antipodals are either two or three in number. There is nuclear fusion and multiplication resulting in polyploidy.
In all the members of the present investigation generally the antipodals persist till late in embryogeny (globular stage).

Secondary multiplication of the antipodal cells, multiplication and subsequent fusion of nuclei resulting in polyploidy and persistence long after fertilization, all point out to a nutritive role for antipodals, even when their haustorial nature is not obvious structurally. Davis (1966) observes that antipodal multiplication in Compositae (Asteraceae) often results in persistent haustorial cells. But Pullaiah (1984) argues that this i s not true always as has been noticed in *Melampodium divaricatum,* where haustorial nature is not evident even though antipodals persist till late in embryogeny.

In the family Asteraceae the endosperm is either *ab initio* cellular or free nuclear. No endosperm haustoria have been reported so far.

In the tribe Heliantheae, *ab initio* cellular endosperm has been reported in *Eclipta erecta* (Bhargava, 1935); *Tridax procumbens,* (Maheswari and Roy, 1952); *Parthenium argentatum* (Esau, 1944, 1946); *Blainvillea rhomboidea* (Sundara Rajan, 1972), *Spilanthes acmella* (Mukherjee and Kothare, 1973; Sundara Rajan, 1974a); *Chrysanthellus indicus* (Deshpande and Kothare, 1973); *Galinsoga parviflora* (Pullaiah, 1918a) and *Parthenium hysterophorous* (Pullaiah 1982a). In all other instances (see Table 3.3) where reports are available, the endosperm is free nuclear.

In the tribe Senecioneae, all the reports (see table 3.6) indicate a free nuclear endosperm except for *Emilia sonchifolia, Gynura nitida* (Pullaiah, 1976) and *Tussilago farfara* (Vernin, 1952). In *Notonia grandiflora* (Pullaiah, 1976), there is no endosperm formation as there is no fertilization.

In the present work, all the members of the tribe Heliantheae show *ab initio* cellular endosperm (except for *Synedrella nodiflora*) while all the members of the tribe Senecioneae show free nuclear endosperm. In *Emilia sonchifolia* included in the present work, the endosperm is free nuclear, while the report of Pullaiah (1976) for the same taxon indicates a cellular endosperm.

In *Synedrella nodiflora* the endosperm is free nuclear, but in some ovules the cell walls in nuclear endosperm are laid down very early (during embryogeny) with the result different ovules with same stage of embryogeny may show free nuclear or cellular stage of endosperm.
Embryogeny in the family is remarkably uniform and always corresponds to *Senecio* variation of the Asterad type. In all the members of the family investigated so far including the present one it is the same.

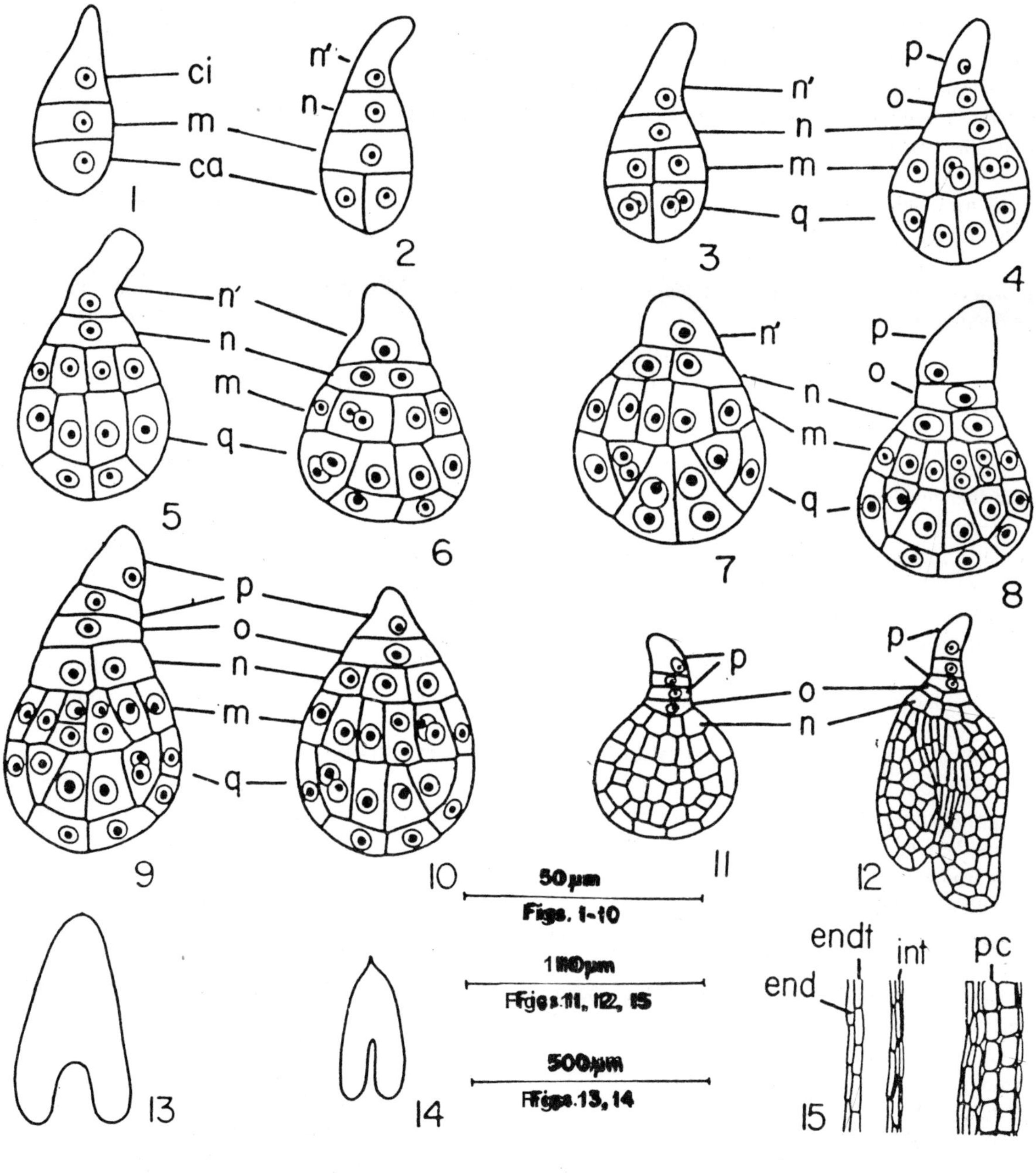

Plate No. 3.21 : *Senecio tenuifolius* (Explanation of Figures)

Figs : 1 Linear proembryo, **2** Asterad embryo, **3-11** Development of the embryo up to the late globular stage, **12** Cordate embryo, **13-14** Dicotyledonous embryos (outline), **15** Structure of the fruit wall in L.S.

(for explanation of various abbreviations, see text on embryogeny)

Apomixis in Asteraceae is of frequent occurrence. While haploid parthenogenesis has been recorded so far only in *Brachycome iberidifolia* (Sharma and Murthy, 1977) the recurrent apomixis (aneuspory, gonial apospory and somatic aposopory) is very common. Most members of Asteraceae in which somatic apospory has been reported are facultative apomicts. Davis (1968) and Pullaiah (1984) have reviewed apomixis in Asteraceae. In the present work somatic apospory with a few nucellar cells at the chalazal region developing into uninucleate embryo sacs is seen in *Spilanthes acmella.* Earlier reports on apospory in the tribe Heliantheae include *Coreopsis bicolar* (Gelin, 1934), *Parthenium argentatum* (Esau, 1944, 1946), *Rudbeckia laciniata* (Battaglia, 1945) etc.

In the tribe Senecioneae, gonial and somatic apospory have been recorded both for *Senecio ludens* and *S.tenuifolius.*

In *Senecio ludens,* gonial apospory leads to the formation of unreduced embryosacs. But no instance of fertilization of the unreduced embryo sac or parthenogenetic development has been noticed.The instances of gonial apospory here are sporadic and consequently, there is very little disturbance is seed setting. In *Senecio ludens* also somatic apospory is sporadic and except for initial signs of development, no aposporous cell develops into an embryo sac.

In *Senecio tenuifolius,* gonial apospory leads to the formation of unreduced embryo sacs. But no instances of further development are noticeable.

Somatic apospory in *Senecio tenuifolius* however is very interesting. It is of two categories. In the non invasive type, the aposporous cell develops into a two nucleate embryo sac. In the invasive type, the aposporous cell invades the nucellar lobe and replaces the normal MMC or tetrad. No abnormality is noticeable at a later stage except that the embryo sac is unreduced. A similar phenomenon has been reported earlier for only *Calotis lappulaceae* (Davis, 1968a). But in *Senecio tenuifolius* however, the unreduced embryo sacs (organised) degenerate, with no further development leading to about 2-3% disturbance is seed setting.

CARPOLOGY AND SYSTEMATICS IN ASTERACEAE

Carpology refers to the study of the structure of the achene (fruit). The structure of the pericarp and to some extent the seed coat are variable and have proved to be useful systematically. Some notable contributions on the carpology of Asteraceae include; structure of seeds and fruits of Heliantheae (Maheswari Devi and Ranjalkar, 1980); *Verbesina enceliodes* (Misra, 1972); *Eclipta erecta* (Pandey, 1976); *Galinsoga parviflora* (Pandey, 1983a); tribe - Heliantheae (pandey, 1983b); *Blumea mollis* (Rajashekar, 1980) etc.

The present work on the members of Heliantheae and Senecioneae shows clearly that the structure of the pericarp is highly useful at the generic level as well as at the species level.

Among the investigated members of the tribe Heliantheae, chemical depositions and their shape in the pericapr wall is a characteristic feature in *Blainvillea rhomboidea, Spilanthes acmella* and *S.paniculata,* while very little deposition is found in *Synedrella nodiflora.* In *Blainvillea rhomboidea* the outer wall of the epidermis is very thick and there are some chemical depositions. In addition to this, large chemical deposits are found deep inside the fruit wall.

Spilanthes acmella and *S.paniculata* can be easily distinguished on the basis of their fruit anatomy. In *S.paniculata,* the fruit wall is made up of only two layers of cells, a broad celled epidermis and an inner layer having chemical deposits. In *Spilanthes acmella,* the fruit wall is distinguishable into an epidermis a layer of hypodermis made up of cells identical to epidermis, two layers of shiny isodiametric cells and an inner layer with large chemical deposits.

Senecio ludens and *S.tenuifolius* can be identified on the basis of following anatomical features of the fruit wall.

	S. tenuifolius	*S. ludens*
1.	Epidermis made up of elongated cells. No hairs	Epidermal cells rectangular, unicellular hairs present.
2.	Hypodermis made up of two layers of collen chyma	Three layers of collenchyma in the hypodermis
3.	Inner most zone of pericarp made up of 2,3 layers of thin walled elongated cells	same

In the investigated members the integument (seed coat) is characterised by the presence of three zones, an outer zone made up of seed coat epidermis and one or two wall layers, a middle zone generally empty representing the cell layers consumed by the endothecium and an inner zone consisting of the remnants of the endothelium and the endosperm. In *Emilia sonchifolia* the middle zone of the seed coat is not empty. It has a few wall layers which are yet to be digested.

SPECIES CATEGORIZATION IN EMILIA

In the genus *Emilia,* two species viz., *E.flammea* (Prakasha Rao *et al* 1979) and *E.sonchifolia* have been worked out embryologically. Postfertilization changes in *E.flammea* are not known yet. On the basis of the available embryological features, the two species can be easily separated.

	E. flammea	*E. sonchifolia*
1.	Anther tetrasporangiate	Anther bisporangiate
2.	Tapetum glandular	Tapetum plasmodial
3.	One row of PMCs	Two rows of PMCs
4.	Dimorphic pollen	All pollen are uniform
5.	Pollen grains accumulate dark contents	No such thing is observed
6.	Hypostase absent	Hypostase present
7.	Antipodals three in number and equal in size	Antipodals three in number and lower most one largest

The above account clearly points to the usefulness of embryological features in species categorization. This is specially true for *Emilia* where only few and overlapping morphological characters are used in species categorization (Gamble, 1948; Saldanha and Nicolson, 1976).

Phyletic appraisal of the tribes Heliantheae and Senecioneae (on embryological grounds)

The Heliantheae and Senecioneae have been accorded different phyletic positions within the family by various synantherologists. Bentham (1873), Bessey (1925), Hutchinson (1916) and Cronquist (1955) opine that Heliantheae is the most primitive tribe and others are derived from it. Small (1919) howeve is of the opinion that the tribe Senecioneae appears to be the most primitive. Most scientists agree that Heliantheae could be the basic tribe due to the presence of paleaceous receptacle, herbaceous involucre and tailless anthers (Stuessy, 1977). On chemical grounds also (Mabry and Bohlman, 1977) Heliantheae is believed to be the basic tribe. But Stuessy (1977) opines , " . . but these data are by no means conclusive". He further observes "other considerations hint, at a more derived status for the tribe". Palynological characters (presence of internal foramina in the ektexine of the pollen wall) and absence of proper generic boundaries (especially in the sub tribe Verbesininae) are not what is expected of the oldest group in the family.

It will be of interest to known whether embryological characters have any solution to offer regarding the primitiveness or otherwise of the tribes Heliantheae and Senecioneae. Unfortunately a vast majority of taxa belonging to both these tribes are yet to be embryologically investigated. This makes it impossible to arrive at a generalization based on the distribution of primitive or advanced embryological characters. Maheswari Devi (1963) opines that embryological features in this family do not aid either in distinguishing one tribe from the other or in tracing the evolutionary tendencies within the family. This is true because embryological characters regarded as advanced or primitive are found in different members of the same tribe. In such a situation, where tribes seem to possess both advanced and primitive characters, the only way to assess the evolutionary status of a tribe is by the percentage distribution of advanced and primitive characters. In terms of embryology however even this assessment cannot be conclusive for a large majority are embryologically unknown. Hence any opinion expressed on the evolutionary status of Heliantheae and Senecioneae on embryological grounds is bound to be tentative.

A perusal of the distribution of embryological characters in Heliantheae show that anthers are both tetrasporangiate (primitive) and bisporangiate (advanced). However there seems to be a preponderance of the latter (unfortunately many of the embryological reports do not mention this simple character, number of sporangia). All the members of Heliantheae in the present investigation are bisoporangiate. Sporogenous tissue is not extensive and tapetum is plasmodial. Endosperm is both nuclear (primitive) and *ab initio* cellular (advanced). Among the members embryologically investigated in Heliantheae 63% have nuclear endosperm, while 37% have cellular endosperm.

In the tribe Senecioneae, the anthers are generally tetrasporangiate except for 2-3 instances of bisporangiate condition. The tapetum is plasmodial except for a solitary instance. Male archesporium eventually becomes two rowed. Endosperm is both nuclear as well as cellular with the former reported in all instances of *Senecio*. In other cases both free nuclear and cellular endosperm have been reported.

A comparison of the embryological features of the tribes Heliantheae and Senecioneae, will indicate that there is a mixture of both primitive and advanced features in the former, whereas in the latter, primitive traits appear to be more frequent. While it is true that embryological data are ambiguous, it is also true that Heliantheae cannot be regarded as more primitive than Senecioneae on embryological grounds.

Unity and diversity of embryological characters in Asteraceae

A biological study of Asteraceae will bring out the fact, that the family indeed symbolizes the dictum **diversity in unity.** Having undergone various gene modifications in the course of evolution, the family has retained a few characters which seem to be immutable. These traits appear to have contributed to the evolutionary success of the family and the genes responsible for these have not altered perhaps ever since the origin of the family. One such morphological character is the capitulum. Within the basic framework of the capitulum (unity) however any number of modifications (diversity) are found in the family.

The embryo development in Asteraceae is another unifying character which is exhibited in all the members without any exception. *Senecio variation* of the Asterad type of embryogeny is of universal occurrence in the family. Even when eggs develop parthenogenetically there is no variation in embryogeny. The genes responsible for embryo development pattern seem to have remained unaltered ever since the origin of the family. However there is diversity in other embryological characters.

The reasons for the occurrence of such unifying characters like, capitulum, and Asterad embryogeny are not far to seek. In traits which are vital for the existence and survival of the family, it looks as if the genes are highly conservative. The possible reasons for this may be :
(a) A single dominant gene which can withstand the pressures of mutation
(b) Group of genes may contribute either collectively or severally towards the unifying characters, so that alterations if any, in one or two genes of the group will not have any effect on the ultimate expression of the trait.

INTERRELATIONSHIPS OF THE FAMILY ASTERACEAE
(on embryological grounds)

Various systematists have accorded different positions to Asteraceae in their scheme of classification. Engler and Prantl (1892) and later Engler (1926) classified the family in the order Campanulatae along with Campanulaceae, Goodeniaceae, Brunoniaceae, Stylidiaceae and Calyceraceae. Bentham and Hooker (1862-83) included the family in the order Asterales along with Valerinae, Dipsaceae and Calycerae. Hutchinson (1926) included Adoxaceae also in the order Asterales. Rendle's (1938) Campanulales includes families Campanulaceae, Goodeniaceae and Compositae.

The families of the order Campanulatae of Engler (1926) have common embryological features like tetrasporangiate anthers, Dicotyledonous type of anther development,mostly glandular tapetum (except in Asteraceae), simultaneous type of cytokinesis in microsporocytes, anatropous, tenuinucellate and unitegmic ovules, Polygonum type of embryo sac development, presence of integumentary tapetum, and *ab initio* cellular endosperm etc.

Rosen (1946), traces the origin of Asteraceae from Lobeliaceae and observes "The forerunners of Stylidiaceae, Goodeniaceae and Compositae (Asteraceae) are to be among like Campanulaceae and Lobeliaceae. According to Leonhardt (1949), the heads of Asteraceae are derived from the thyrsoid type of inflorescence of Campanulaceae. The other families of the order like Stylidiaceae, Goodeniaceae and Brunoniaceae originated from Lobeliaceae at various levels and must have given rise to Asteraceae. According to Rosen (1946) Goodeniaceae and Brunoniaceae may have originated from primitive Goodeniaceous ancestors, which must have had their origin from Lobeliaceae.

According toe Maheswari Devi (1983), the families of the order Campanulatae of Engler (1926) differ from Asteraceae in having more than one row of archesporial cells in the anther, glandular anther tapetum, smooth exine of the pollen grains, three antipodals which degenerate before fertilization, cellular endosperm with both chalazal and micropylar haustoria, presence of starch as reserve food in the endosperm (in most members of Goodeniaceae and Lobeliaceae) and Solanad type of embryo development. Thus, Maheswari Devi (1963) opines that the embryological features are not in favour of retaining the family Asteraceae either in the Campanulatae of Engler (1926) or in the Campanulales of Rendle (1938).

The inclusion of the family in Asterales of Bentham and Hooker (1962-63) and Hutchinson (1926) together with Dipsacaceae and Valerianaceae seems to be justified on embryological grounds (Mahesswari Devi, 1963). In these two families the anther tapetum is plasmodial. Dipsacaceae and Asteraceae share integumentary tapetum. The endosperm is cellular in some members of Asteraceae. Vernin (1952) proposed that Compositae (Asteraceae) could be placed in Asterales along with Dipsacaceae, Valerianaceae and Calyceraceae.
It has to be pointed out however that embryological data to not provide conclusive evidence of inclusion of

Asteraceae either in Campanulatae or Asterales. In such a situation it seems best to obtain data from other sources and embryological data should be used only as a collateral evidence. But the available evidence (on embryological grounds) however points out that Asteraceae is more likely to be included in Asterales rather than in Campanulatae. Chemical evidence on the other hand (see General Introduction) does not seem to favour either of the above mentioned views.

Bessey (1915), Gundersen (1950) etc., have divided Asteraceae into a number of separate sub families. Bessey (1915) has elevated all the 14 tribes to the level of families (Heliantheae, Ambrosiaceae, Heleniaceae, Arctotidaceae, Calendulaceae, Inulaceae, Asteraceae, Vernoninaceae, Eupatoriaceae, Anthemidaceae, Senecionidaceae, Carduaceae, Mutisiaceae and Lactucaceae). Gundersen (1950) elevated the tribe Cichorieae to Cichoriaceae and included it in the Asterales along with Calyceraceae. Johansen (1950) opines "some embryologists have seggregated tribes within the family; while others have split it into a number of separate families. If the latter procedure is adopted it is believed that the embryology of the entire group can be established on a more satisfactory basis". But there does not seem to be any embryological ground to support Johansen's (1950) view, for not only the entire family has random distribution of primitive and advanced embryological characters, even the tribes are not homogenous, in terms of their embryology. The only unifying feature is embryogeny, but it is common to all the tribes justifying the retention of Asteraceae as a single family.

LITERATURE CITED

* AFZELIUS, K. 1924. Embryologische and zytologische Studien in *Senecio* and verwandten Gattungen. *Acta. Hort. Berg.* **8** : 123-219.

* ___________ 1936. Apomixis in der Gattung *Arnica. Svensk bot. Tidskr.* **30** : 572-679.

AHLSTRAND, L. 1978. Embryology of *Urisinia* (Compositae). *Bot Notiser.* **132** : 109-116.

___________ 1979a. Embryology of Arctotidae - Arctotinae (Compositae). *Bot Notiser* **132** : 377-380.

ANDERSON, L.C. 1970. Embryology of *Chrysothamnus* (Astereae, Compositae). *Madrano.* **20** : 337-342.

* BAMBACIONI, V. 1941. Ricuerche morfologiche sulle Lauraceae. Embriologia della *Umbellaria californica a del Laurus canariensis. Ann. Di. Bot.* (Roma) **22** : 99-116.

BANCHETTI, C. 1961. The development of the female gametophyte of *Senecio leucanthemifolia* var. *pinnatifidus. Caryologis.* **14** : 303-311.

BANERGI, I. 1940A. A contribution to the life history of *Tridax procumbens. J. Bombay Nat. Hist. Soc.* **42** : 89-99.

___________ 1940b. A contribution to the morphology and cytology of *Carthamus tinctorius. Proc. Nat. Inst. Sci. India.* **6** : 73-86.

___________ 1940c. A contribution to the life hisotory of *Blumea laciniata. J. Indian bot. Soc.* **21** : 295-307.

___________ 1942. A contribution to the life history of *Blumea laciniata. J. Indian bot. Soc.* **21** : 295-307.

___________ and S. Pal, 1959. A contribution to the life history of *Synedrella nodiflora. J. Linn. Soc. Land. Bot.* **55** : 810-817.

* BATTAGLIA, E. 1945. Fenomeni citologici nuovi nella embriologesche (Semigamia) e nella microsporogenesi (doppio nucelo di restituzione) di *Rudbeckia laciniata. N. G. Bot. Ital.* **52** : 34-38.

___________ 1951. Development of the tetrasporic Embryo sac of *Chrysanthemum viscosum. Bot. Gaz.* **112** : 490-494.

BENTHAM, G. 1873. Notes on the classification, history and geographical distribution of Compositae. *Jour. Linn. Soc. (Bot.)* Lond. **13** : 335-557.

___________ and HOOKER, J.D. 1862-63. *General Plantarum*, London.

BERGMAN, B. 1944. A contribution to the knowledge of the embryo sac mother cell and its development in two apomicts. *Sevensk. bot. Tidsr.* **38** : 249-259.

* BERUTI, R. 1961. Lo svilupo del gametofito femminile di *Hypochoeris aetensis.* Caryologia. **14** : 319-325.

BESSEY, E.A. 1915. The Phylogenetic Taxonomy of flowering Plants. *Annal. Missour. Bot.Gdn.* Vol. II Nos. **1-2.**

BARGEN, L. N. 1972 Embryology and Achene morphology of endemic Canarian species of *Chrysanthemum* subgenus *Argyranthemum* (Asteraceae). *Norw. J. Bot.* **19** : 149-170.

BHARGAVA, H.R. 1935. Contribution to the morphology of *Eclipta erecta. Proc. Indian Acad. Sci.* **B. 1** : 325-339.

* CARANA, E. 1915a. Ricerche sull embriogonesi delle Asteraceae. *Ann. di. Bot.* (Roma) **13** : 251-301.

CHAMBERLAIN, C.J. 1895. Contributions to the embryology of *Aster novae anglica. Bot. Gaz.* **20** : 205-212.

CHENNAVEERAIAH, M.S. & PATIL, R.M. 1971. Apomixis in *Blumea*. *Phytomorphology*. **21** : 71-76.

CHOPRA, S. 1968a. Development of the Female gametophyte in Niger *(Guizotia abyssinica) Curr. Sci.* **38**, (**18**) : 392-393.

____________ 1972. Morphological studies in Niger *(Guizotia abyssinica)* with special reference to the effect of 2, 4-D and gamma rays. Ph. D. Thesis, Agra. Univ.

____________ & SINGH, R.P. 1974. Effect of gamma rays and 2, 4-D on ovule, female gametophyte, seed and fruit development in *Guizotia abyssinica*. *Phytomorphology*. **26** : 240-249.

____________ and ____________ 1976. Effect of gamma rays and 2, 4-D on ovule, female gametophyte seed and fruit development in *Guizotia abyssinica*. *Phytomorphology*. **26** : 240-249.

CICHAN, M.A. & BARABARA, F. PALSER. 1882. Development of normal and seedless achenes in *Cichorium intybus* (Compositae). *Amer. J. Bot.* **69** : 885-895.

COOPER, G.O. 1936. Cytological investigations of *Erechtites hieracifolia. bot. Gaz.* **98** : 348-355.

CRONQUIST, A. 1955. Phylogeny and Taxonomy of the Compositae. *Ann. Midl. Nat.* **53** : 478-511.

* DAHLGREN, K.V.O. 1920. Zur Embryologie der Kompositen mit besonderer Beruckischtigung der Endospormbildung. *Zeit. F. Bot.* **12** : 481-516.

* ____________ 1924. Studien uber die Endospermbildung der Kompositen. *Svensk. Bot. Tidskr.* **18** : 177-203.

DAVIS, G.L. 1961a. The life history of *Podolepis jaceoides*. 1. Microsporgenesis and male gametogenesis. *Phytomorphology*. **11** : 86-87.

____________ 1961b. The life history of *Podolepis jaceoides*. 2. Megasporogenesis, female gametophyte and embryogeny. *Phytomorphology*. **11** : 206-219.

____________ 1961c. Abnormal embryo sacs in *Calotis cuneifolia. Aust. J. Sci.* **23** : 413-414.

____________ 1961d. The occurrence of synergid haustoira in *Cotula australis. Aust. J. Sci.* **24** : 296.

____________ 1962a. Embryological studies in the Compositae. 1. Sporogenesis, gametogenesis and embryogeny in *Cotula australis. Aust. J.Bot.* **10** : 1-12.

____________ 1963a. Embryological studies in the Compositae. 3. Sporogenesis gametogenesis and early embryogeny in *Minuria denticulata. Proc.Linn.Soc.* N.S.W. **88** : 35-40.

____________ 1963b. Generative apospory and diploid parthenogenesis in *Brachycome ciliaris* var. *languinosa. Aust. J.Sci.* **36** : 90.

____________ 1964a. Somatic apospory and polyembroony in *Minuria integerrrima. Nature.* **204** : 94.

____________ 1964b. The embryology of *Minuria integerrima* : a somatic apomict. *Phytomorphology*. **14** : 231-239.

____________ 1964c. Embryological studies in the Compositae. 4. Sporogenesis, gametogenesis and embryogeny in *Brachycome ciliaris. Aust. J. Bot.* **12** : 142-151.

____________ 1964d. Development of the female gametophyte of *Minuria cunninghamii. aust. J.Bot.* **12** : 152-156.

____________ 1966. Systematic embryology of the Angiosperms. John Wiely and Sons, Inc., New York.

___________ 1967. Apomixis in Compositae. *Phytomorphlogy.* **17** : 270-277.

___________ 1968a. Apomixis and abnormal anther development in *Calotis lappulaceae* (Compositae) *Aust. J. Bot.* **16** : 1-17.

___________ 1968b. The embryology of *Epaltes australis* (Compositae). *Proc. Linn.* N.S.W. **93** : 184-192.

DESAI SAROJ, G. 1981a. Apomixis in *Liggera alata.* 68th Indian Science Congress. Part III. Sec. *Bot. Abstract.* P.51.

___________ 1981b. Apomixis in *Laggera aurita.* Fourth All India *Bot. Conf.* pp. 75.

___________ 1982a. Embryological studies in few members of Compositae. Ph. D. Thesis, Univ. Nagpur.

___________ 1982b. Reduced and unreduced embryo sacs in *Solidago purpurea.* 69th Indian Sci. Congr. Part III. Sec. Bot. pp.87.

DESHPANDE, P.K. 1960a. Morphology of the endosperm in *Caseulia axillaris. Curr. Sci.* **29** : 56-57.

___________ 1960b. Studies in the family Compositae. Male and Female gametophytes of *Parthenium hysterophours. J. Biol. Sci.* **3** : 26-29.

___________ 1961b. Fertilisation and development of endosperm, embryo and seed coat in *Eclipta prostrata. Bull. Bot. Soc.* **41** : 505-509.

___________ 1962b. Contribution to the embryology of *Caseulia axillaris. J. Indian Bot. Soc.* **41** : 540-549.

___________ 1962c. A reinvestigation of the endosperm in *Tridax procumbens. Cur. Sci.* **31** : 113-114.

___________ 1962d. Studies in the life history of *Glossocardia bosvallea. Bull. Bot. Sci.* **31** : 113-114.

___________ 1964a. A contribution to the life history of *Volutarella* ramosa (= *V. divaricata*) *J. Indian Bot. Soc.* **43** : 141-148.

___________ 1964b. A contribution to the embryology of *Bidens biternata. J. Indian. Bot. Soc.* **43** : 149-157.

___________ 1970. Compositae. In *Comparative embryology of Angiosperms.* Indian National Science Academy, New Delhi. 325-33.

___________ and M.M. KOTHARE, 1971. A contribution to the embryology of *Adenostemma lavenia.* Proc. 59th Indian Sci. Congr. Part III : 331-332.

___________ and ___________ 1973. A contribution to the life history of *Chrysanthellum indicum* Proc. 60th Indian Sci. Congr. Part III : 327.

___________ and ___________ 1976. Embryology of *Cyathocline purpurea. J. Indian Bot. Soc.* **55** : 205-212.

DIETERT, R.A. 1938. The morphology of *Artemisia tridentata, Lloydia.* **1** : 3-74.

DUTTA, S. 1939. A note on the development of female gametophyte in some Indian Compositae. *Curr. Sci.* **8** :471-472.

ENGLER, A. 1926. *Die Naturlichen Pflazenfamilien.* Lepig.

ESAU, K. 1944. Apomixis in Guayale. *Proc. Nat. Acad. Sci.* (Wash) **30** : 352-355.

___________ 1946. Morphology of reproduction in Guayule and certain other species of *Parthenium. Heilgardia.* **77** : 61-101.

* FAGERLIND, F. 1941. Die Embryosackentwicklung bei *Tanacetum valgare* and einigen *Chrysanthemum. Arten. Svensk. Bot. Tidskr.* **35** : 157-176.

* ___________ 1947b. Macrosporogenesis and Embryosackbildung bei agamospermischen *Taraxacum* biotypen. *Svensk. Bot. Tidskr.* **41** : 365-390.

GAMBLE, G.S. 1948. Flora of Madras Presidency Vol. II.

GATES, R.R. 1920. A preliminary account of the meiotic phenomena in pollen mother cells and tapetum of lettuce *Lactuca sativa. Proc. Roy. Soc. London* **B 91** : 216-223.

GELIN, O.E.V. 1936. Zur Embryologie and zytologie von *Berkheya bergiana* and *B.aldami. Svensk. Bot. Tidskr.* **30** : 324-328.

GILL, B.S. and M.IQBAL. 1981. Embryological studies in *Pterotheca falconeri. Curr. Sci.* **50** : 781-782.

GODINEAU, J.C. 1973. Le Sac Embryonnaire des Angiosperms. Morphogenese et infrastructure. *Bull. Soc. Bot. Fra. Coll. Morphol.*

* GOLDFLUS, M. 1898. Sur la structure et les functions de 1 assise epitheliate et des antipodes chez les Composees. *Journ. de. Bot.* **12** : 374-384.

GORMUGLIO, G., ARENA, M., VILLARI, R. AND ROSITTO, M.1983. Embryological studies in *Senecio ambigus* and *S.gibbosus Phyton* (Horn), **23 (1)** : 39-42.

* GUIGNARD, L. 1881. Sur le role des antipodes. *Bull. Soc. Bot. de France* : 197-202.

* __________ 1882. Recherches sur le sac embryonnaire des phanerogames Angiosperms. *Ann.Sci.Nat.Bot.* **(6) 13** : 136-199.

GUNDERSON, A. 1950. Families of Dicotyledons. Waltham, Mass.

HARLING, G. 1950. Embryological studies in the Compositae. 1 Anthemideae-Anthemidinae. *Acta.Hort Berg.* **15** : 135-168.

___________ 1951a. Embryological studies in the Compositae. 2. Anthemideae-chrysantheminae. *Acta.Hort Berg.* **16** : 1-56.

___________ 1951b. Embryological studies in the Compositae. 3. Asterae. *Acta. Hort. Berg.* **16** : 73-120.

___________ and U. HOLEMBERG, 1961. The development of the embryo sac in *Sanvitalia procumbens. Bot. Notiser.* : 353-360.

* HOFMESITER, W. 1849. Die Entshehung des Embryo der Phanerogamen. Lepzig.

* HOLMGREN, I. 1915. Die Entwicklung des Embryosackes bei *Anthemis tincotoria. Svensk. Bot. Tidskr.* **9**: 171-183.

* ___________ 1919. Zytoloische Studien uber die Corthflanzung Die Den Gattungen *Erigeron* and *Eupatorium.* Kgl. *Svensk. Vet. Akad. Hendl.* **59** : 1-118 Diss. Stockholm.

HOWE, T.D. 1964. Development of Embryo sac in *Ratibida tagetes. Am. J. Bot.* **53** : 678.

___________ 1975. The female gametophyte of three species of *Grindelia* and *Prionopsis ciliata* (Compositae). *Am. J. Bot.* **63** : 273-279.

HUTCHINSON, J. 1916. Notes on African Compositae II Kew Bulletin XXXI.

___________ 1926. The Families of flowering Plants, Vol. I. Dicotyledons.

JOHANSEN, D.A. 1950. *Plant Embryology,* Waltham, Mass.

* JUEL, H.O. 1900. Beitrage zur Kenntnis der Tetradenbilgung, *Juhrb, F. Wiss. Bot.* **35** : 626-659.

KAPIL, R.N. and SETHI, S.B. 1962a. Gametogenesis and seed development in *Ainsliaea aptera.*

Phytomorphology **12** : 222-234.

KAUL, V. 1972. Embryology and development of fruit in Chichorieae II. *Hypochoeris radicata.* Proc. 59th Indian Sci. Cong. Part III. Abstr. No. 150. Page 450.

___________ 1973. Cytology and Embryology of Indian Cichorieae (Compositae). Ph. D. Thesis, Univ. of Rajasthan, Jaipur.

___________ and SINGH, D. 1971. Development of fruit in *Sonchus arvensis.* Proc. 58th Indian Sci. Cong. Part III. Abstr. No. 150. Page 450.

___________ DANTHAN. A.S.R. and SINGH, D.1975. Embryological studies on the genus Sonchus. J.Indian Bot. Soc. 54 : 238-245.

___________ and SINGH, D. 1982. Cytoembryological studies on Indian Taraxacum. Microsporangium, microsporogenesis and microgametogenesis. *Cytologia.* **47** : 125-135.

KOTHARE, M.M. 1972. Embryology of the Family Compositae. Ph. D. Thesis, University of Nagpur.

LALITHA, D. 1981. *Embryological and Palynological studies in Compositae.* Ph. D. Thesis, Bangalore University, Bangalore.

___________ 1986. Embryology of *Pluchea tomentosa. J. Indian bot. Soc.* **65** : 47-52.

LAVILLE, P. 1911. Observations sur la development de l'ovarie en fruit chez les Composees. *Bull. Soc. Bot. France* 58.

MAHESWARI, P. and ROY. S.K. 1952. The embryo sac and embryo of *Tridax Procumbens. Phytomorphology.* **2** : 245-252.

___________ and SRINIVASAN, A.R. 1944. Contribution to the embryology of *Rudbeckia bicolor. New Phytol.* **44** : 135-142.

MASHESWARI DEVI, H. 1957. Embryological studies in the Compositae. 3. *Gerbera jamesonii. Proc. indian Acad. Sci.* B. **46** : 68-74.

___________ 1959. Embryological studies in Compositae and Gentianales. D.Sc. Thesis, Andhra University.

___________ 1963. Embryological studies in Compositae 4. Heliantheae. Proc. Indian Acad. Sci. B. **58** : 274-290.

___________ and PULLAIAH, T. 1976a. Embryological investigations in the Melampodiinae; I *Melampodium divaricatum. Phytomorphology.* **26** : 77-86.

___________ and ___________ 1976c. Embryology of safflower (*Carthamus tinctorius*). *The Botanique.* **7** : 63-70.

___________ and ___________ 1977a. Embryological abnormalities in *Carthamus tincotorius. Acta. Bot. Ind.* **5** : 1-8.

___________ and ___________ 1977b. Embryology of *Eupatorium odoratum,* an ampomict. In : Recent trends and contacts between Cytogenetics, Embryology and Morphology. Today and Tomorrow. pp. 307-314.

___________ and ___________ 1978. Structure and development of Pollen, ovule, female gametophyte and seed in Helenieae (*Hymenantherum tenuifolium* and *Gaillardia lorenziana*). In : Physiology of sexual reproduction in Flowering Plants. pp. 413-417. Kalyani Publishers, New Delhi.

___________ and ___________ 1980. Embryology of *Melampodium cinereum* (Heliantheae; Compositae) Proc. 67th Indian Science Congress. Part III. Abstract. 98 pp. 37.

__________ and RANJALKAR, M.M. Development and structural anatomy of seeds and fruits in Heliantheae (Asteraceae). In : Biochemistry, Developmental and structural anatomy of Angiosperms : A symposium. pp. 59-7.

MARTIN, R.W. and SMITH, F.H. 1955. Megagametophyte development in *Chrysanthemum leucanthemum. Bot. Gaz.* **116** : 243-249.

MARTINOLI, G. 1939. Contributo all embriologia delle Asteraceae. 1-3 *N.G. Bot. Ital.* **46** : 259-296.

__________ 1942. Contributo all embriologia delle Asteraceae. 6 *N.G. Bot. Ital.* **49** : 311-326.

McDONALD, C.C. 1927. A study of seed development in three species of *Erigeron* with special reference to apogamy. *Bull. Torrey Bot. Cl.* **56** : 479-497.

MESTRE, J.C. 1957a. The presence of two ovules in the ovary of some Cynarees. Observation on *Centaurea collina* and *Carduus defloratus. Bull. Soc. Bot. France.* **104** : 37-40.

__________ 1963-64. Recherches d embryogenie comparee : *Les rapports phylogenetiques des Composees.* Diss. Univ. Paris.

MISRA, S. 1957. Floral morphology of the family Compositae. 1 The flower and gametophyte of *Flaveria repanda. J. Indian Bot. Soc.* **36** : 503-512.

__________ 1965. Floral morphology of the family compositae. 3. Embryology of *Siegesbeckia orientalis. Aust. J. Bot.* **13** : 1-10.

__________ 1972. Floral morphology of the Composiate IV. Tribe *Vernonieae; Vernonia anthelmentica Bot. Mag.* (Tokyo). **85** : 187-199.

MITRA, J.N. 1947. A contribution to the embryology of some Compositae. *J.Indian Bot. Soc.* **26** : 105-123..

MOTTIER. D.M. 1893. On the Embryo sac and Embryo of *Senecio aureus. Bot. Gaz.* **18** : 245-253.

MUKHERJEE, P.K. and KOTHARE, M.M. 1973. Life history of *Spilanthes acmella.* Proc. Indian Sci. Cong. Part III. Abstr. Page 330.

__________ and DESAI, S.G. 1980. Apomixis in *Eupatorium triplinervis.* Third All India. Bot. Conf. pp. 61.

MURTHY Y.S. and SHARMA, H.P. 1977. Embryological studies in Compositae. II *Bellis perennis.* A casual apomict. In : Recent trends and contacts between Cytogenetics, Embryology and Morphology. Ed. Dnnansagar. V.R. *et. al.* 257-260.

NEWCOMB, W., 1972. The development of the Embryo sac of sunflower, *Helianthus annus;* before and after fertilisation. Ph.D Dissertation, University of askatchewan, Saskatoon, Saskatchewan.

__________ 1973a. The development of the Embryo sac of sunflower *Helianthus annuus;* before fertilisation. *Can. J. Bot* **51** : 863-878.

__________1973b. The development of the Embryo sac of sunflower *Helianthus annuus;* after fertilization. *Can. J. Bot.* **51** : 879-890.

__________ and STEEVES, T.A. 1971. *Helianthus annuus;* Embryologenesis, Embryo sac wall projections before and after fertilization. *Bot. Gaz.* **132** : 367-371.

NORRIS, H.W. 1892. Development of ovule in *Grindelia squarrosa. Amer. Natur.* **26** : 703-704.

OSAWA, J. 1913. Studies on the cytology of some species of *Taraxacum. Ark. Zellforsch.* **10** : 450-469.

PADMANABHAN, D. 1962. A reinvestigation of the endosperm and endothelium in *Tridax procumbens. Phytomorphology.* **12** : 356-361.

PALM, B. 1914a. ber die Embryosackentwicklung eingier Kompositen *Svensk. Bot. Tidskr.* **8** : 447-453.

____________ 1934. Ein neur embryosack typus bei *Rubdeckia hirta. Bot. Notiser.* 453-427.

PALM, B.T. 1931. Antipodals in *Zinnia, Trans. Illinois State Acad. Sci.* **124**, 143-147.

PANDEY, A.K. 1976. Development of seed and fruit in *Eclipta erecta. Geobios.* **3**,194-195.

____________ 1977. Development and Structure of seeds and fruits in some Compositae. Ph. D. Thesis, Kanpur Univ. Kanpur.

____________ 1980. Structure and behaviour of endothelium in some Compositae. *Acta. Bot. Indica.* **8** : 50-56.

____________ 1983a. Development and structure of seeds and fruits in *Galinsoga parviflora. J. Indian Bot. Soc.* (Supp) 62 : 54.

____________ 1983b. Anatomy of seeds and fruits in some Heliantheae (Compositae). J. Indian Bot. Soc. (Supp) **26** : 54.

____________ and CHOPRA, S. 1979. Development of seed and fruit in *Gerbera jamesonii.* Geophytology. **9** : 171-174.

____________ and SINGH, R.P. 1973. Structure of male gamates in some Compositae. *Plant Science.* **5** : 148-150.

____________ and ____________ 1976. Development of female gametophyte in *Hymenantherum tenuifolium. Curr. Sci.* **45** : 29-30.

____________ and ____________ 1978. Development of seeds and fruits in Compositae, Cichoreae. *Phytomorphology.* **28** : 198-206.

____________ SUMAN CHOPRA and SINGH R.P. 1986. Development and Structure of seeds and fruits in Compositae - Tribe Helineae *J. Indian Bot. Soc.* 65 : 416-421.

PODDUBNAJA ARNOLD, V.A. 1931. Ein versuch der Anwendungung der Embryologischen. Methode beider Losung eingier systematischer Fragen. I. Vergleichende embryologische zytologische untersuchungen uber die gruppe Cynareae. Fam. Compositae. Beigh. Bot. Ztbl. **48A**. 141-237.

POPHAM, R.A. 1938. A contribution to the life history of *Galinsoga ciliata. Bot. Gaz.* **99** : 543-555.

PARKASA RAO, P.S., SUNDARI, K.T. and NARAYANA, L.L. 1979. Some embryological features of *Emilia flammea. Curr. Sci.* **48** : 129-131.

PULLAIAH, T. 1976. Embryological investigations in Compositae with some observations on Basellaceae. Ph. D. Thesis. Andhra University.

__________ 1978a. Studies in the Embryology of Compositae III. The tribe Asterae. *Bot. Mag. Tokyo.* **91** : 197-205.

__________ 1978b. Embryology of *Tithonia. Phytomorphology.* **28** : 437-444.

__________ 1979a. Embryology of *Adenostemma, Elephantopus* and *Vernonia* (Compositae). *Bot. Notiser.* 132 : 51-56.

__________ 1979b. Studies in the Embryology of Compositae IV. The Tribe Inuleae. *Am. J. Bot.* **6** : 1119-1127.

__________ 1981a. Studies in the Embryology of Heliantheae (Compositae). *Plant Syst. Evol.* **137** : 203-214.

__________ 1981b. Embryology of *Acanthospermum hispidum. Curr. Sci.* 50 (1) 992-994.

__________ 1981c. Embryology of *Coreopsidinae.* (Heliantheae : Compositae). Fourth All India Bot. Conf. pp. 56.

__________ 1982a. Embryology Seed Coat and Fruit wall of *Parthenium hysterophorus*. *Jap. J. Bot.* **58 (7)**.

__________ 1982b. Embryology of *Sonchus oleraceus*. *Curr. Sci.* **51** : 708-710.

__________ 1982c. Studies in the Embryology of Compositae. The Tribe : Eupatorieae. *Ind. J. Bot.* 5.

__________ 1982d. Embryology of *Achillae squarrosa*. *J. Indian Bot. Soc.* 61.

RAGHAVAN, T.S. & VENKATASUBBAN, K.R. 1941. Contribution to the cytology of *Tridax procumbens*. *Proc. Indian Acad. Sci.* B. **13** : 85-108.

RAJSEKHAR, G. 1980. The development of embryo sac. endosperm and the structure of fruit wall in *Blumea mollis*. *Curr. Sci.* **49** : 156-158.

RANGASWAMY, V. and PULLAIAH, T. 1986. Studies in the embryology of *Senecio candicans* Dc. (Compositae) *J. Indian Bot. Soc.* **65**, 209-212.

RENZONI-CELA, G. 1963. Ricerche citoembriologiche distribuzione geografica di *Cirsium casabonae*. *N.G. Bot. Ital.* **70** : 493-504.

__________ 1970. (Studies on the genus *Centaurea* (Astereaceae) : Embryology of *Centaurea cineraria* var *veneris*. *N.G. Bot. Ital.* 104 : 467-468.

RENDLE, A.B. 1938. Classification of Flowering Plants, Vol. II> Dicotyledons.

ROSEN, W. 1946. Further notes on the embryology of Goodeniaceae. Meddel, Fr. Goteburgs. *Bot. Tragid.* 16 : 235-49.

ROSENBERG, O. 1906. Uber die Embryobildung in der Gattung *Hieracium*. *Bere dtsch. bot. Ges.* 24 : 157-161.

__________ 1907. Experimental and cytological studies in the *Hieracia*. 2b Cytological studies on the apogamy in *Hieracium*. *Bot. Tidsk.* **125** : 143-170.

SCHNARF, K. 1931. *Vergleichende Embryologie der. Angiospermen*. Berlin.

SALDHANHA, C.J. and NICOLSON, D.H. 1976. *Flora of Hassan District*. Amerind Publishers. India.

SEHGAL, C.B. 1966. Morphological and Embryological studies on *Erigeron bonariensis* (= *E.linifolius*). *Beitr. Bio. Pfl.* **42** : 161-183.

SHARMA, H.P. 1975. Morphology and Experimental studies in some members of Asteraceae. Ph. D. Thesis Meerut Univ.

__________ and MURTHY, Y.S. 1977. Embryological studies in the Compositae 3. Tribe Astereae. *Brachycome iberidifolia*. A casual apomict. *J. Indian Bot. Soc.* **56** : 35-44.

SINGH, R.P. and PANDEY, A.K. 1985. Development and structure of seeds and fruits in Compositae. Cynareae. *Phytomorphology*. **34** (1-4), 1-9.

SMALL, J. 1919. The Origin and Development of Compositae. *New Phytologist*, reprint No. 11. William Wesley and sons. London.

SNOW, E. 1945. Floral morphology of *Chrysothamnus nausoosuspeciosus*. *Bot. Gaz.* **106** : 451-462.

SOUEGES, R. 1920a. Embryogenie des Composees. Les premiers stades du development de 1 *embryon chez le Senecio vulgaris C.R. Acad. Sci.* Paris, **171** : 254-256.

__________ 1948. Embryogenie et classification Essai d'un systeme embryogenique. Partie speciale : Ire Periode du systeme. Paris.

STORK, M.A. 1920. Studies in the genus *Taraxacum*. *Bull. Torrey. Bot. Cl.* **47** : 199-210.

STUESSY, T.F. 1977. Heliantheae - systematic review. In : *The Biology and Chemistry of the Compositae.* Vol I, II. (Eds. Heywood, V.H., Harborne, J.B. and Turner, B.L.) Academic Press, N.Y.

SUBRAMANYAM, K. 1951. Interrelationships of Campanulate. *J.Mysore University* B. **12** : 5-13.

SULLIVAN, V.I. 1976. Diploidy, polyploidy and agamospermy among species of *Eupatorium* (Compositae). *Can. J. Bot.* **54** : 1907-1917.

SUNDARA RAJAN, S. 1978a. Embryological studies in Compositae. Floral morphology, sporogenesis and gametogenesis in *Emilia sonchifolia. Curr. Sci.* **37** : 26-28.

____________ 1968b. Observations on female gametogenesis in *Synedrella nodiflora. Curr. Sci.* **37** : 385-386.

____________ 1972. Embryological studies in Compositae 3. A contribution to the embryology of *Blainvillea rhombodiea. Proc. Indian Acad. Sci.* **B 75** : 167-176.

____________ 1974a. Embryological studies in Compositae 4. A contribution to the life history of *Spilanthes acmella* (= *S. calva*). *Proc. Indian Acad. Sci.* **B. 79** : 267-282.

____________ 1974b. Embryological studies in Compositae 5. On the Embryology of *Eclipta alba. Curr. Sci.* **43** : 56-57.

TIAGI, B and TAIMNI, S. 1960. Embryo sac development in *Vernonia cinerascens* and seed development in *V. cinerea. Curr. Sci.* **29** : 406.

____________ and ____________ 1963. Floral morphology and embryology of *Vernonia cinerascens* and *V. cinerea.* Agra Univ. *J. Res. (Sci).* **12** : 123-137.

VAIDYA, P.B. 1916. Male and female gametophytes of *Acanthospermum hispidum. J. biol. Sci. Bombay.* **4** : 11-15.

____________ 1973. Embryology in *Acanthospermum hispidum. J. Univ. Poona. Sci. Tech.* **44** : 5-10.

VENKATESWARLU, J. 1939. A note on the structure and development of ovule and embryo sac in two species of *Launaea. Curr. Sci.* **8** : 556-57.

____________ 1941b. Vascular supply in the ovules of some Compositae. 1 *Launaea pinnatifida. Proc. Indian Acad. Sci.* **B. 51** : 38-46.

____________ and MAHESWARI DEVI, H. 1955a. Embryological studies in Compositae. 1. *Launea pinnatifida. Proc. Indian Acad. Sci.* **B. 41** : 38-46.

____________ and ____________ 1955b. Embryological studies in Compositae. 2. Helenieae. *Proc. Nat. Inst. Sci. India.* **B. 21** : 149 - 161.

* VERNIN, J. 1952. Contribution a letude du development de 1 albumen et de I 'embryon chez les Composees. Thesis. Paris.

* VESQUE, J. 1978. Development du sac embryonnaire des phanerogames angiospermes. *Ann. Sci. Nat. Bot.* **(6) 6** : 237-285.

VIDAYATHI, O. 1973. Cytoembryological investigations in some Compositae from the tribes Arctotideae and Calenduleae. Diss. Univ. Moscow.

WALTER, R and KUTA, E. 1971. Cytological and embryological studies in *Sonchus : 1 Sonchus asper* and *S. oleraceus. Acta. Biol. Cracov. Ser. Bot.* **14** : 103-109.

* **Original not seen**

4

CHEMOSYSTEMATICS

Plant systematics may be said to be the scientific study of the kinds and diversity of organisms and the relationships among them (Raveb *et al,* 1976). It is a broad field of science which uses characters and data from many disciplines. In former times, much systematic work was based on morphological characters and the study was mainly carried out on examination of preserved herbarium specimens. These studies involved an examination of the morphology of relatively small number of plants.

Present day systematics however lays emphasis on the examination of large number from living populations together with herbarium study. This method of study often called **Biosystematics**, attempts to study as much of the biology of the plant as possible and utilizes these data to clarify taxonomic and evolutionary relationships of the organisms involved.

Modern plant systematics has a very wide data base. Information from such diverse scientific disciplines like Physics, Chemistry, Mathematics, Statistics etc., besides the traditional biological data are made use of in constructing phylogeny and establishment of a meaningful system of classification.

One of the major tools of modern systematics are phyto chemical data which aim to provide chemical basis for the diversity in plants. According to Birch (1974), as both morphological and chemical features are determined by the genetic make up of an organism, the structure of the molecule must be as much a character as any other. Thus it is scarcely surprising that cytological as well as chemical data have provided valuable complementary information about the placement of groups within the taxonomic system.

Cronquist (1973), opines that the morphological characters on which the different systems of classification are based are nothing but external expression of the biochemical complexity. The phytochemical and cytological bases of systematics may therefore be regarded as more basic, fundamental and privileged than other classes of characters (Heywood, 1966).

The application of chemical data together with cytology has been recognised as a useful tool for understanding the genetic variation occurring within plant populations. (More *et al* 1970, Moore, 1972).

Eventhough modern chemosystematics is only a few decades old, the idea of potential importance of chemical evidence in taxonomy is fairly old. Application of chemical data to taxonomy has been traced back to 1699, when Petiver suggested its use for systematic purposes. Even Linnaeus (1751), quoted by Hegnauer in 1962) opined that plants similar in form have similar chemical constituents. Rochleder as early as 1847, published a treatise on phytochemistry. Somc of the earlier taxonomists who have commented on the systematic utility of chemical constituents include DeCandolle, Hallier and Molisch (Hegnauer 1969).

The work of Abbott (1886), represents one of the earliest attempts to correlate chemistry with the phylogenetic level of development. She opined that all saponin containing plants are closely allied and possess some structural similarities (Radford *et al,* 1974). Greshof in 1909 (cited by Gibbs, 1958), suggested that chemical characters should be included in a natural classification. Gibbs (1956-57) established the relationship of some families of dicotyledonous plants and later, He (1963) brought out a historical review of chemotaxonomy. Some of the workers in the past few decades who have made significant contributions to chemosystematics include Alstan and Turner (1963), Swain (1963), Gibbs (1958), Hegnauer (1962), Turner (1967), Fairbrothers (1968), Harborne (1964, 1967a, 1975, 1982), Markham (1982) etc.

Three outstanding reviews on phytochemistry (appeared in early 1960s) which have highlighted the use of chemical characters in finding solutions to taxonomic problems and in analysis of the introgression are :

1. Chemotaxonomie Pflanzen (Hegnauer, 1962-65),
2. Biochemical systematics (Alstan and Turner, 1963a), and
3. Chemical plant Taxonomy (Swain 1963).

In recent years, chemosystematics has been recognised as an indispensible tool of modern taxonomy, not only because of the understanding of the significance of chemical compounds in systematics, but also because of the availability of rapid methods of physical analysis and elucidation of structure of organic compounds with the help of sophisticated techniques.

What chemical compounds are of use in systematics and how to satisfactorily classify the vast array of chemical compounds present in plants is much debated. According to Radford *et al* (1974) "Virtually every chemical component found in plants including both the primary metabolites (involved in vital functions in the cell) and the secondary products of metabolism (accumulated in the cells but not considered to be involved in vital processes) has been used in chemical comparison of taxa at various taxonomic levels".

Regarding the classification of chemical compounds of use in systematics there does not seem to be a satisfactory classification yet. Mentzer (1966), recognised three large classes of substances which can further be subdivided. These are

- i. Primary or basic constituents (proteins, nuclei acids and their derivatives, chlorophylls, polysaccharides etc).
- ii. Secondary constituents (Amino acids, betalins, flavonoids etc). and
- iii. Miscellaneous substances (Crystals, raphides, cystoliths, starch grains etc).

Turner (1967) recognised only two groups :

- i. Macromolecular compounds (proteins, nucleic acids etc) and
- ii. Micromolecular compounds (Amino acids, phenolic compounds etc.,)

The application of macromolecules and micromolecules to taxonomic problems involves basically different approaches (Seigler, 1981).

Studies of macromolecules involve the primary products of DNA. Changes in amino acid sequence within the protein indicate alteration in the base pair sequence in the DNA molecule. Amino acid sequence in proteins, especially in cytochrome *c* indicate that these data provide valuable information about phylogeny and relationships at the higher taxonomic levels like families, orders etc. (Seigler, 1981). However very little information is available on the sequence of amino acids in plant cytochromes. It has been made evident that sequences of amino acids in genera of the same family are more similar to each other than to those of other families, and that families thought to be closely related by morphological evidence generally resemble each other more closely than less related families (Fairbrothers *et al* 1975).

Some of the recent investigations have established that micromolecular chemical (Secondary metabolities) data can also provide valuable insight into evolutionary process. According to Turner (1967, 1969b) chemical studies on secondary metabolities have been proved to be useful in resolving many problems of speciation and evolution, but in contrast to protein sequencing data having generally been applied to the study of lower taxonomic categories. However they may also be of value at higher taxonomic levels (Seigler, 1981).

In order to understand the role played by secondary metabolites in systematic considerations, it is necessary to find out how and why they arose in plants. Many of the proteinaceous materials in plants are enzymatic. In primitive organisms, these are involved mainly in the synthesis of primary metabolites. As these organisms evolved, genetic material and its derived proteins duplicated and increased both in terms of quantity and redundancy. Mutations occurring at this level changed the functioning of the enzymatic proteins and consequently the products. These products called secondary metabolites were of a less critical nature and less widely distributed due to the random occurrence of mutations. This underscores the fact that secondary metabolites are direct gene products and the ability or inability to produce them can be inherited. As a result, organisms producing a kind of secondary metabolite may be regarded as related or have arisen from a common source.

Determination of the homologous orogin of similar compounds in several systematic groups is one of the basic problems inherent in the taxonomic application of secondary compounds (Hegnauer 1967). Thre are two ways in which taxa accumulate the secondary metabolites. These are,

1. Taxa which are basically different might accumulate the same compound if they have been subject to same mutational pressure, which may be regarded as convergent evolution and
2. Taxa accumulate the secondary metabolities following the same pathway which is a true indicator of phyletic association. It is therefore suggested that the mere presence of a secondary metabolite in two taxa need not be regarded as indicating a phyletic relationship if other characters do not warrant such a treatment (Seigler, 1974).

Of what use are secondary metabolities to the plants? Are they merely products of altered genes produced due to random mutation and being only of diagnostic value? Secondary metabolites are distributed in all parts of the plant both in lower and higher plants. But their function and diversity is yet to be satisfactorily explained. Classically, the secondary metabolites have been viewed as waste or excretion products (Mothes, 1966). But some of the recent researches indicate that they may have many coevolutionary, defensive and attractive roles (Whittaker and Feeny 1971, Freeland and Janzen 1974, van Emden, 1973). They may be also of primary metabolic importance (Seigler and Price, 1976; Robinson, 1974; Swain, 1976; Jones, 1979). Flavonoids, particularly anthochlor pigments act as honey guides thus making the plants conspicuous to insect pollinators. Some of the flavonoids are known to be controlling the growth in pea plants (Galston, 1969).

The fact that secondary metabolites are not altogether redundant for the plant can be supported by the observation that they (Secondary metabolites) have survived the selective pressure over the long years of evolution. Their widespread distribution and immense diversity is another evidence of their necessity to plants. There is a great need for intensive work on secondary metabolites with reference to the significance of their development and environmental variation. Table No.4.1 gives a list of major groups of secondary compounds in plants.

Among the secondary metabolites, the ones that exhibit great diversity and much useful in systematic considerations are the phenolic compounds. The term phenolic compound encompasses a wide variety of plant substances, which possess in common an aromatic ring bearing one or more hydroxyl substitute. Phenolic compounds usually occur in vacuole and are water soluble as they frequently combine with sugar to form glycosides. All the phenolic compounds show intense absorption in the UV region of the spectrum due to their aromatic nature. In addition to this, all phenolic compounds exhibit bathochromic shifts in their spectra in the presence of alkalies.

Among the various phenolic compounds, flavonoids are perhaps the most ubiquitous in land plants and most exploited for taxonomic purposes. Bate Smith (1962) regards them as chemically previleged. Alston and Irwin (1961) believe that variation in flavonoid substances offers a greater potential for taxonomic work than amino acids. Harborne (1976b), remarks that flavonoids can be used as taxonomic markers as they posses "The necessary requirements for chemical charactaristics to be useful in plant Taxonomy". These requirements are :

(a) Structural variability
(b) Chemical stability
(c) Wide spread distribution in the plant kingdom and
(d) Easy and rapid identification

Majority of the flavonoid studies for systematic purposes are based on chromatographic techniques (paper, TLC, or GLC). Chromatographic evidence obtained in the form of spots on chromatogram can be used as a corroborative evidence for systematic purposes (Rao, 1980). Grant (1968) and Gupta (1978) opine that chromatographic spots are excellent markers and are much more important than mere chromosome numbers. For the purposes of systematic study, the compound in question is extracted with a suitable solvent and separated using the technique of either chromatography or electrophoresis. The presence of similar flavonoid spots on the chromatogram may serve as a preliminary evidence of their relationhip. But as has been pointed out already, when other data do not support such a presumption, the mere presence of identical spots on the chromatogram should not be taken as a definite evidence of relationship, as the compound in question may accumulate in taxa following entirely different biosynthetic pathways. At the lower taxonomic ranks however (genus, species etc), presence of identical spots may be regarded as indicative of affinity as the taxa under study are already known to be related. This would facilitate the study of interspecific or intraspecific variation. Flavonoids have relatively low molecular weight and are non polymeric and products of metabolism. They are not actively concerned in cellular metabolic processes (Bate Smith, 1963). These characters together with their immense structural variability due to genetically controlled substitution of OH and OCH_3, at different carbon positions on the flavonoid skeleton and their non susceptibility to environmental changes enhance the importance of flavonoids as taxonomic markers. Harborne (1984), recognises some ten classes of compounds (see table 4.2) in flavonoids. Their colour properties in visible and UV light are given in Table 4.3 Among the functions attributed to flavonoids the important ones are disease resistance and attracting pollinators (Harborne 1967a), inhibition of growth of saprophytic organisms (Swain 1973), determination of feeding preference in animals, particularly insects (Bate Smith, 1963; Zielske *et al*, 1972 and Doskotch *et al* 1973). Isman and Duffey (1981), have indicated that flavonoids may offer natural resistance to plants possessing them.

A correlation exists between the distribution pattern of flavonoids and evolution from woody habit to herbaceous habit in flowering plants. Woody plant families predominatly contain flavonols which are infrequent or replaced by flavones in the predominatly herbaceous families. This suggests that during evolution, the ability to oxidize the 3' position in the flavonone precursor is lost (Harborne, 1966, 1967a, Crowden *et al*, 1969).

Flavonoid distribution pattern has been of systematic value mainly in the following respects. These are :

(a) In determining the evolutionary trends within the family as primitive species have simpler flavonoid patterns. But this may not be true always. As seen in several groups of species from the related genera

Parthenium, Hymenoxys and *Ambrosia* of Asteraceae, more highly evolved members have simplified patterns of secondary compounds (Mabry, 1974).

(b) In confirming the reclassification within the families and

(c) In providing a choice between alternative system of classification.

There are several instances where flavonoid compounds have been studied for various taxonomic purposes like phylogenetic affinities, tracing the ancestry of natural hybrids, species interrelationships etc. Some of these are - separation of species in Iridaceae (Riley and Bryant, 1961); study of interspecific hybrids in *Baptisea* (Alston and Turner, 1963b); tracing ancestry of amphidiploid *Viola* (Stebbins *et al* 1963); evolution in the *Asplenium* complex (Smith and Levin, 1963); species interrelationship in *Lotus* (Harney and Grant 1964, 1965; and Grant and Wheltter 1966); *Tragopogon* (Brehm and Ownbey, 1965, Belzer and Ownbey, 1971) and *Avena* (Rajhanthy *3et al* 1971); systematic studies in *Citrullus vulgaris* (kaur *et al* 1973) and *Garcinea* (Chennaveriah and Razdan, 1980) etc.

The use of flavonoids in taxonomic research has been reported from a wide range of taxa, such as *Pinus* (Erdtman 1963); Centrospermae (Mabry, 1966); *Malus* and *Pyrus* (Williams 1966); Gesneriaceae (Harborne 1966, 1967b); Leguminosae (Alston and Turner, 1963; Alston 1966); Eucryphiaceae (Bate Smith and Swain, 1966) etc.

A few other instance where flavonoids have helped in systematic considerations are as follows - Bignoniaceae and Gesneriaceae (Harborne, 1967c); Plumbaginaceae (Harborne, 1967b); *Primula* (Harborne 1968); tribe; Genistae of Leguminosae (Harborne 1969), *Potentilla* spp (Asker and Frost, 1970); *Briza* (Moore *et al* 1970); Barley (Frost and Holm, 1971, Frost *et al* 1979); *Potentilla* (Asker and Frost, 1972); *Asplenium* complex (Smith and Harborne, 1971); oats (Asker and Frost, 1973); Umbelliferae (Harborne, 1971b); Carprifoliaceae (Bohm and Glennie 1971); *Avena* (Itiyama and Grant, 1972); *Galax aphylla* (Harborne and Williams 1973); *Ulmus* (Bate smith and Richens, 1973); *Populus acuminata* (Crawford 1974); phylogenetic implication of Julinaceae (Young 1976); Polemoniaceae (Smith *et al* 1977); *Saxifraga* (Bohm and Wilins, 1978) and intraspecific variation in *Chenopodium fremontis* (Crawford and Mabry, 1978).

Crawford (1978) has reviewed the use of flavonoids in the biosystematic studies of many monocotyledonous families.

Ellagaitannis (a class of flavonoids) have a very limited distribution. This has been used to shift *Daphniphyllum* from Euphorbiaceae to Daphiniphyllaceae (Okuda *et al* 1980).

Some of the recent contributions on flavonoid based systematics include those on numerical chemotaxonomy of *Bauhinia* (Nageshwar *et al* 1984); anthochlor pigments from the petals of *Musseanda* (Harborne *et al* 1983); survey of leaf flavonoids in Portulacaceae (Narayana, 1986); flavonoids as taxonomic markers in some Cocosoid palms (William *et al* 1983); chemotaxonomy of *Ceratophyllum* (Sundari *et al* 1982); chemotaxonomic study of Umbelliferae (Santa Ram *et al* 1983); chemosystematics of *Ipomea* and some other related data (Nair *et al* 1986); chemotaxonomy of Vitaceae of Bihar (Singh *et al* 1986) etc.

Flavonoid spots obtained on the chromotogram (TLC) have been subjected to statistical analysis by many workers resulting in numerical assessment of phyletic relationship. Various numerical parameters like *paired affinity index, coefficient of similarity, matching coefficient, biochemical distance* and *correlation coefficient* have been adopted to analyse the flavonoid spot pattern on the chromatogram and interpret it in terms of phylogenetic relationship. There have been numerous contributions on numerical chemosystematics. Some of the noteworthy contributions which began with the pioneering work of Sokal and Sneath (1963) and

Ellison *et al* (1962) are in *Brassica* (Dass and Nybom, 1967); *Saxifraga* (Jaworska and Nybom 1968); *Secale* (Dedio *et al* 1969b); *Aegilops* (Kaltsikes and Dedio, 1970a and b); *Avena* (Dass, 1972); *Triticinae* (Bose, 1972); *Curcumerinae* (Dass *et al* 1974); *Petunia* (Natarella and Sink 1974); *Anona* (Dass *et al* 1975); *Hordeum* (Frost *et al* 1975, 1979); *Setaria* (Gupta and Singh, 1977); *Blumea* (Singh and Jorapur 1975;____________ Arditti and Fischh 1977; Bopiah, 1986) and Campanales (Leela Rao, 1983).

Spot pattern obtained on TLC plate can be satisfactorily interpreted where the difference between the two taxa are large and qualitative in nature. In cases where the differences are small or quantitative in nature, it is generally felt that more critical methods of analysis and evaluation are necessary. Many workers like Lorens, Schultz and Schaefter (1964), Mathews (1966), Wiemarch (1972) have pointed out the merits and demerits of numerical chemosystematics. Dass *et al* (1976) have discussed the relative merits of various numerical parameters chiefly using spot intensity and spot area on TLC and have suggested that biochemical distance calculated without squaring the spot areas are most reliable.

Asteraceae has been most extensively investigated from the point of view of chemosystematics based on secondary metabolites. The family is chemically distinct Inulin type fructans, seed oils containing characteristic fatty acids, bitter sesquiterpene lactones, pentacyclic triterpene alcohols, accumulation of derivatives of caffeic acid, flavones and flavonols, essential oils and acetylenic compounds (except for Senecioneae), constitute the chemical make up of the family. According to Hegnauer (1977) "The combined occurrence of sesquiterpene lactones, acetylenic compounds and inulin type fructans is almost as characteristic of the Compositae as are their head like inflorescences". Triterpenes and flavonoids are present in almost every member of the family. According to Harborne (1977), flavonoid pigments are universally distributed in Compositae and are potentially promising as taxonomic markers.

Eventhough Asteraceae is one of the families where extensive and intensive investigations have been carried out on flavonoids, considering the size of the family, the available data are really meager. Harborne (1977) comments "considering the vast size of the Compositae (Asteraceae), our information on the flavonoids in the family is very sparse. Relatively few species have been examined perhaps some 400 taxa representing 100 genera -----". Eversince the publication of the book, The Biology and Chemistry of the Compositae (Eds. Heywood *et al* 1977), chemical data on the family have been pouring in. Even then the data are still inadequate for a detailed interpretation. The available data however clearly point out the trends of chemical diversity in the family. A considerable range of flavonoid and of glycosidic or other combinations have been encountered in different tribes and substribes of the family. The common flavonols, kaemferol and quercetin and the common flavones, apienin and luteolin are widely distributed in the family. Surprisingly the leaf flavonoids are not what is excepted in a highly evolved family (Harborne, 1977).

Previous work on flavonoids in Astercear are numerous. Two reviews viz. The Biology and Chemistry of the Compositae Vols. I and II (Eds. Heywood, Harborne and Turner, 1977) and - The Flavonoids, Advances in Research (Eds. Harborne and Mabry, 1982) provide an excellent account of almost all flavonoid researches in the family. Some of the principal works which have a bearing on systematics are cited here.

Flavonoid structure in Asteraceae began perhaps as far back as 1902, with the studies of A.G. Perkin on the flavonol quercetagetin of *Tagetes erecta*. But as Harborne (1977) points out, much of the flavonoid data available has arisen accidentally as a by product of investigations quite unrelated to plant taxonomy. Bate Smith (1962), in a survey of dicotyledonous Angiosperms has analysed the flavonoids of 39 species belonging to 28 genera. The following are some of the noteworthy contributions to the study of flavonoids of Asteraceae - studies on Anthocyanins of *Dimorphoteca* (Ball *et al* 1972), chemosystematic aspects of flavonoid distribution in 22 taxa of *Helenium* (Beirner, 1973); flavonoids of *Lasthenia* (Bohm et al 1974); variations in the chromatographic patterns in *Tragopogon* (Brehm and Ownbey, 1965); flavonoid chemistry of *Coreopis mutica* (Crawford, 1970); anthochlor pigments in *Coreopsis grandiflora* (Geissman and Heaton, 1943); correlation between flavonoid chemistry and plant geography in *Senecio radicans* complex (Glannie et al 1971); leaf flavonoids and systematics in *Matricaria* and *Tripleurospermum* (Greger, 1975); flavonoids and the systematics of Anthemidae (Greger, 1969); flavonoid pattern in the *Chrysanthemum* complex, chemosystematics of *Pyrohappus* (Northington, 1974); isolation of flavonoids in *Polymnia uvedalia* (Herz and Bhat, 1973); constituents of *Cosmos bipinnatus* (Nakaoki, 1936); isolation of quercetaget in from *Parthenium tomentosum*; constituents of *Bidens frondosa* (Romussi and Pagani, 1970); isolation of Hymenoxin, a flavonoid from *Helianthus anguistifolius* (Waddell, 1973); flavonoid and other phenolics of *Cichorium* (Rees et al 1984); flavonoids of *Artemesia sublessingiana* (Rubkhovskaya *et al* 1985); flavonoids of *Senecio aureus* (Pegay, 1986); flavonoids of *Anaphalis veluntina* (Ikramov *et al* 1986); polyoxygenated flavones from *Ageratum conyzoides* (Vyas *et al* 1986); chemotaxonomy of wormwood - *Artemesia obsinthum* (Chivala *et al* 1983); chemosystematic study of phenolics in *Sonchus* (Mansour *et al* 1983)l the flavonoids of *Arnica frigida* and *A.lousieana* (Merfort *et al* 1986), Polyphenolic compounds of *Galinsoga ciliata* (Szepczynaka *et al,* 1984); flavonoids of *Spilanthes oleraceous* (Verykokidou vitsaropoulos, 1983); leaf flavonoid chemistry and the taxonomy of *Coreopsis* sections *Pugiopappus* and *Euleptosyne* (Crawford *et al* 1984); chemosystematics of *Blumea* (Dakshini and Prithipal Singh, 1977) and systematic implications of flavonoids and sesquiterpene lactones in *Vernonia* (Mabry 1975). Flavonoid composition correlating with geographical distribution has been reported in three species of *Hymenoxys, H. scaposa, H. acaulis* and *H.ivesiana* (Seeligman and Alston 1967). The three species contain more than 30 flavonoids. The pattern of distribution of these compounds are correlated more strongly with population positions along an east west gradient extending from Arizona to Texa. It was observed that small and isolated populations of island taxa have fewer and simple flavonoid compounds than their mainland counterparts which are not isolated (Mabry 1973).

Ellison *et al* (1962) in a very significant chemosystematic work on *Bahia* have shown as to how crude biochemical profiles can be used for comparison within a genus. Using several species of *Bahia,* they have described a method of representing the chemical affinities of each species in comparison with others on a polygonal graph. The patterns of these polygons can be looked upon as indicative of biochemical similarities or differences. Based on their observations, Ellison *et al* (1962) have indicated that the genus *Bahia* consists of two phyletic series, the *alternifolieae* and *oppositifoliae.*

Gelnnie *et al* (1971) have correlated the flavonoid chemistry in 44 clones from 25 species of *Senecio,* mainly those belonging to the succulent *S.radicans* complex. They have opined that flavonoid patterns did not show any relationship either with morphology or cytological characters. Instead they showed correlation with plant geography. They have shown that taxa from Madagaskar, the Canary islands and Kenyà and markedly different in their flavonoid complement from south and south west African plants.
Bain and Denford (1985), have studied the flavonoid variation in the *Senecio streptanthifolius* complex. The complex includes *S.streptanthifolius* and four related species *S.pauper culus, S. multilobatus S. carnus* and *S.neomexicanus.* They have divided *S.streptanthifolius* into two subgroups.

Belzer and Ownbey (1971) have studied the inheritence of flavonoid patterns in several species of *Tragopogon* and their F_1 and F_2 hybrids. Inheritence was generally additive in the F_1 plants while considerable seggregation and recombination occurrred in the F_2 plants.

Robinson *et al* (1978) have correlated the chemical constituents in *Lagascea, Allosipermum, Calea,* etc., with subtribal classification of the tribe Heliantheae.

The genus *Bidens* has been extensively investigated for its anthochlor pigments. Analysis of anthochlor pigments in terms of pollination ecology, grouping of taxa in *Bidens pilosa* complex and natural hybridization studies have been conducted to understand and analyse the genome diversity. Scogin and Zakar (1976) have shown in an analysis of floral UV absorption area, that the anthochlor pigments are present in areas correspponding to the UV absorption region. In a study of 15 species of *Bidens* they have shown that there are three groups based on UV absorption/reflection. These are

(a) Anthochlors absent in the ligule,
(b) Anthocholors present only at the basal 50% of the ligule and
(c) Anthochlors present all along the length of the ligule.

This grouping seems to correspond well with the taxonomic grouping of the investigated species in sections, like, Meduseae, Platycarpea, Psilocarpaea and Heterodonta.
In a report on the rayed and rayless populations of *Bidens pilosa* from South India, Sundara Rajan (1977) has suggested as to how the elimination of rays has helped in the rayless population getting established in the area.

The review of the previous chemosystematic work on Asteraceae clearly indicates that in spite of the extensive work, much needs to be still done in this area. In addition to this, chemosystematic work on Indian Asteraceae is woefully inadequate. Except for the chemosystematic work on *Blumea* (Dakshini and Singh 1977), there does not seem to be any other report on flavonoid profiles and systematics of Indian Asteraceae. In view of this, the present investigation is undertaken.

The present work encompasses 51 taxa of the tribes Eupatorieae (4 taxa), Heliantheae (26 taxa) and Senecioneae (21 taxa). Very few taxa (4) are studied in Eupatorieae only to assess the spot pattern of the locally available members.

In the tribe Helaintheae (26 taxa) the purpose is to analyse

(a) Correlation with substribal classification with the help of locally available members.
(b) Study of possible interspecific hybridization in *Galinsoga.*
(c) Study of intraspecific variation and adaptive potentiality in *Bridens pilosa, Eclipta prostrata, Spilanthes* and *Synedrella nodiflora.*

In the tribe Senecioneae (21 taxa) four genra represented in the local flora (*Senecio, Gynura, Emilia* and *Notonia*) have been studied to understand the intrageneric relationship. In the genus *Senecio* (13 taxa), a detailed numerical assessment is worked out mainly to understand the interspecific relationship, generic affinity and correlation of flavonoid spots with morphological traits. In the other genera flavonoid spot pattern has been analysed in relation to species categorization.
In most of the instances flavonoid spot pattern has corroborated the observed morphological findings and appears to complement them.

In the present work. Identification of individual flavonoids is not attempted nor are they characterized chemically, since the main idea of the work is to analyse the spot pattern for systematic purposes. The findings of the present work indicate that even without the identification and characterization of individual flavonoids, the spot pattern on TLC plate gives enough evidencde for a basic chemosystematic work.

MATERIALS AND METHODS

The plants were collected from various locations in Karnataka (see map), except for *Notonia walkeri* which was collected from Nilgiri Hills, Tamilnadu.
Leaves were picked from the healthy plants belonging to the same age group, for the extraction of flavonoids.

Extraction

The leaves were first cleared in water and cut into small bits, air dried and stored. Flavonoids were extracted following the method recommended by Harborne (1973a). The leaf tissue was powdered (0.2gm) and hydrolysed with 25ml of 2N Hydrochloric acid for about 45 minutes in a boiling water bath. The solution was allowed to cool and then filtered. The filterate was extracted twice with 25ml (each time) of ethyl acetate. The two fractions of the ethyl acetate extract were pooled together and left overnight for evaporation and concentrated to dryness. The flavonoids which remain as residue adhere to the sides of the glass vial. The residue was dissolved in 1ml of ethanol and a suitable aliquot was used for chromatographic studies.

Preparation of TLC plates

The method described by Stahl (1969) was followed in the preparation of chromatographic plates. Cellulose microcrystalline of 'Loba chemie', India was used for thin layer chromatography, 15 grams of cellulose powder was mixed with 90ml of double distilled water and homogenized into a slurry in a blender. The slurry was poured off into an applicator (Desaga type) and the thickness was adjusted to 0.3mm.

Glass plates of 20cm x 20cm and 3mm thickness were used for coating with the cellulose slurry. These were arranged on the plastic template in a straight line. The applicator was placed at the left and of the template and the cellulose slurry was poured into the applicator tube. The lever of the applicator tube was turned by 180^0, and the cylinder was moved evenly with a uniform speed over the glass plates spreading the cellulose slurry over (the glass plates) with a thickness of 3mm. The coated glass plates were allowed to dry for about 24 hours at room temperature.

Loading

The coated glass plates were marked into a 17cm x 17cm square for the purpose of migration of the solvent and the sample.

A spotting guide was used to load the sample uniformly on to the chromatographic plate. Approximately 2 ml of flavonoid extract was used for spotting. The extract was spotted with the help of capillary tube about one cm., from the boundary of the plate. The spots were allowed to dry for about 1/2 an hour. The front lines were marked on the plates in both the directions and 4 plates were stacked together using spacers at each corner. The outermost plates was tied by rubber bands. In this method 4-8 plates could be developed simultaneously (Balsgard system - Nybom, 1966).

Developing the chromatogram

The chromatographic chamber was filled to a height of about 0.5cm from the bottom with the developing solvent. The chamber was allowed to stand for about 30 minutes to make it saturated with solvent vapours. The chromatographic plates were introduced into the chamber and were allowed to get saturated with the solvent vapour before the solvent was allowed to run on the chromatographic plates. After 30 minutes, the chromatographic plates were dipped into the solvent and the time needed for the solvent to run the distance of 17cm was noted. The plates were then removed, thoroughly air dried and reintroduced into another chromatographic chamber saturated with the vapour of the solvent used for the second run. The direction of the second run of the solvent on the chromatographic plate was at right angles to the first run.

Solvent

Of the several solvent combinations tried, the following, gave a very good separation of flavonoids.

First run	Chloroform	Acetic acid	Water
	50%	45%	5%

Second run	Acetic acid	Water
	50%	50%

After the second run, the plates were removed from the chamber separated out from the stack and air dried.

Detection of Flavonoids

The dried chromatographic plates were exposed ammonia fumes in order to detect the phenolic spots. Spraying of freshly prepared mixture of 1% $A1C1_3$ + 10% NaOH in water was also tried to detect the spots on the chromatogram. Between the two methods, exposure to ammonia fumes was found to be better.

After exposure to ammonia fumes, the plates were observed under long wave UV light (360nm)

The boundaries of the various spots were marked on the plate and the colour noted. A tracing sheet was kept on the chromatographic plate and the boundaries of the spots were copied.

For each taxon, at least six plates were developed in order to minimize experimental errors and to obtain an average of the Rf values of the spots. Each distinct colour of the spot at any particular Rf value was regarded as a distinct phenolic compound and the total number of spots on the chromatographic plate represents the total number of different phenolic compounds for a particular taxon. Rf value was calculated separately for both directions of the solvent.

Master chromatogram

The phenolic spots of two or more taxa belonging to a genus were interpolated on to a common chromatographic diagram called Master chromatogram or composite chromatogram (Composite key).

On the basis of Rf value and colour under UV light, spots occupying identical positions on the chromatogram were given the same number indicating that they represents the same compound. The area of each spot was recorded with the help of a planimetre for the purpose of numerical assessment. These are represented in each chromatogram by means of histograms.

Numerical assessment

Data obtained from chromatograms have been used to interpret them in terms of homology of phenolic compounds amongst the various taxa studied. Numerical assessment has been made at the level of genus in most instances and in the case of the tribe Heliantheae, the chromatographic data of the taxa have been used to assess the subtribal classification.

In order to assess the phyletic relationship between the taxa, the following similarity measures of numerical taxonomy have been applied. These are :

(a) Matching coefficient and Paired affinity index
(b) Coefficient of similarity and negative correlation
(c) Biochemical distance.

The first two measures have been used to indicate and assess the homology between the phenolic compounds of the taxa and these relate to the qualitative variations depending on the presence or absence of identical spots. The biochemical distance however relates to quantitative variations and depends on the intensity (area) of the spots. These numerical methods of assessment have been described by Sokal and Sneath 91963) and used by Dass and Nybom (1967), Dedio *et al.*, (1969) etc.

Matching coefficient

This deals with the comparison of identical spots (positive match) between the two taxa and also deals with the absence of certain characters (negative match) in the two taxa taken for comparison. Consequently, this can be used when the taxa used for comparison are large. There must at least be three taxa involved to identify the negative matches. Matching coefficient is expressed as follows.

$$\textit{Matching coefficient} = \frac{m+n}{m+n+d}$$

where, m = number of positive matches (substances in question occuring in both the taxonomic units).
n = number of negative matches (substances in question missing in both the taxonomic units)
d = number of differences (substances occuring in only one of the taxonomic units)

Coefficient of similarity

This deals with the positive matches i.e., number of characters (identical spots) common for both the taxonomic units. This can be employed even when the taxonomic units available for comparison are only two. Here of course, the number of negative matches will not be known and the assessment of phyletic relationship has to depend on only positive matches. Coefficient of similarity is expressed as follows :

$$\textit{Coefficient of similarity} = \frac{m}{m+d}$$

Negative correlation

This is a method to assess the relationship based on the degree of difference between the two taxa in question. This is calculated as follows.

$$\frac{d}{m+d}$$

Here, a higher value obtained indicates a none too close relationship, while a lesser value indicates a closer relationship.

Biochemical distance

A study of the chromatograms will not only reveal a qualitative variation of various phenolics, but also a quantitative variation in terms of intensity as indicated by the area of the spots. In order to calculate the similarity measures between the taxa, the quantitative variation i.e., relative amount of substance in each spot has to estimated. For this purpose, the expression of taxonomic distance described by Sokal and Sneath (1963) and used by Dass *et al* (1967) as *Biochemical distance* has been calculated. This can be expressed as follows :

Biochemical distance = $n/A_1 - A_2$.
Where n = total number of different substances studied.
A_1 and A_2 = areas of the corresponding spots of the taxonomic units A and B in mm^2.
Biochemical distance can be used to indicate the relationship between any two taxa in a group or a genus.

Paired affinity index

This has been calculated only for the genus *Senecio* (Senecioneae) as more number of taxa (12) are studied. Paired affinity index (PAI) is useful in illustrating the degree of chromatographic similarity of difference between species in a large group. PAI is calculated (Ellison *et al* 1962) as follows.

$$\text{PAI} = \frac{\text{Spots common for species A and B}}{\text{Total spots for A and B.}} \times 100$$

PAI is calculated as shown above for each pair of species involved in the study and the resulting percentages are plotted on the appropriate arms of polygonal graph (Hutchinson, 1936). Each polygonal graph shows the comparative quantitative relationship of a single species to all others. PA indices are noted from 0 to 100 with zero at the centre of the circle and 100 at the tips of radii. One value for each species will be always 100, since it represents the species under consideration.

The advantage of polygonal graph in presenting the data, is that, at a glance it shows the pattern of biochemical affinity of any one species with all others of group. According to Ellison *et al* (1962), two polygonal graphs similar in shape (though not in area) indicate a closer affinity between them.

Similarity Matrix

Similarity matrices are constructed computing the data from matching coefficient, coefficient of similarity, negative correlation and biochemical distance.

Group affinity values (GA)

GA has been worked out only for the species of *Senecio*. It is obtained by summing up the values obtained in matching coefficient, negative correlation or paired affinity indices (Ellison *et al* 1962). For instance the summation of PA indices of one taxon with each of the others under consideration provides a numerical value which is called group affinity value. In the same way GA can be calculated for matching coefficient and negative correlation. In negative correlation however, the GA values are actually disaffinity values for they are negative. A higher value of GA in negative correlation indicates lesser relationship and *vice versa*.
For the 12 species of *Senecio* the maximum GA value possible is 1200. This indicates 100% affinity of any one taxon with all others. The minimum possible GA value is 100, which indicates that the taxon in question has no spots common with other taxa considered for comparison.

Isolation values

This has been calculated only for the species of *Senecio*.
The presence of distinctive compounds within the species or specific spots are of potential value in systematic considerations as they provide a clue as to how far the taxon in question is specialized or isolated from the group. This is expressed numerically in terms of *Isolation value* (*Iv*). The total number of species specific spots in relation to the number of spots gives the *Iv*. Isolation values can be calculated by two different methods Iv_n and Iv_s (Ellison *et al* 1962).

$$Iv_n = \frac{\text{Number of species specific spots}}{\text{Total number of spots for all the taxa under consideration}} \times 100$$

$$Iv_s = \frac{\text{Number of species specific spots}}{\text{Total number of spots in the taxon under consideration}} \times 100$$

Iv_n deals with the percentage of distinctive spots with reference to the grand total of all different spots for all species of the groups. When the group consists of large number of species however the Iv_n decreases in usefulness, if there is no close relationship between species. Under such circumstances Iv_s values which give the percentage of unique spots of a species with reference to the total spots of that species are more useful (Ellison *et al.*, 1962). Both Iv_n and Iv_s values are calculated for all the species of *Senecio* investigated.

Determination of total phenolics

The quantity of total phenolics has been estimated for the two forms of (normal leaf and small leaf) *Synedrella nodiflora* and *Calyptocarpus vialis* in order to relate the qualitative and quantitative distribution of phenolics. Extraction and quantification of phenolics involves two steps.

Procedure : Part I

Preparation of standard graph of Ferulic acid : Ferulic acid stock solution (300 mg/ml) is prepared and serial dilutions (in steps of 20 mg) are obtained.

The reaction mixture consists of 1 ml of ferulic acid + 7.5ml of distilled water + 1ml of 35% sodium carbonate. To this mixture, 0.5ml of folin denis reagent (available in the market) is added and optical density (OD) is read at 720nm using a Bauch and Lamb spectrophotometer (spectronic 20). Readings are taken before 2 minutes from the time of adding folin denis reagent. Results are plotted on a graph sheet to obtain a standard graph of ferulic acid.

Procedure : Part II

Extraction and quantification of phenolics from plant tissues : One gram each of fresh material is taken from 3 different parts (stem, leaf and inflorescence, with open ray florets) of 3 plant materials - *Synedrella nodiflora* (normal leaf); *S.nodiflora* (small leaf) and *calyptocarpus vialis.* These parts are cut into small pieces, washed thoroughly in distilled water and kept in beakers. To these, little (about 5ml) amount of 2 N HCl is added and the material pressed into the liquid. Later on, 15ml of 2 N HCl is added to start hydrolysis. The beakers are kept in a hot water bath for about 30 minutes. At the end of this period, the beakers are taken out and cooled to room temperature. The solution is filtered. The filterate is taken in a separating funnel. To this is added equal volume of ethyl ether while gently shaking the funnel. When the solution settles, two layers are visible in which the upper layer contains phenolics. This layer is taken out of the funnel and allow to evaporate. On evaporation of ether phenolics remain as a brown residue. This residue is dissolved in 5ml of 90% ethyl alcohol. 1ml of this alcoholic extract of phenolics is taken for quantification.

The reaction mixture for spectrophotometric analysis consists of 1ml extract + 1ml of 35% $Na_2 Co_3$ + 7.5ml of distilled water + 0.5ml of folin denis reagent. OD is recorded within 2 minutes of the addition of folin denis reagent using a spectorphotometre (spectronic 20) at 720nm.
Concentration of the different phenolic extracts are determined with the help of standard graph of ferulic acid.

OBSERVATIONS

The following is a detailed description of the flavonoid spot profiles of the taxa investigated.

1. **Ageratum conyzoides** (Fig. 4.1, Table 4.4). There are a total of 7 spots, of which 4 spots are grey in colour, 2 greenish yellow in colour and one light yellow in colour. Spot nos. 4 and 5 are common between this taxon and *Ageratum houstonianum.* Spot No.4 is the largest with an area of 276 Sq.mm.

2. **Ageratum houstonianum** (Fig. 4.2, Table 4.5). The TLC plate reveals a total of 9 spots of which 2 spots are light yellow, 3 light green, 2 greenish yellow, one yellow and one light purple in colour. Spot nos. 4 and 5 are shared with *Ageratum conyzoides.* In contrast to *A.conyzoides,* here spot no.4 is only 74 sq. mm in area.

3. **Adenostemma lavenia** (Fig. 4.3, Table 4.6). There are 8 spots in all. Of these, seven are light yellow in colour and only spot no.2 is green in colour. Spot no.1 with an area of 236 sq. mm is the largest.

4. **Chromolaena odorata** (Fig. 4.4, Table 4.7). There are 13 spots. Of these, 6 spots are green, 5 light yellow, one grey and one greenish yellow in colour. Spot no. 3 with an area of 239 sq.mm. is the largest.

5. **Acanthospermum hispidum** (Fig. 4.5, Table 4.8). There are only four spots, with spot no.2 being the largest with an area of 244 sq.mm. Of the four spots, two are light yellow and two are greenish yellow in colour.

6. **Polymnia grandis** (Fig. 4.6, Table 4.9). There are four spots widely separated on the TLC plate. Of these, two are light yellow and two grey in colour. Spot no.2 with an area of 113 sq.mm is the largest.

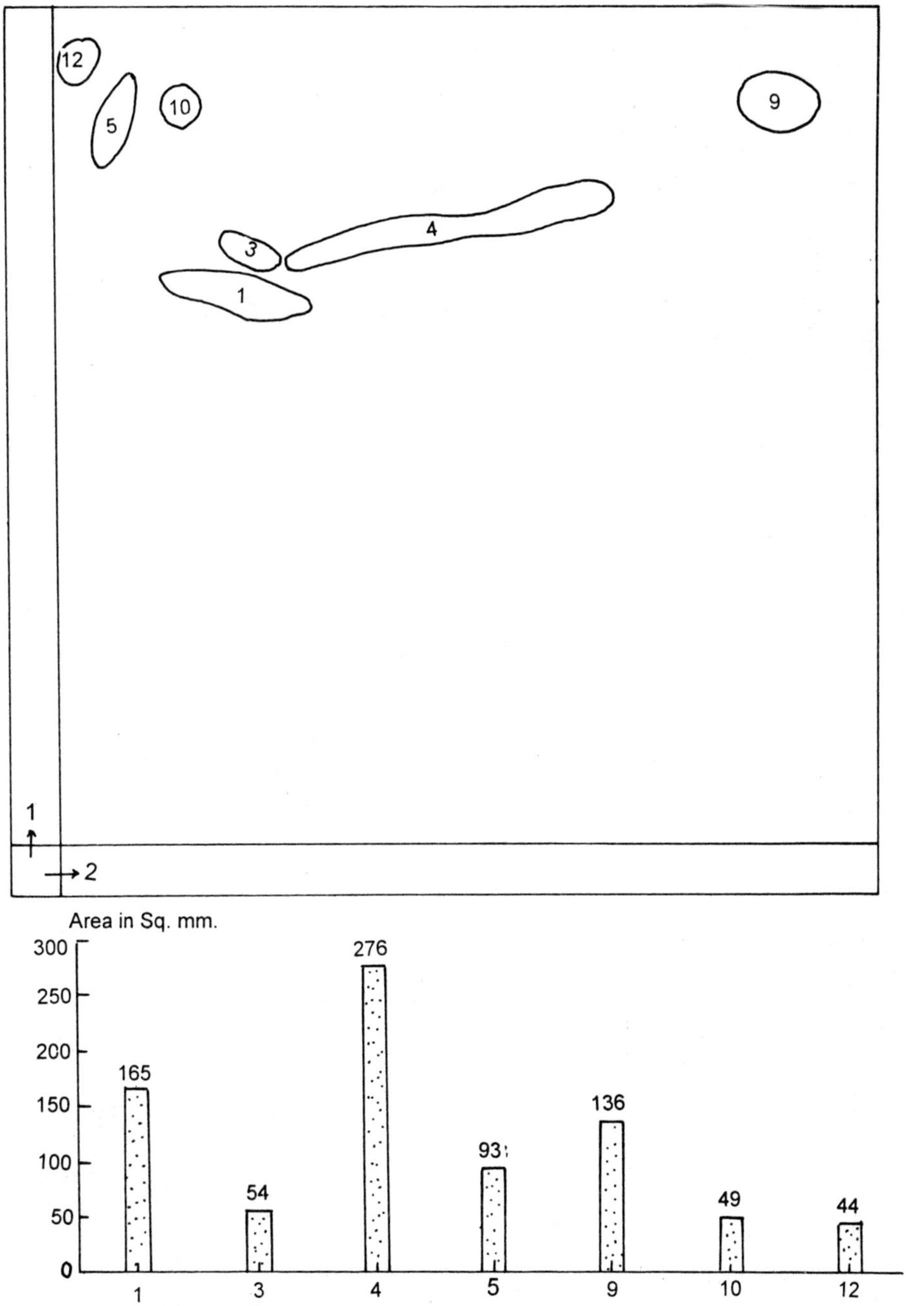

Plate 4.1 Phenolic spot patterns in *Ageratum conyzoides*

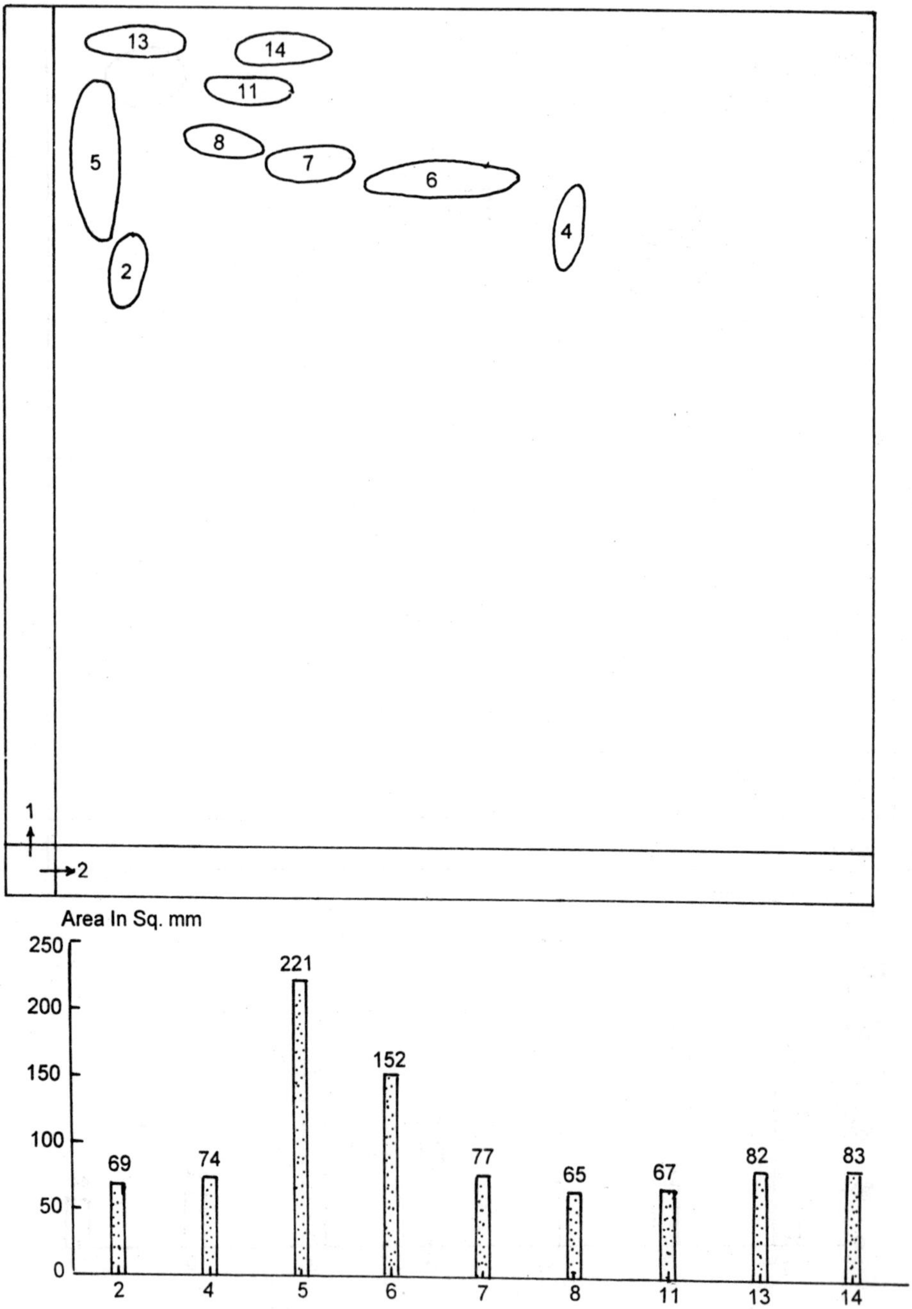

Plate 4.2 Phenolic spot patterns in *Ageratum houstonianum*

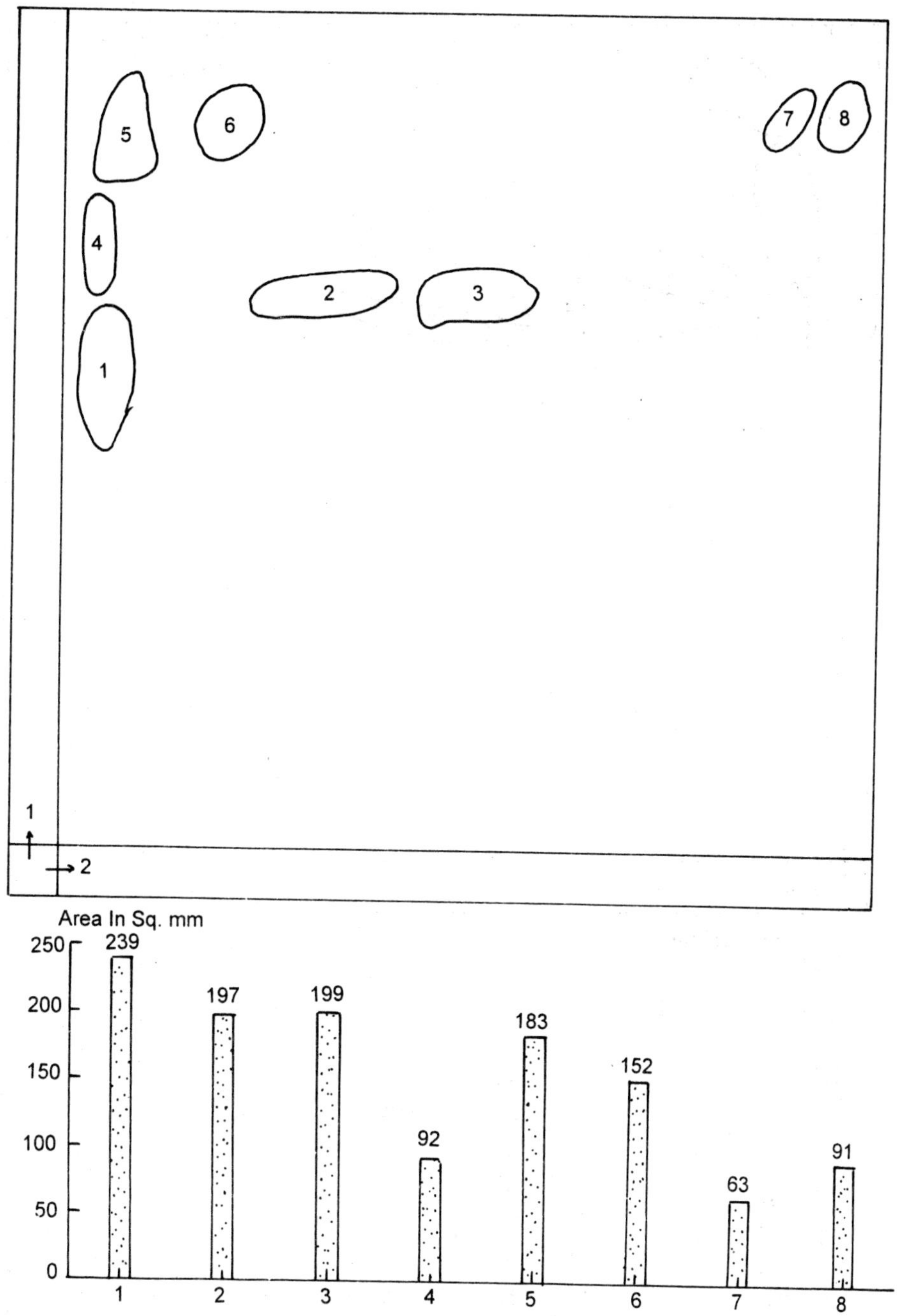

Plate 4.3 Phenolic spot patterns in *Adenostemma lavenia*

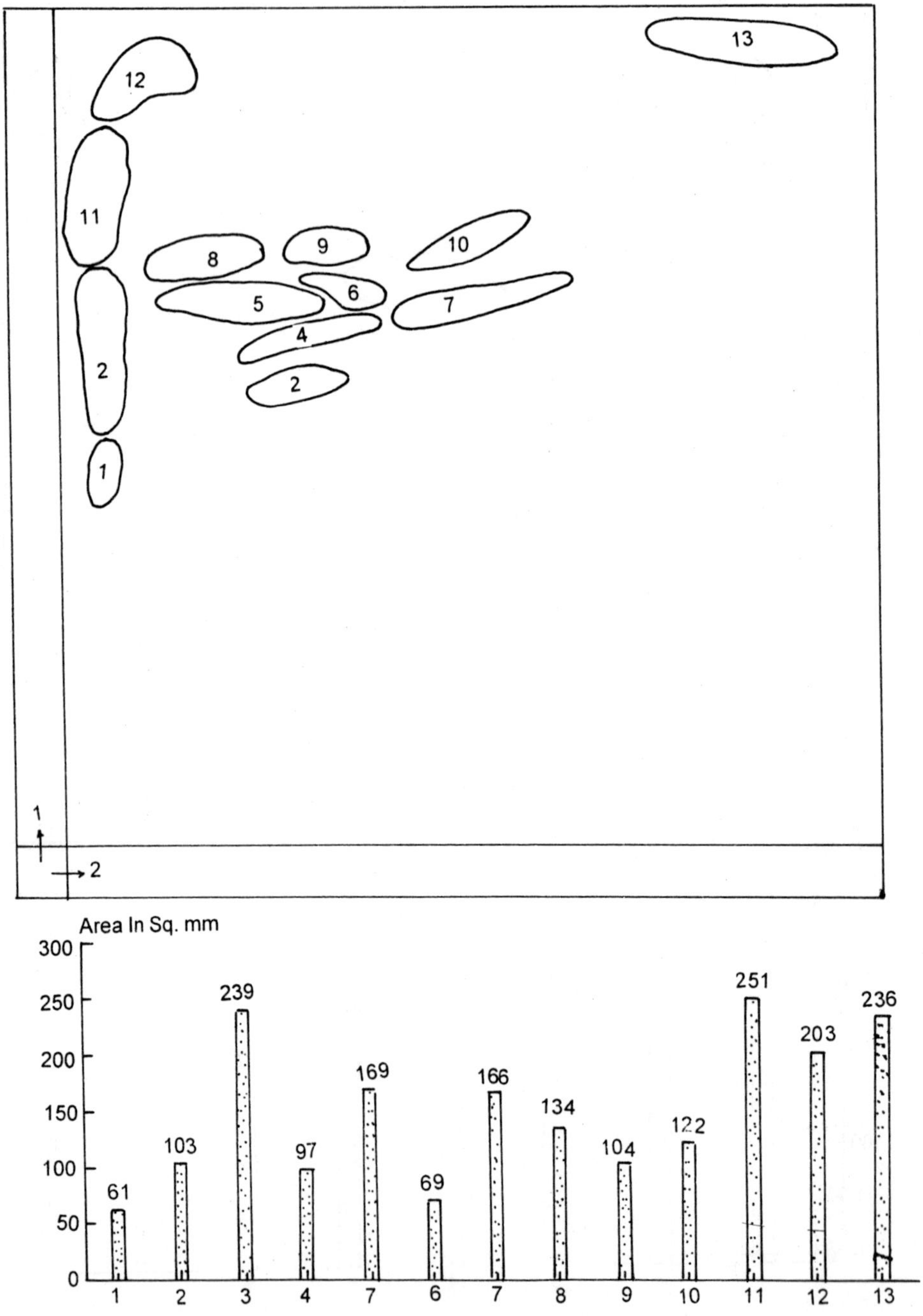

Plate 4.4 Phenolic spot patterns in *Chromolaena odorata*

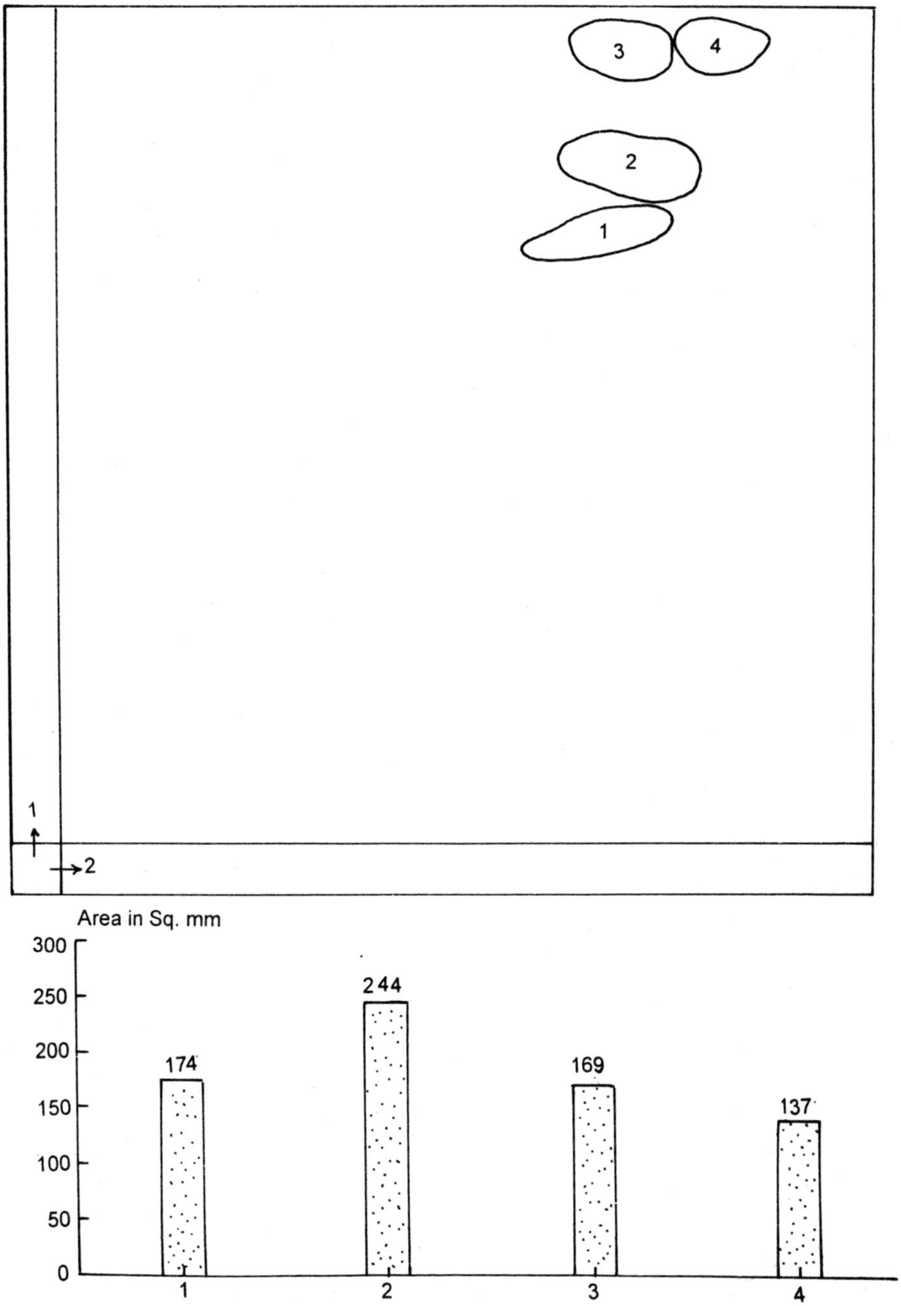

Plate 4.5 Phenolic spot patterns in *Acanthospermum hispidum*

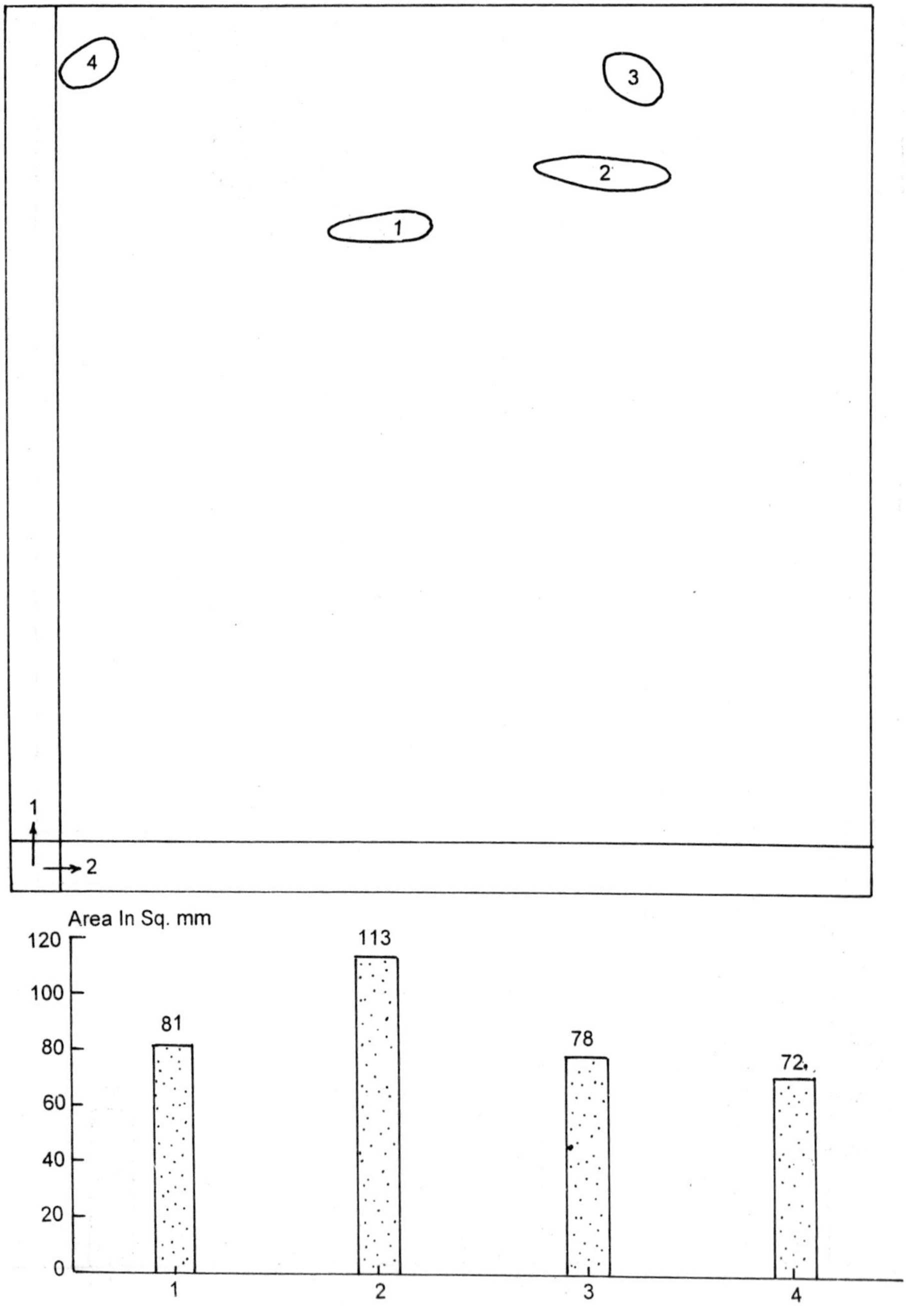

Plate 4.6 Phenolic spot patterns in *Polymnia grandis*

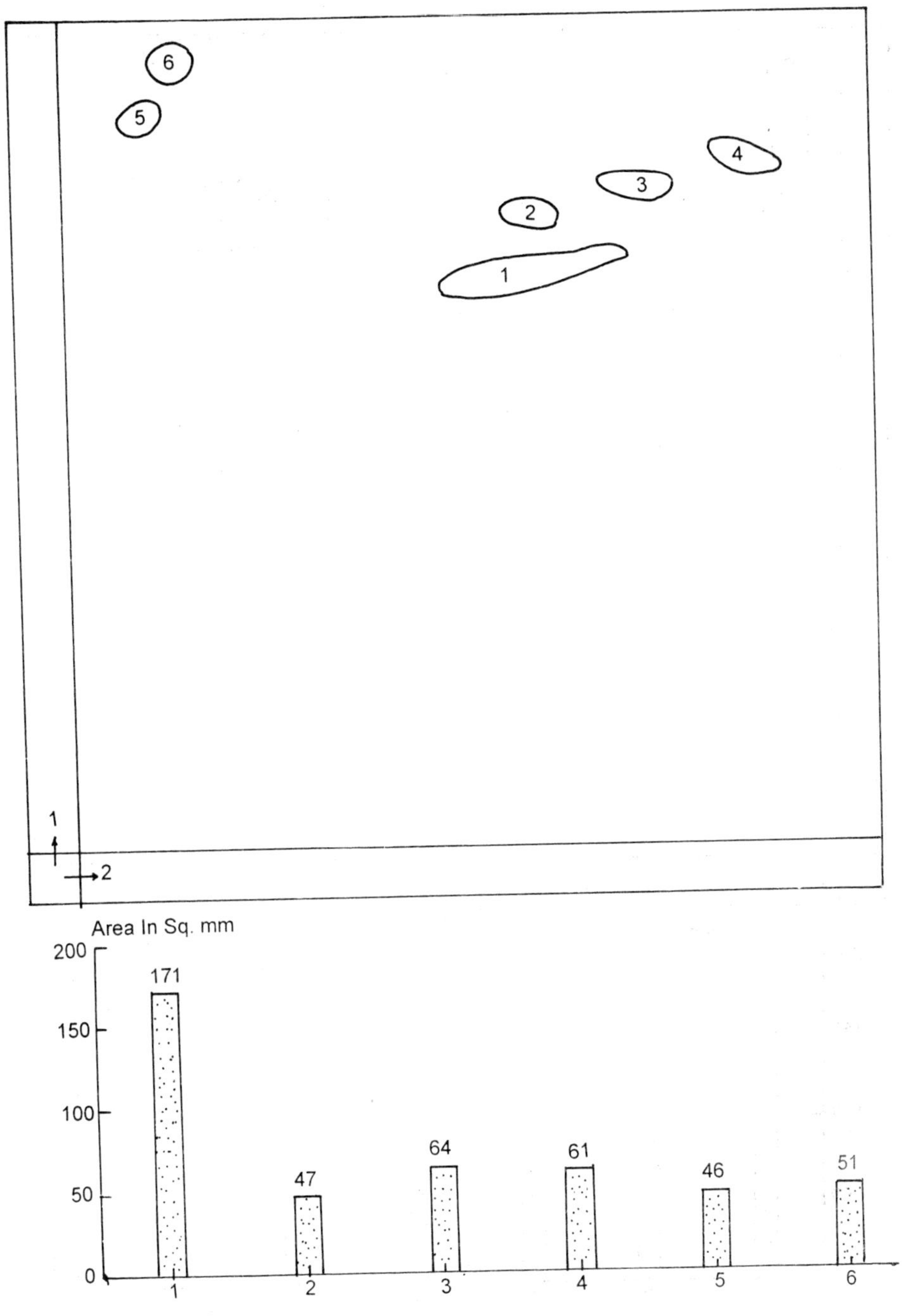

Plate 4.7 Phenolic spot patterns in *Siegesbeckia orientalis*

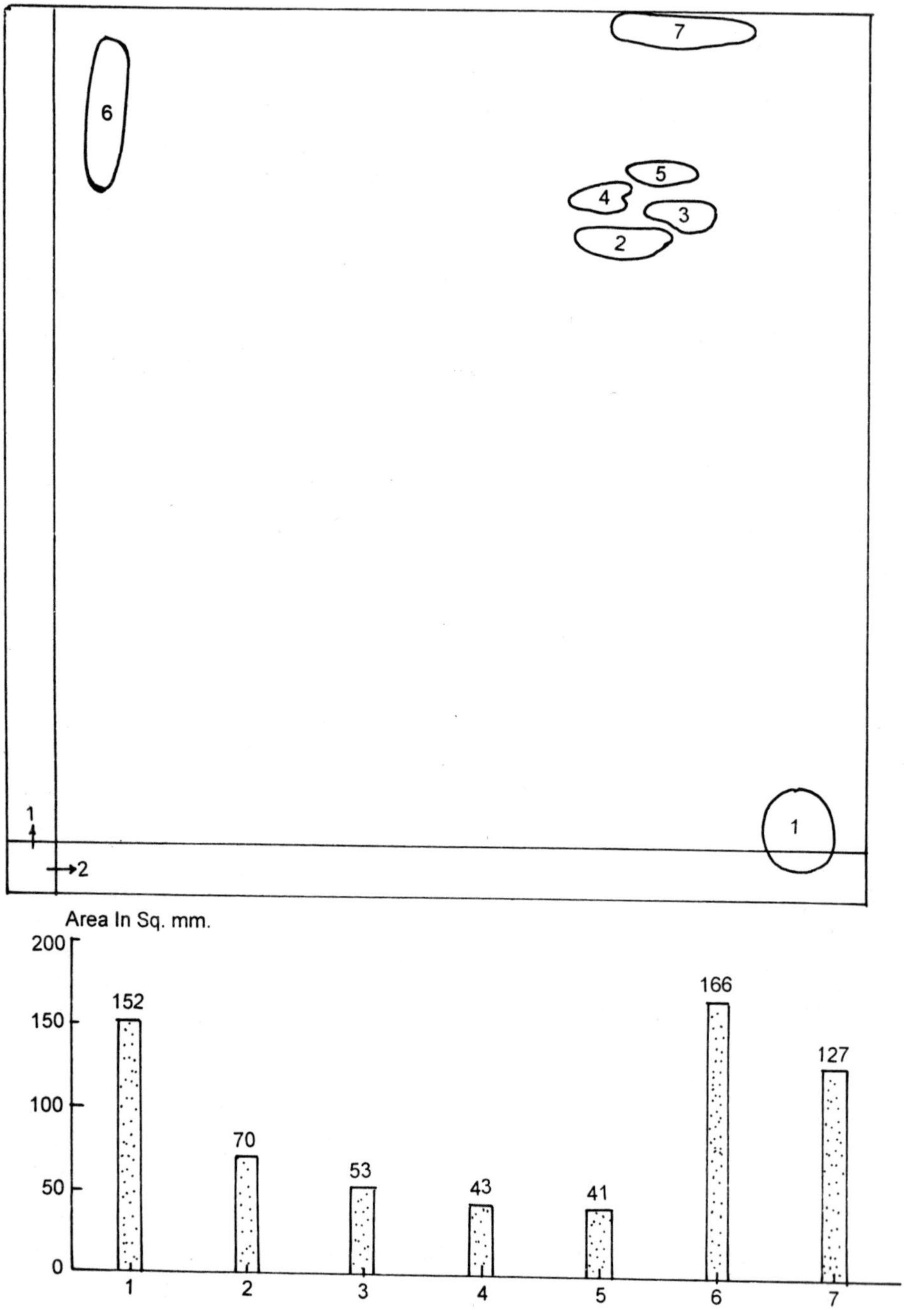

Plate 4.8 Phenolic spot patterns in *Blainvillea acmella* (B = rhombodea)

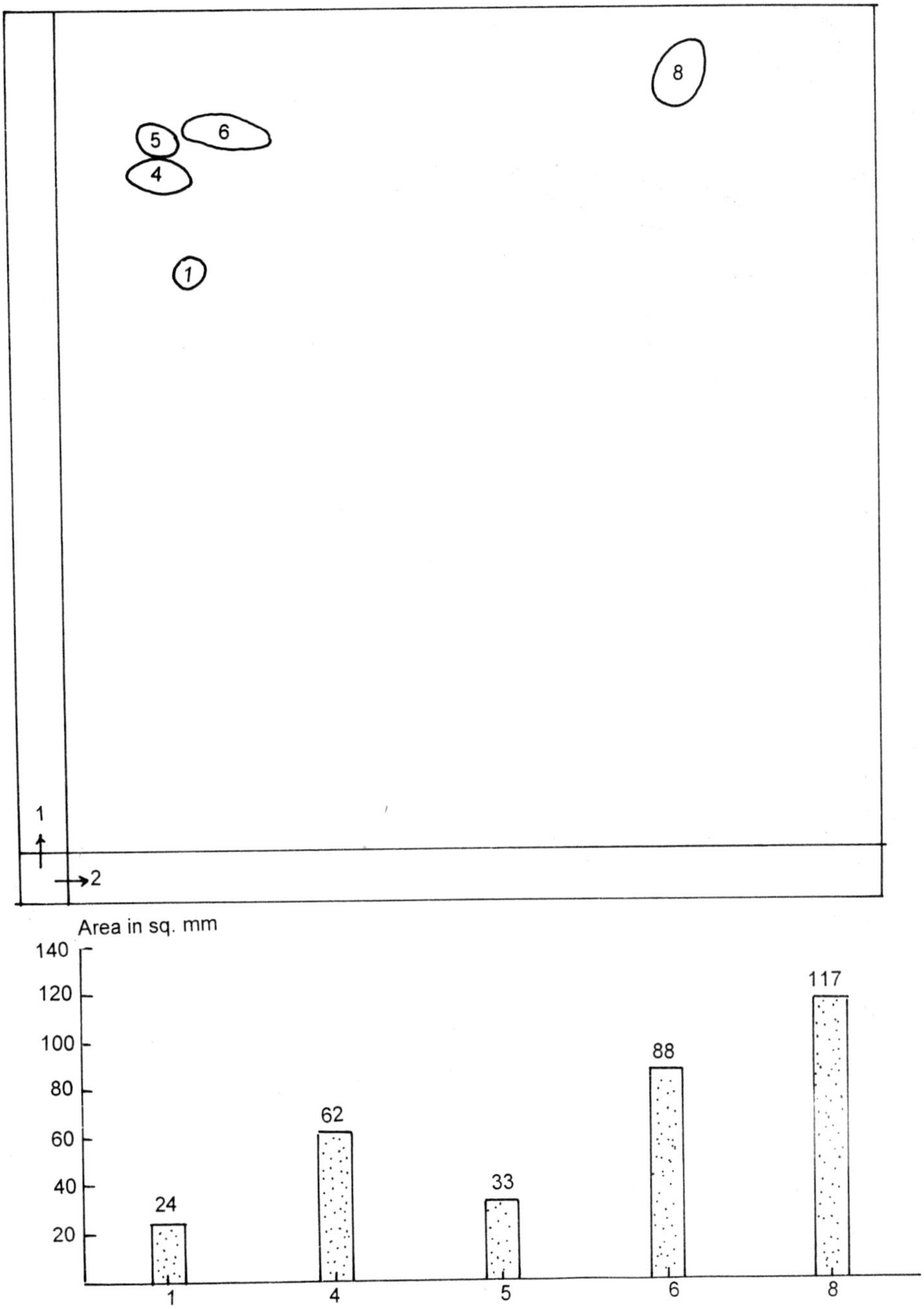

Plate 4.9 Phenolic spot patterns in *Eclipta prostrata* (Normal leaf)

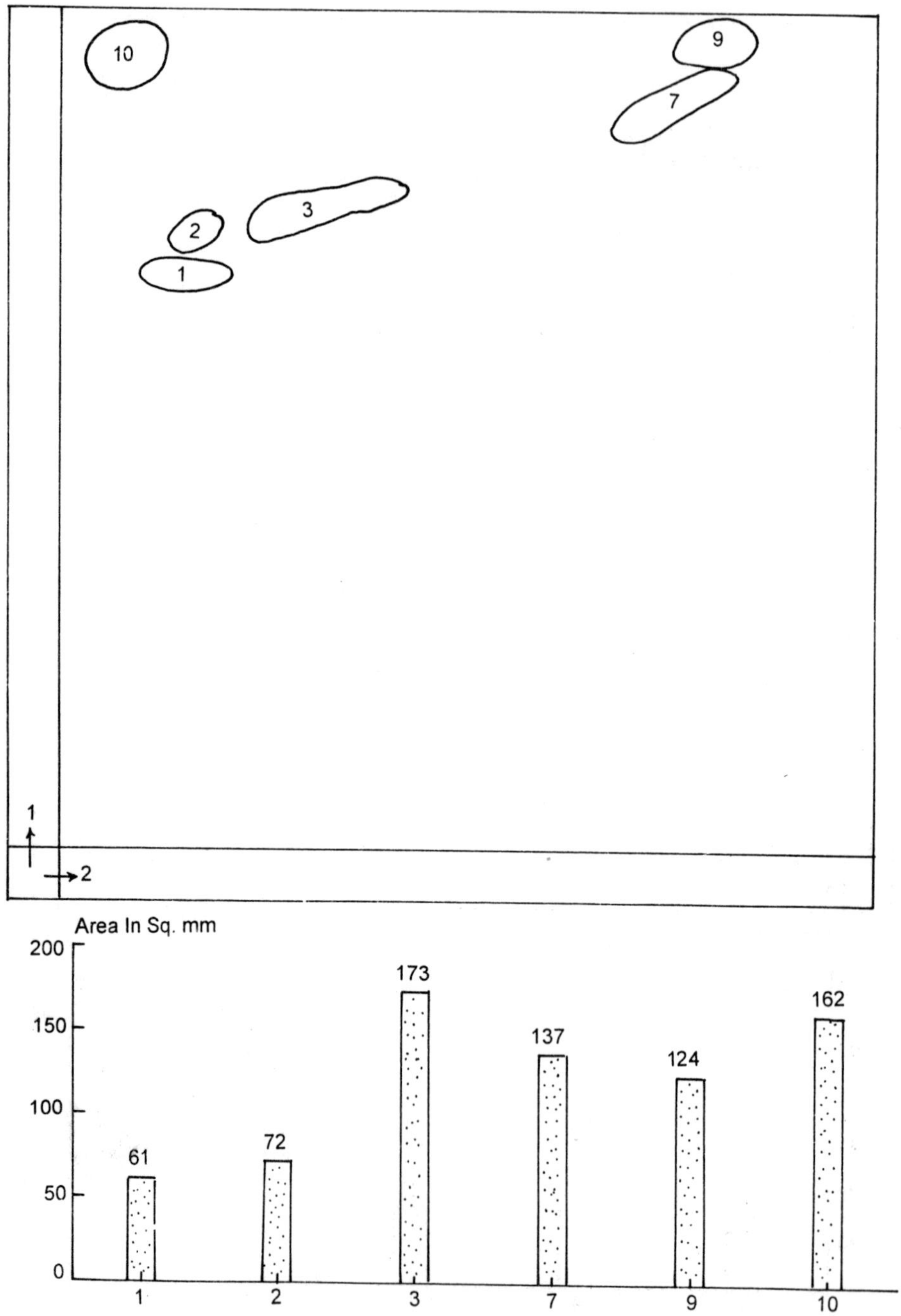

Plate 4.10 Phenolic spot patterns in *Eclipta prostrata* (Small leaf)

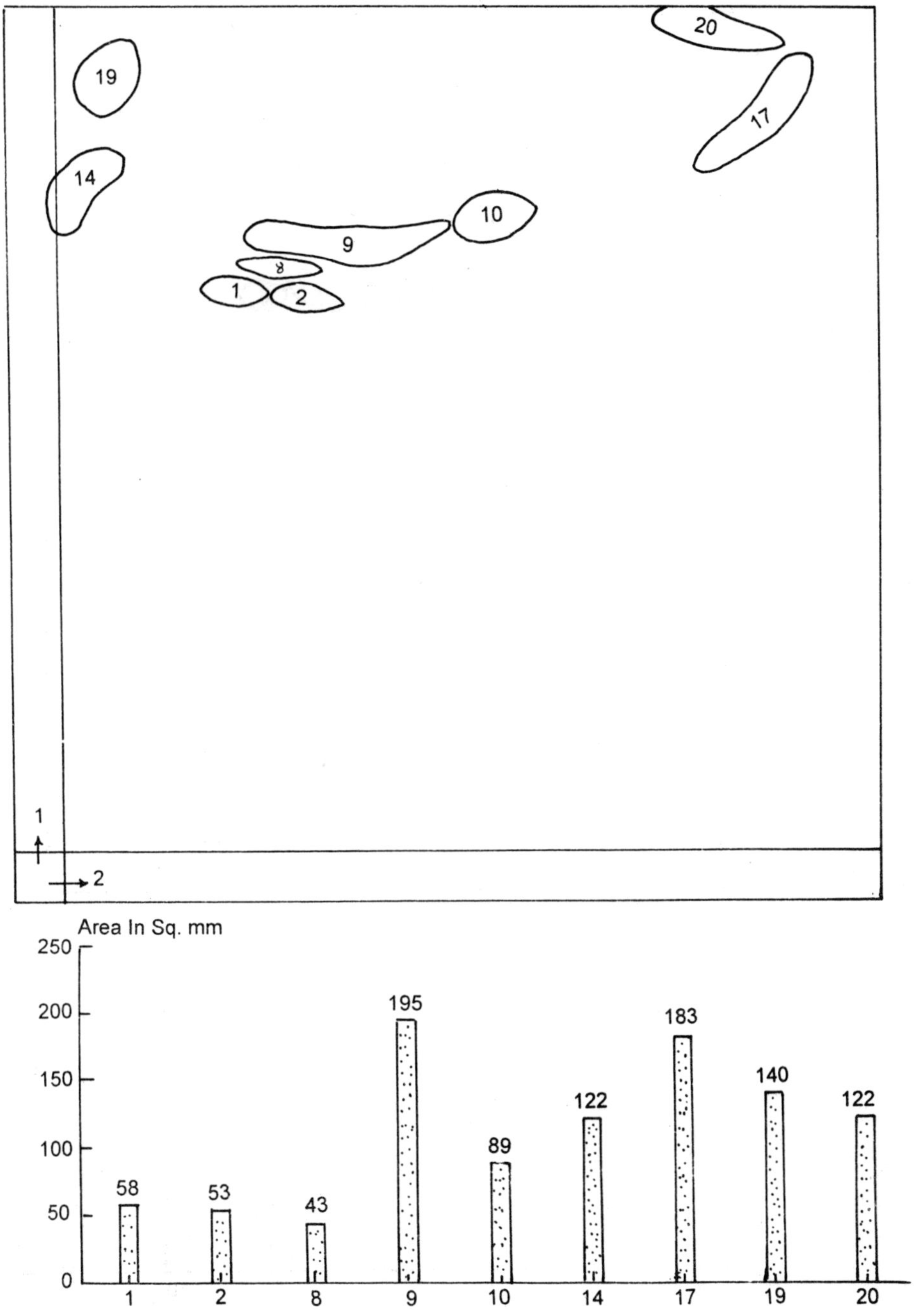

Plate 4.11 Phenolic spot patterns in *Synedrella nodiflora* (Normal leaf)

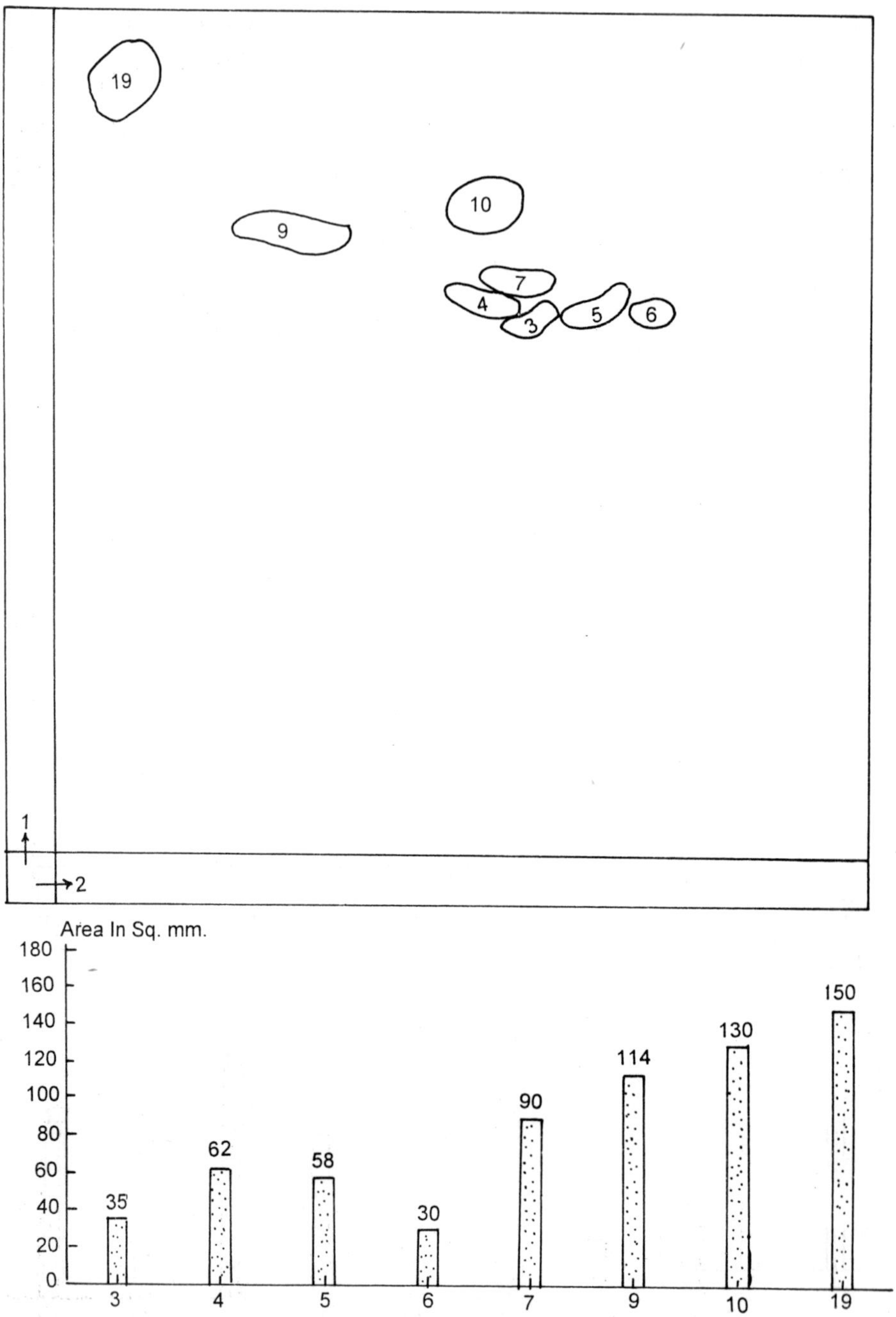

Plate 4.12 Phenolic spot patterns in *Synedrella nodiflora* (Small leaf)

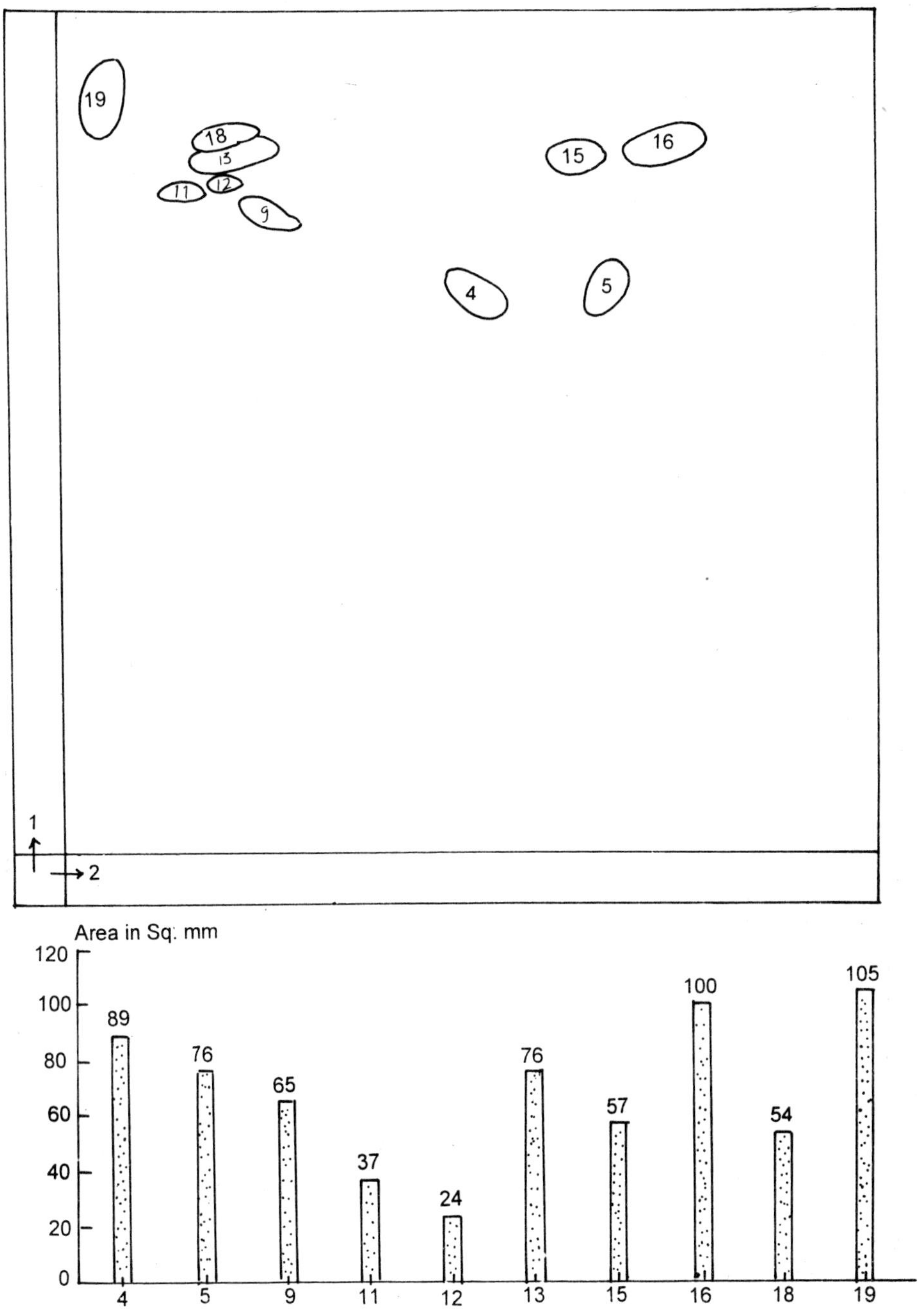

Plate 4.13 Phenolic spot patterns in *Calyptocarpus vialis*

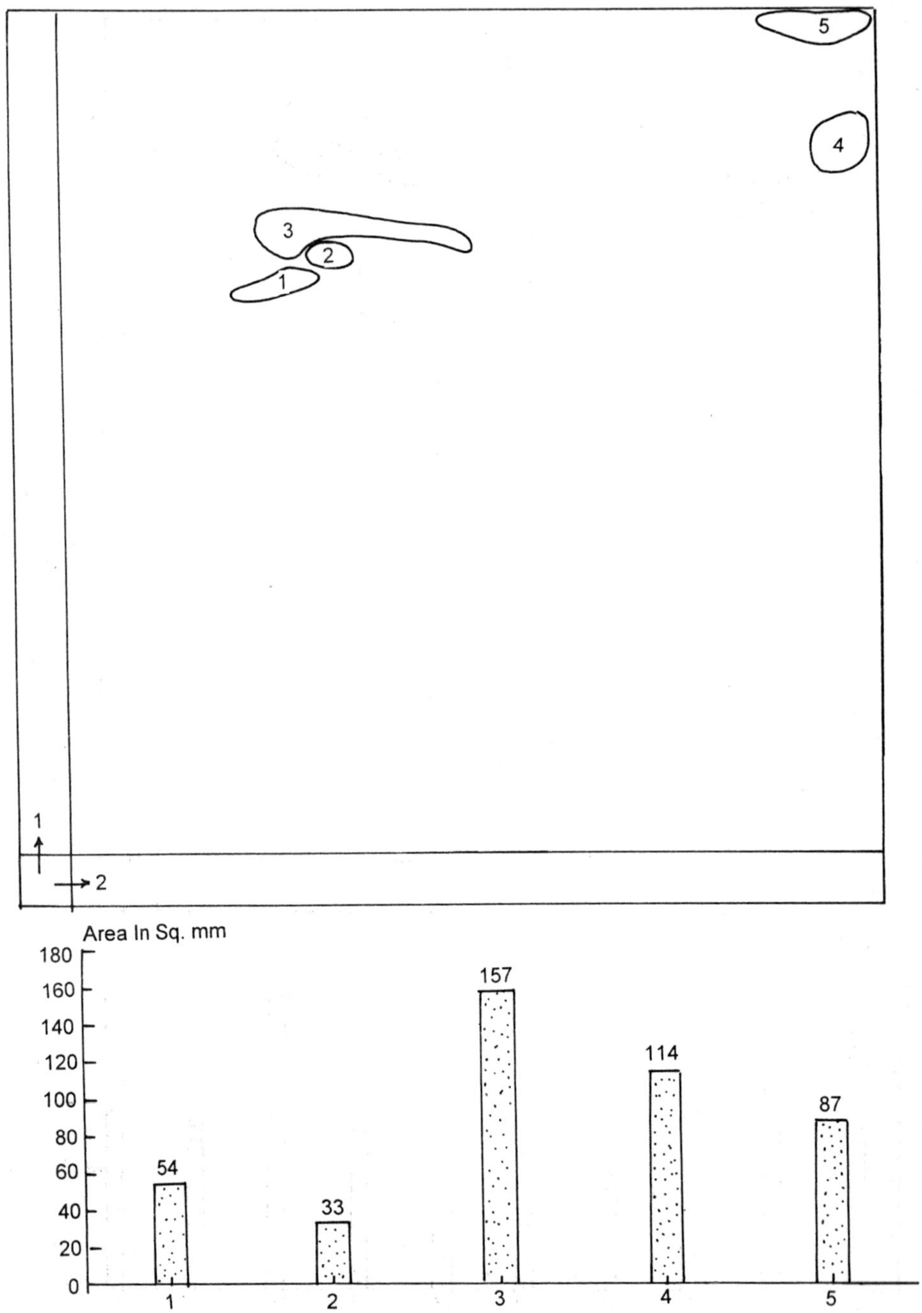

Plate 4.14 Phenolic spot patterns in *Lagascea mollis*

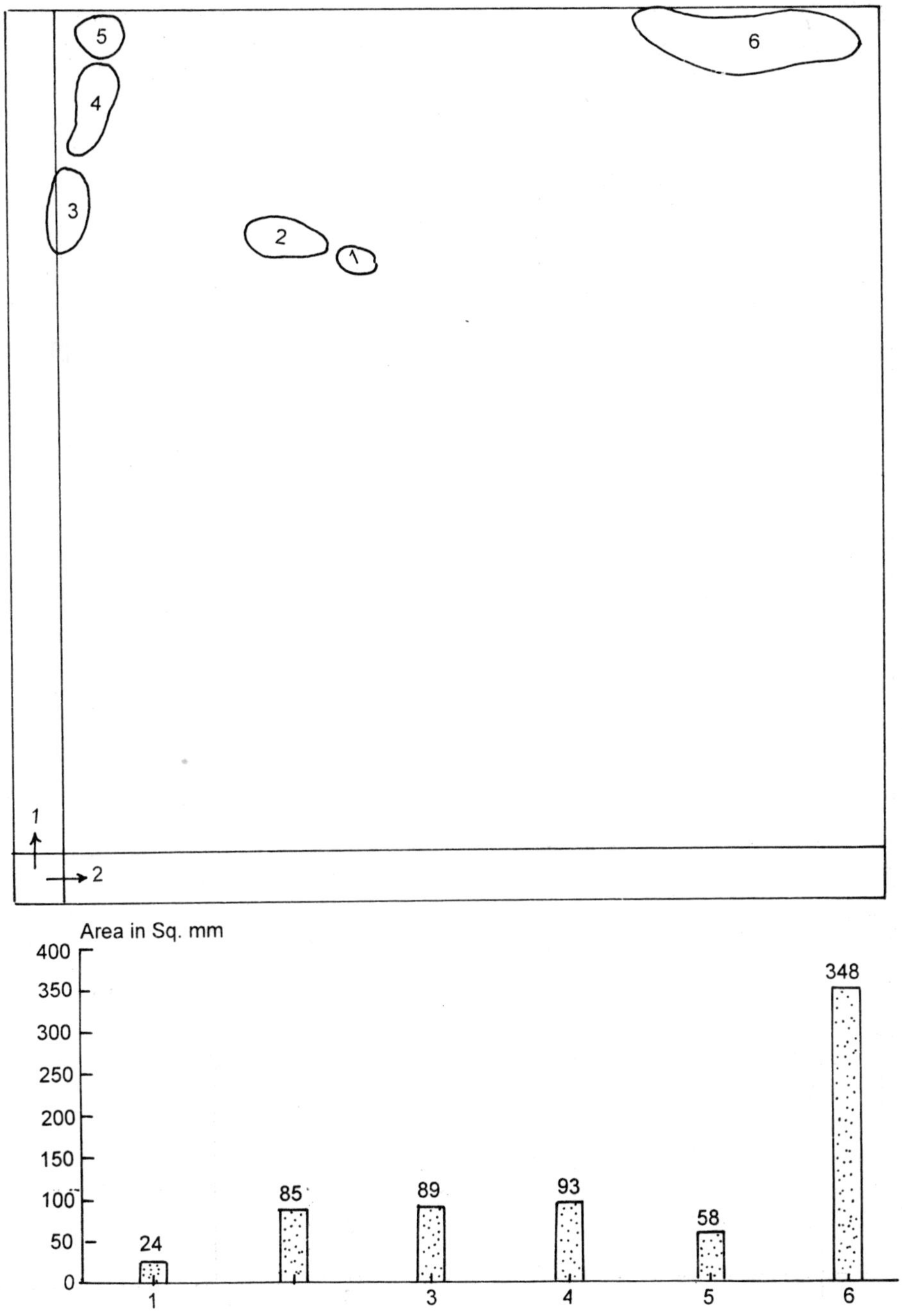

Plate 4.15 Phenolic spot patterns in *Sclerocarpus africaneus*

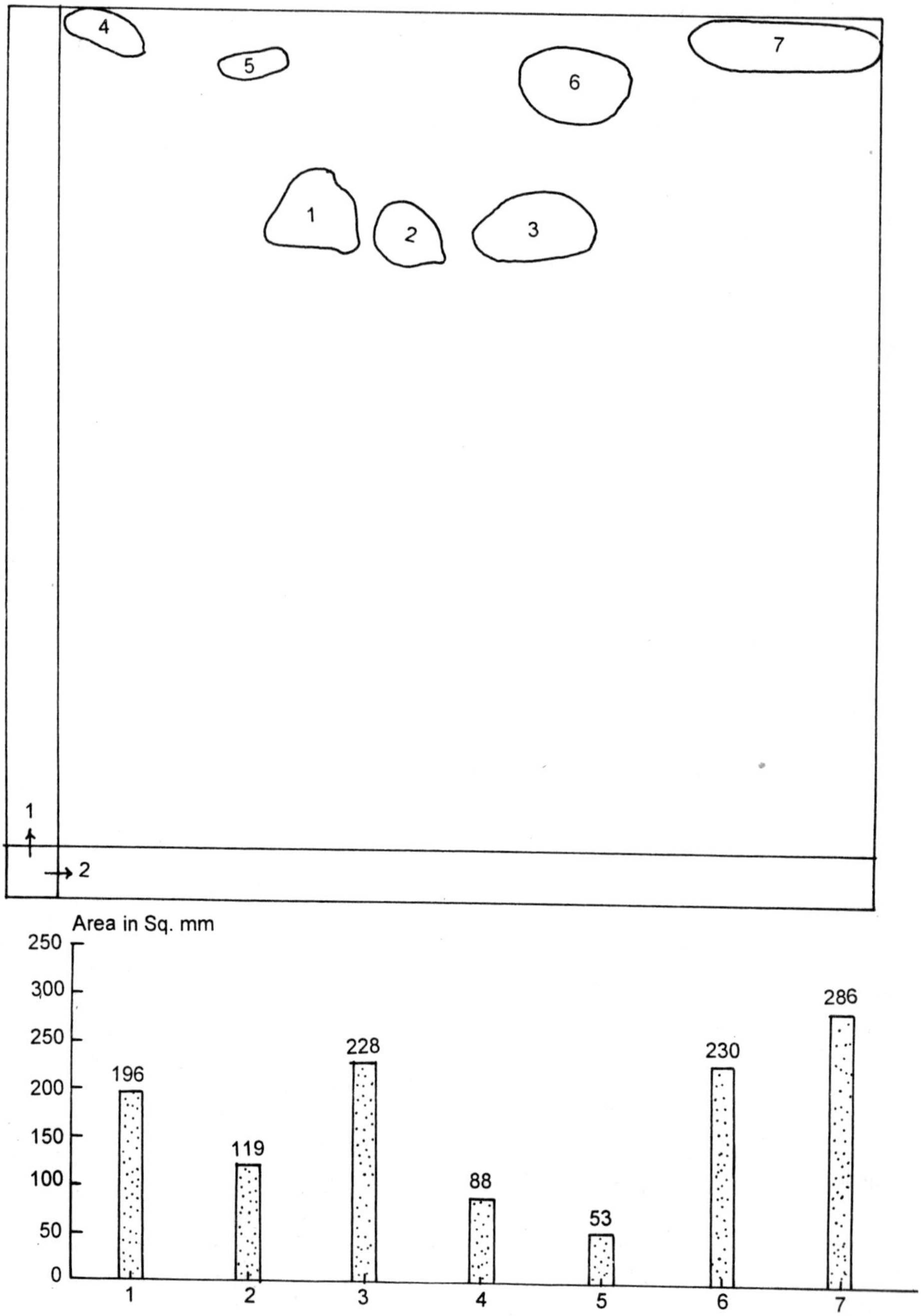

Plate 4.16 Phenolic spot patterns in *Helianthus rigidus*

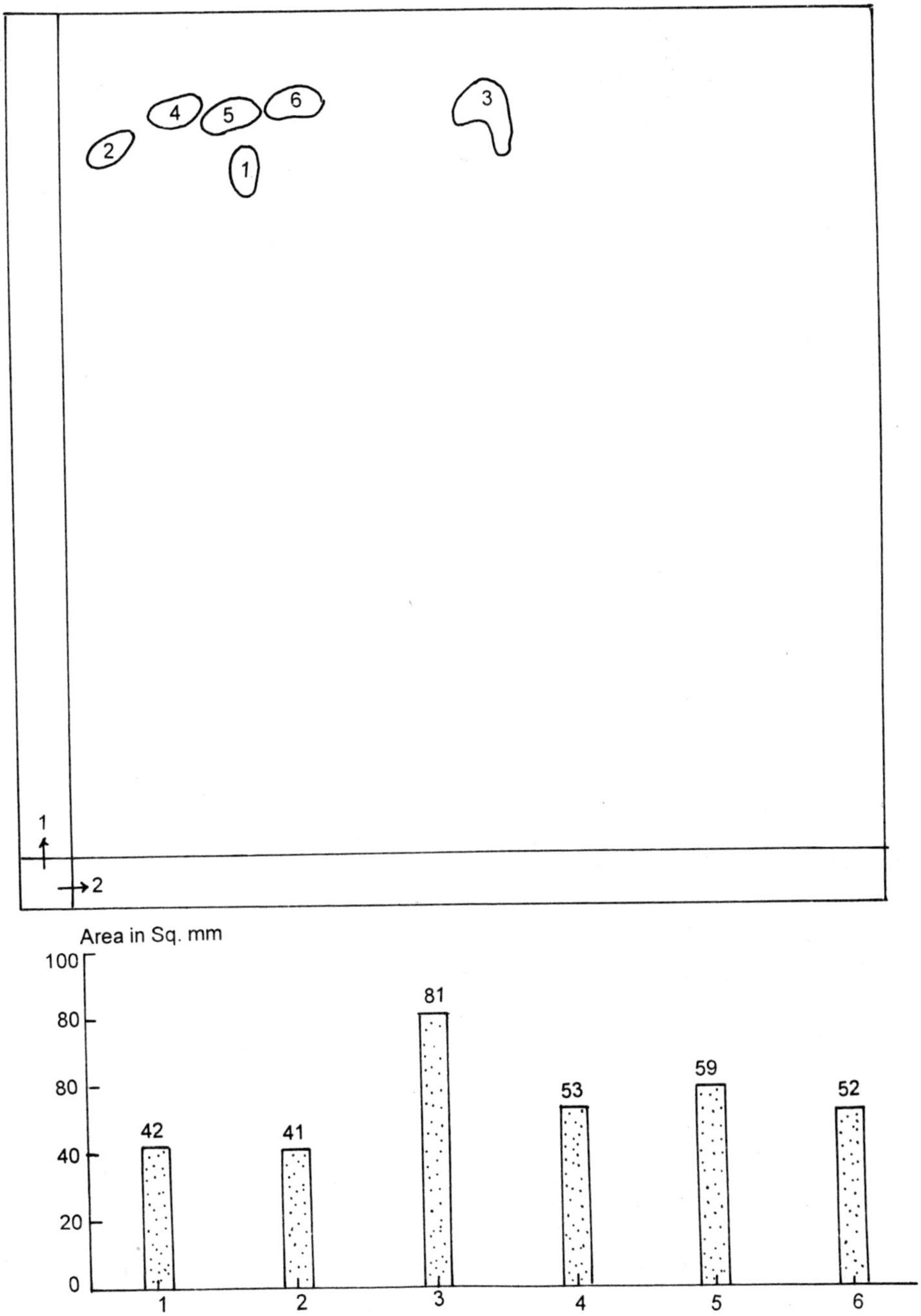

Plate 4.17 Phenolic spot patterns in *Rudbeckia nitida*

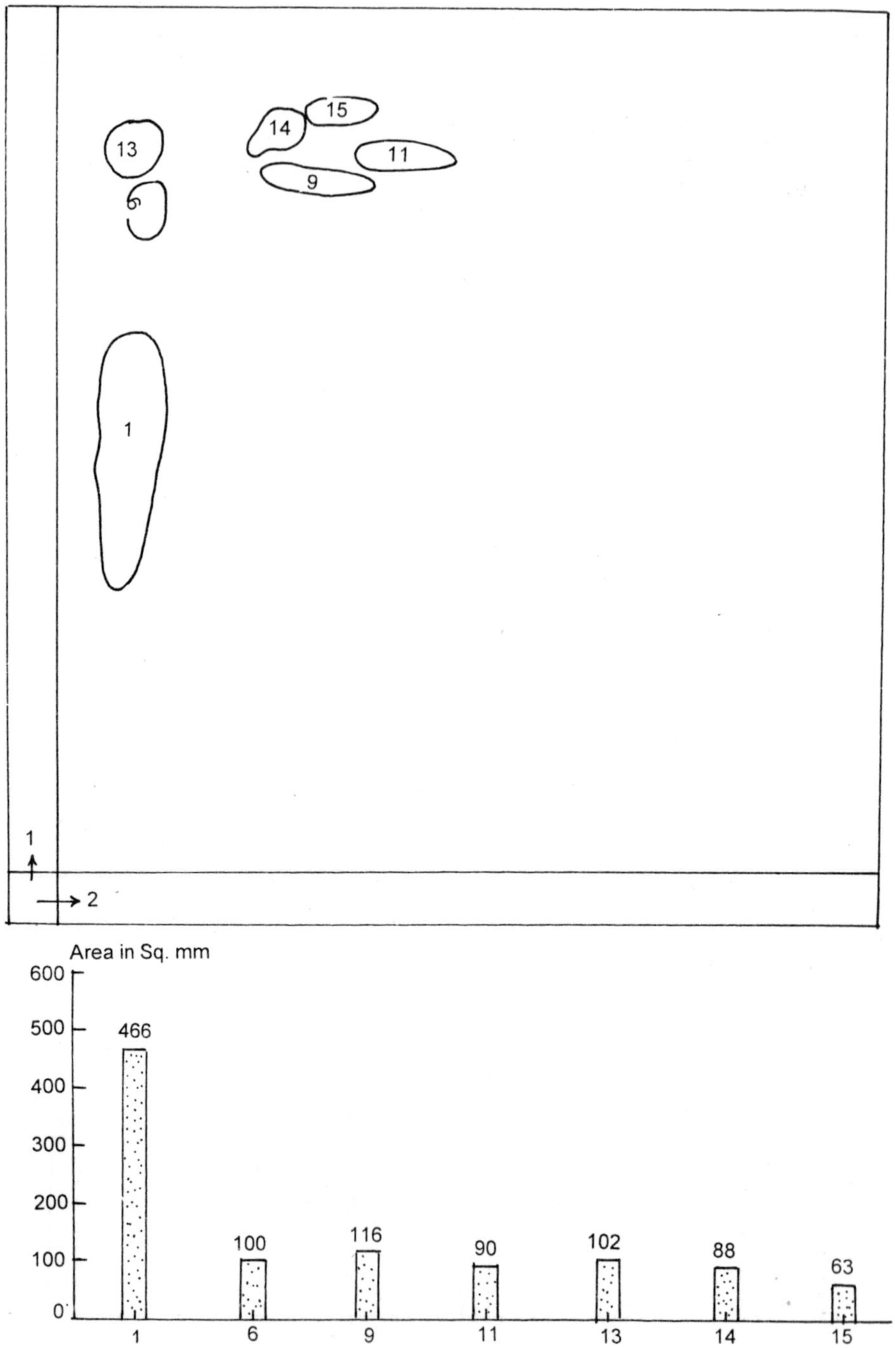

Plate 4.18 Phenolic spot patterns in *Bidens pilosa* (Normal ray)

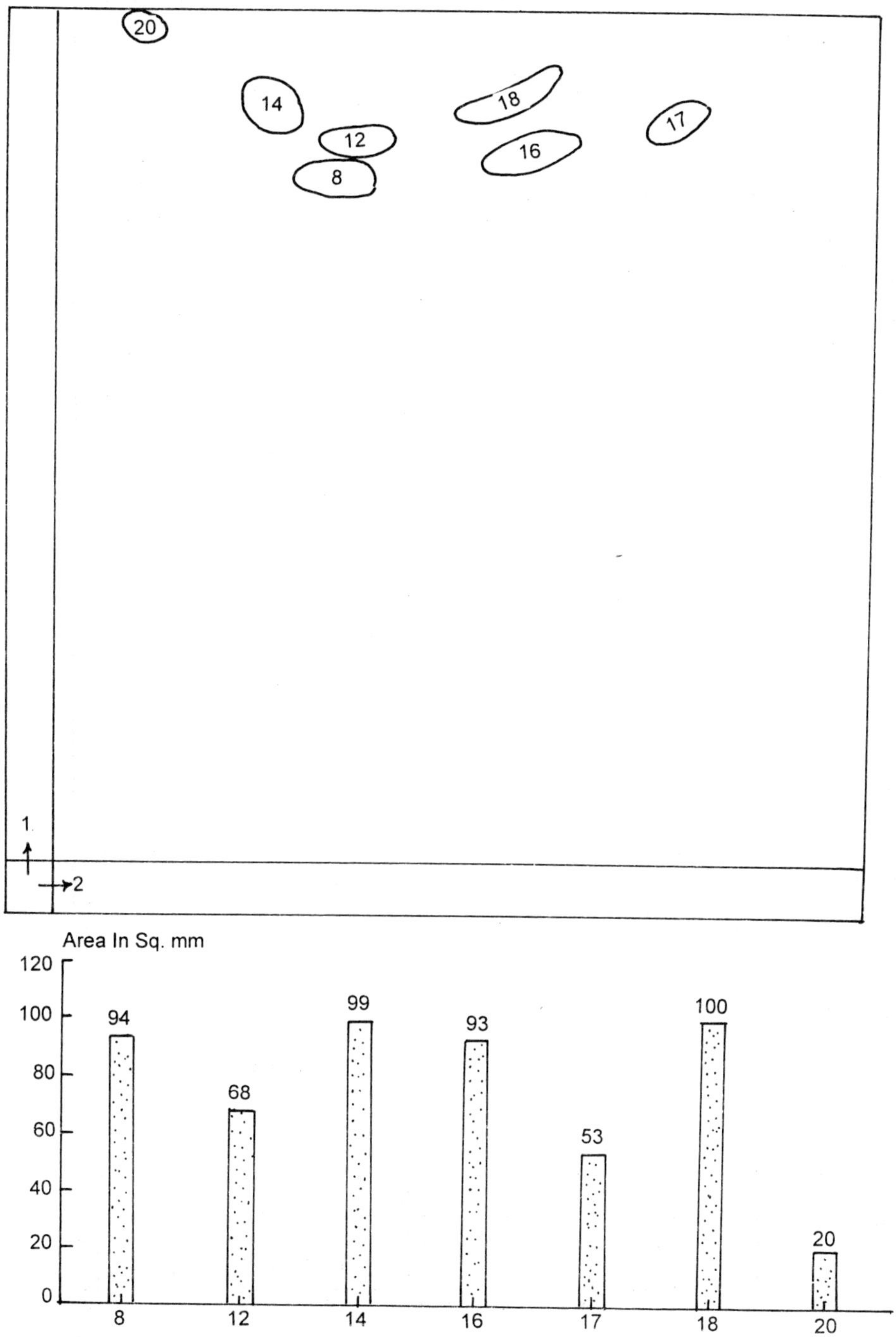

Plate 4.19 Phenolic spot patterns in *Bidens pilosa* (Intermediate)

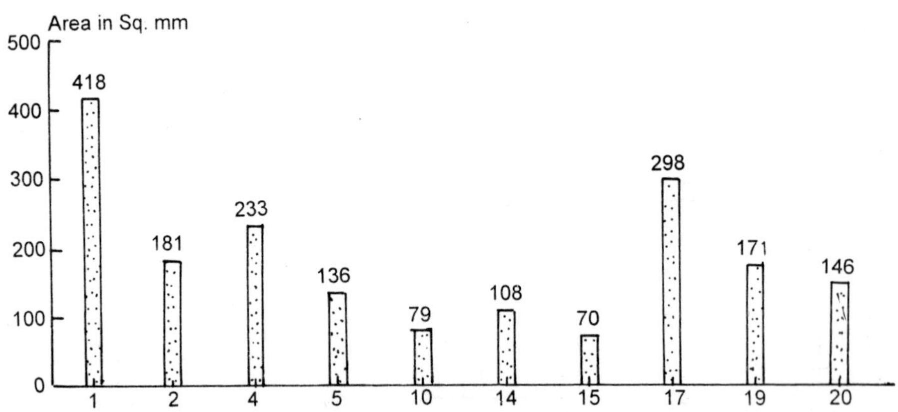

Plate 4.20 Phenolic spot patterns in *Bidens pilosa* (Rayless)

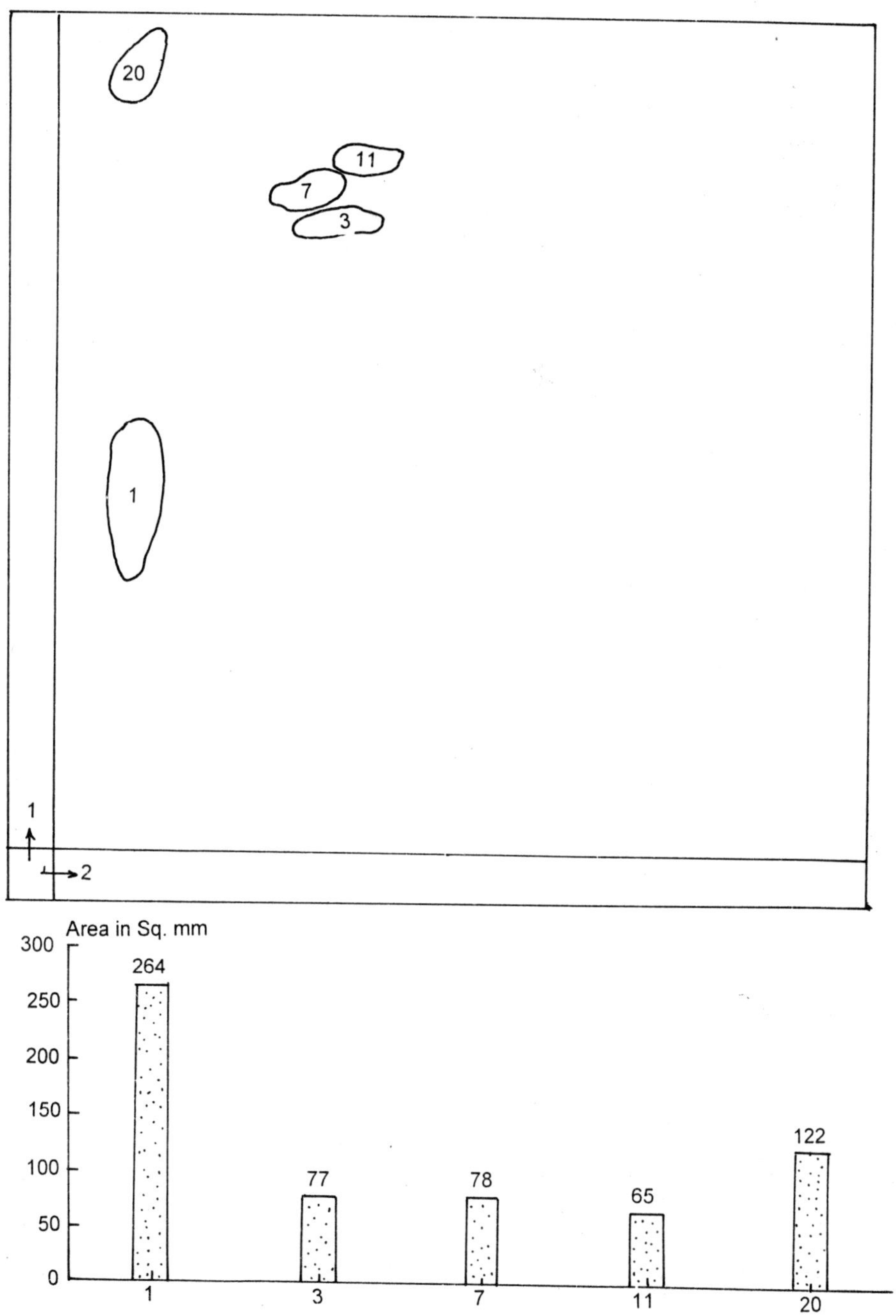

Plate 4.21 Phenolic spot patterns in *Bidens humilis*

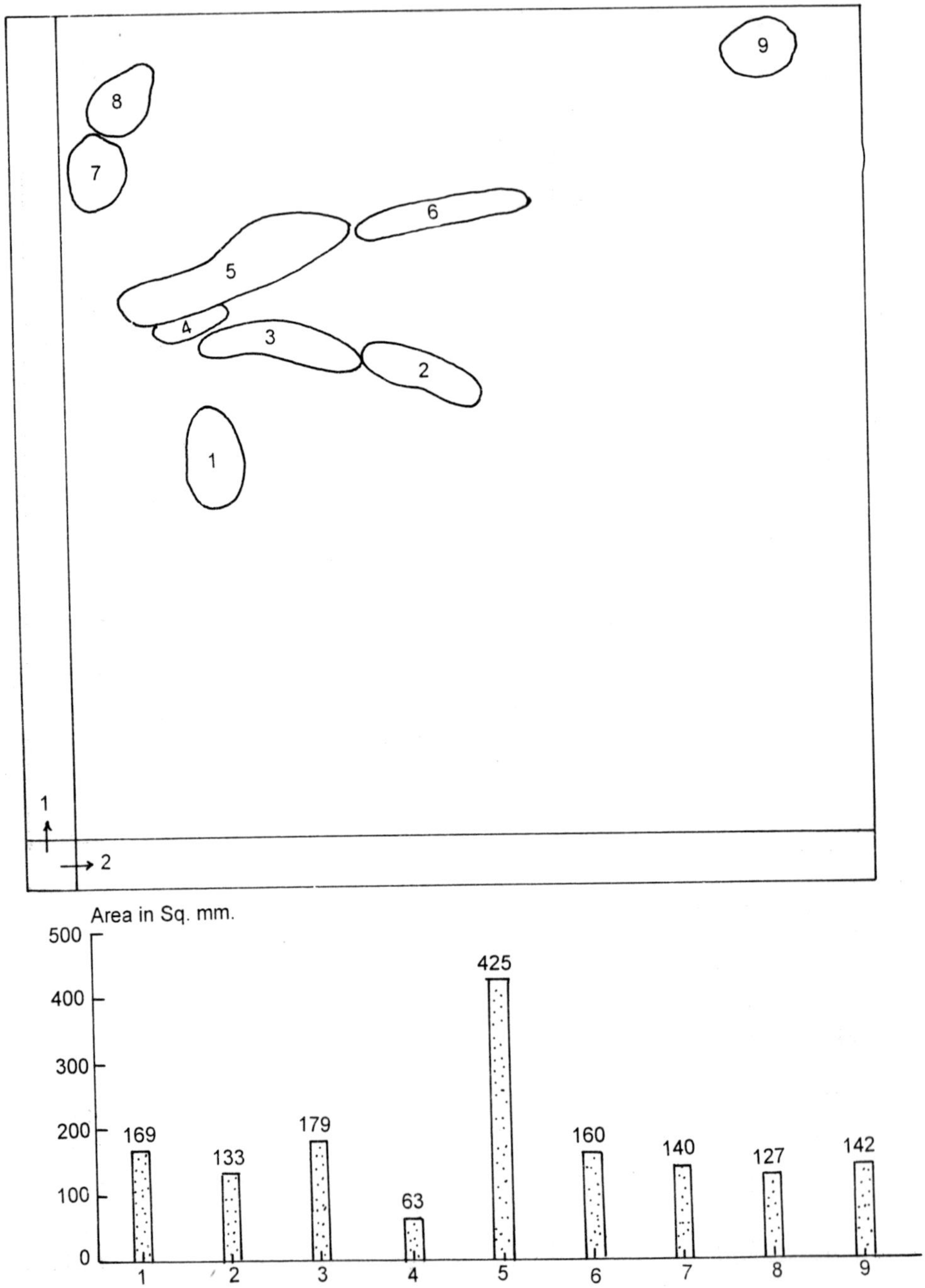

Plate 4.22 Phenolic spot patterns in *Moonia heterophylla*

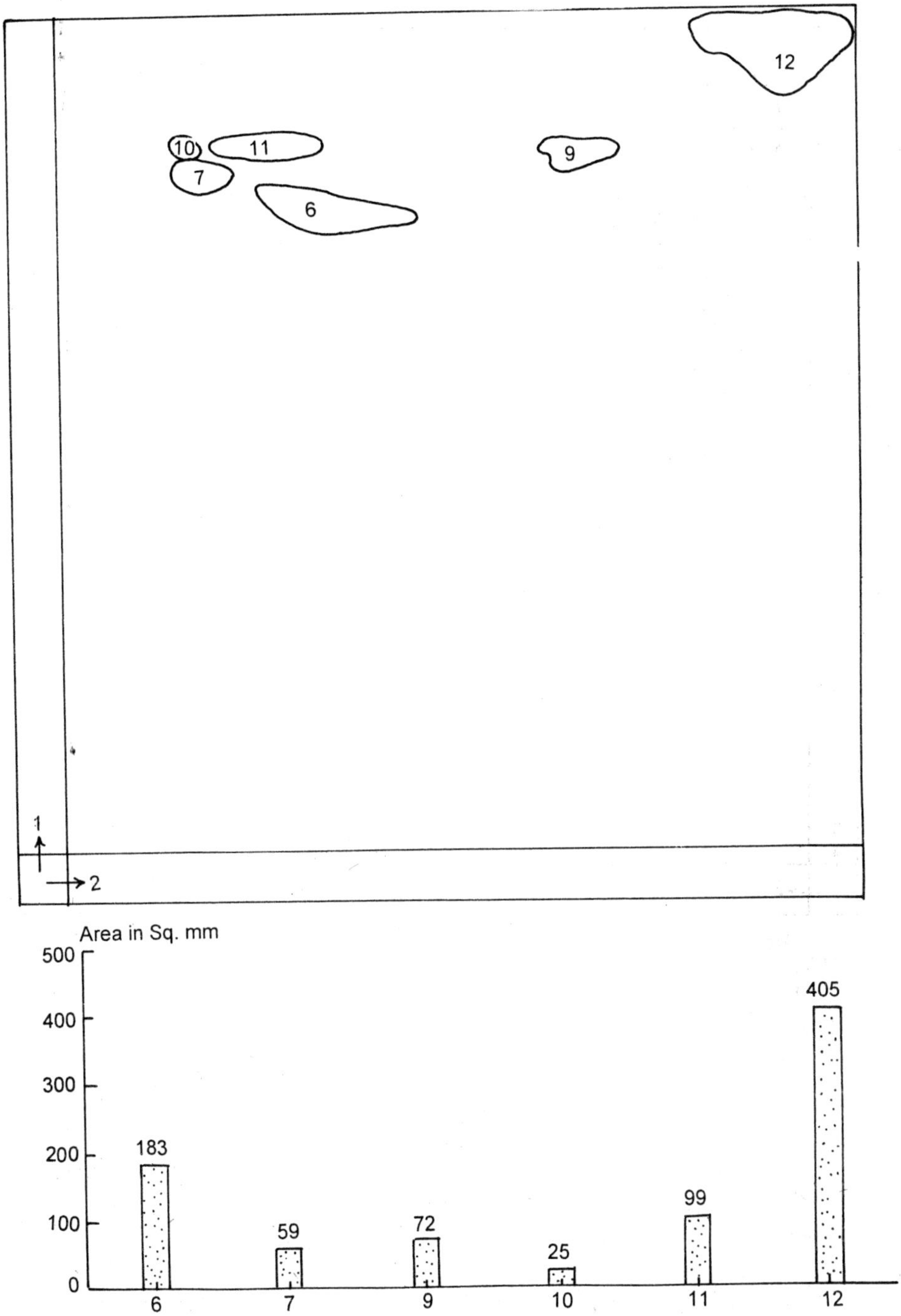

Plate 4.23 Phenolic spot patterns in *Galinsoga parviflora*

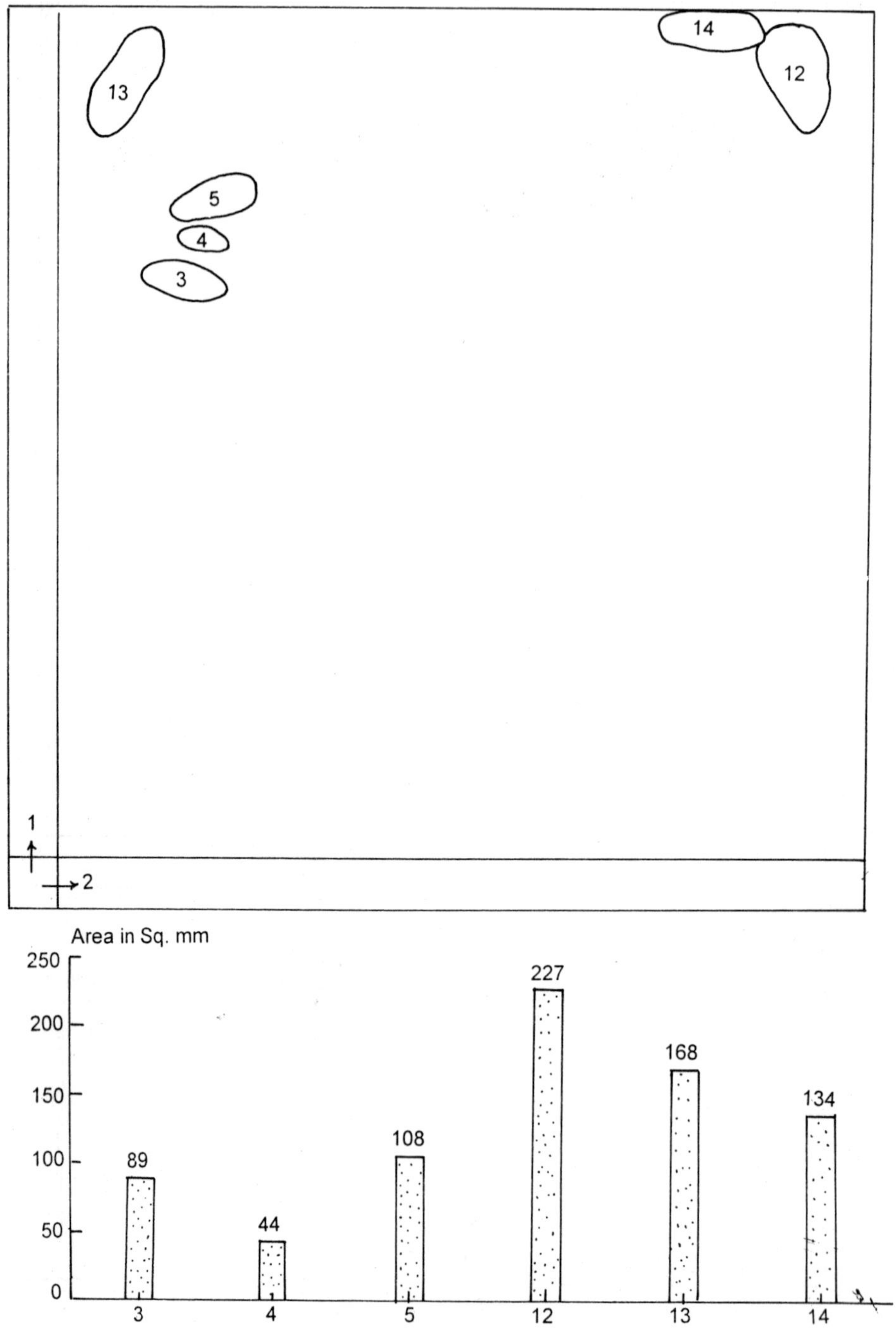

Plate 4.24 Phenolic spot patterns in *Galin soga* (Intermediate)

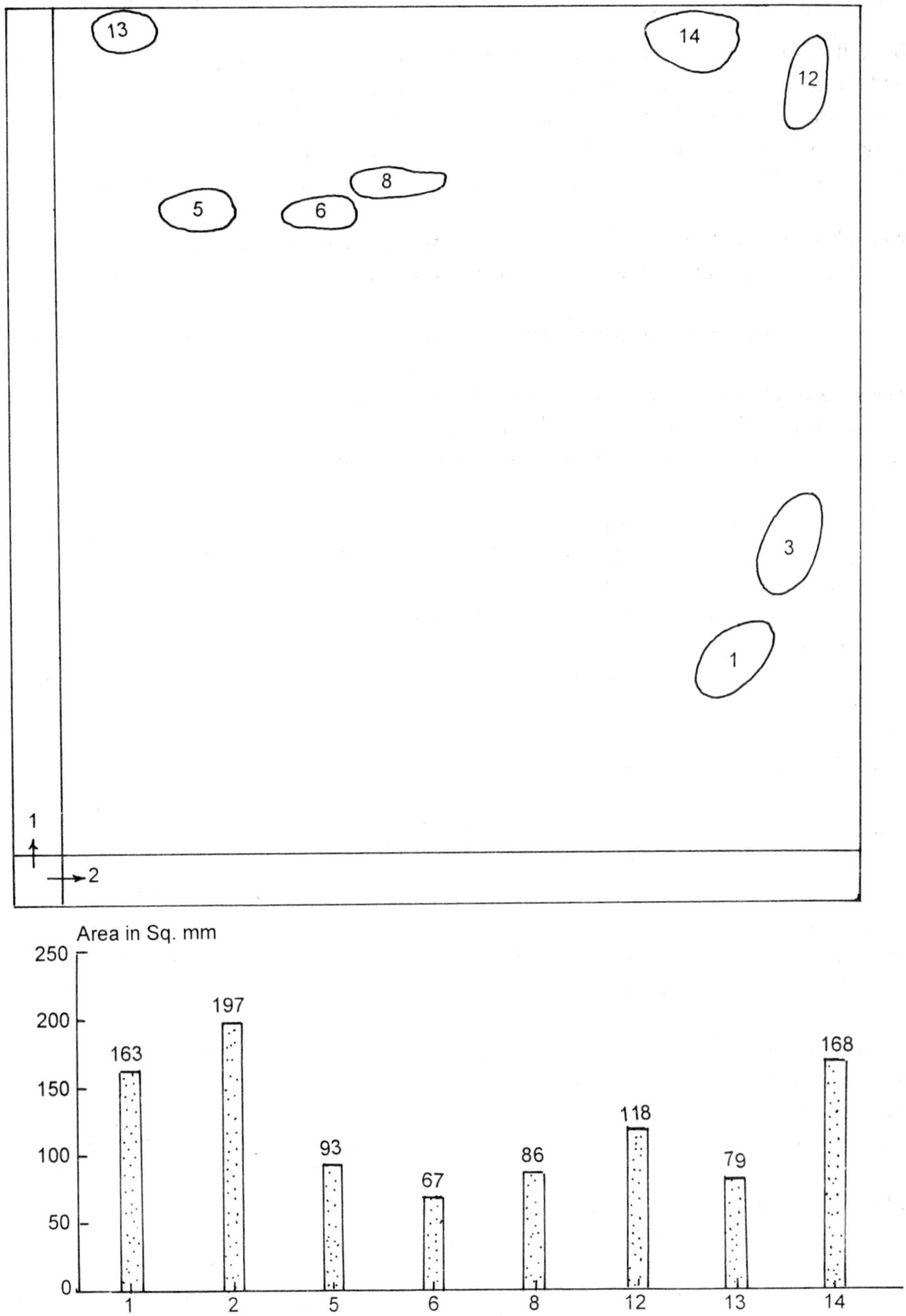

Plate 4.25 Phenolic spot patterns in *Galinsoga quadriradiata* (= G ciliata)

7. Siegesbeckia orientalis (Fig. 4.7, Table 4.10). There are six spots on the chromatogram, with spot no.1, being the largest with an area of 171 sq.mm. Of these spots, three are grey and three are light yellow in colour.

8. Blainvillea acmella (Fig. 4.8, Table 4.11). There are seven spots, Of which 3 are grey, 3 greenish yellow and one light yellow in colour. Spot no.6 is the largest with an area of 166 sq.mm.

9. Eclipta prostrata (normal leaf) (Fig. 4.9, Table 4.12). This taxon is the one that is normally found distributed all over Karnataka. The chromatogram reveals a total of 5 spots, of which 2 are grey, 2 greenish yellow and one light yellow in colour. The spots generally do not exceed 100 sq.mm in area except for spot no.8 (113 sq.mm).

10. Eclipta prostrata (Small leaf) (Fig. 4.10, Table 4.13). The chromatogram shows six spots of which three are light yellow, one greenish yellow, one grey and one violet in colour. Spot no.3 is the largest in area (173 sq.mm).
Spot no.1 is common between the normal and small leaf populations.

11. Synedrella nodiflora (normal leaf) (Fig. 4.11, Table 4.14). There are nine spots on the chromatogram, of which three are greenish yellow, two grey, two light yellow, one green and one greenish yellow in colour. The greenish yellow spot (no.9) is the largest with an area of 195 sq.mm.

12. Synedrella nodiflora (small leaf) (Fig. 4.12, Table 4.15). There are a total of 8 spots, of which 2 are green, 2 dark green, one greenish yellow, one grey, one light violet and one grey in colour. Spot no. 19 which is grey in colour is the largest with an area of 150 sq. mm.
Between the small leaf and normal leaf *Synedrella nodiflora*, spot nos. 9, 10 and 19 are common. Of the total 17 spots encountered in both the taxa, 14 spots are different. Spot no.4 found in small leaf taxon seems to be unique in having a light violet colour, while the spots of other colours are found in both the taxa.

13. Calyptocarpus vialis (Fig. 4.13, Table 4.16). The chromatogram reveals a total of 10 spots, of which 5 are greenish yellow, 3 grey,one dark green and one light violet in colour. All the spots are less than 100 sq.mm in area except for spot no.19 which is 105 sq.mm. in area.
This taxon shares only two spots (9 and 19) with normal leaf *Synedrella nodiflora* while with the small leaf *Synedrella nodiflora*, there are three common spots (4, 9 and 19). Spot no.9 seems to be common for all the three taxa, while spot no.4 is found only in small leaf *Synedrella nodiflora* and *Calyptocarpus vialis*.
Total phenolic contents have been estimated for the two forms of *Synedrella nodiflora* and *Calyptocarpus vialis*. The distribution of phenolics (mg/g) in different parts of the plantbody (leaf, stem and inflorescence) is represented in the form of histograms in Fig. 4.55.

14. Lagascea mollis (Fig. 4.14, Table 4.17). There are five spots. Of these, three are grey and two green in colour. Spot no. 3 with an area of 154 sq.mm is the largest.

15. Sclerocarpus africanus (Fig. 4.15, Table 4.18). There are six spots. Of these, three are green, two grey and one light green in colour. Spot no.6, green in colour and with an area of 348 sq.mm is the largest.

16. Helianthus rigidus (Fig. 4.16, Table 19). There are 7 spots in all. Of these, three are light green, two light yellow, one grey and one green in colour. Spot no.7 with an area of 286 sq.mm is the largest.

17. Rudbeckia nitida (Fig. 4.17 Table 4.20). There are six spots on the chromatogram. Of these spots, three are light green, two green and one light yellow in colour. All the spots are less than 100 sq.mm in area with spot no.3 being the largest (81 sq.mm)

Bidens : In the genus *Bidens*, three forms belonging to *B.pilosa* are studied together with *B.humilis*. In the genus (including *B.pilosa* and *B.humilis*) there are a total of 20 spots. The distribution is as follows :

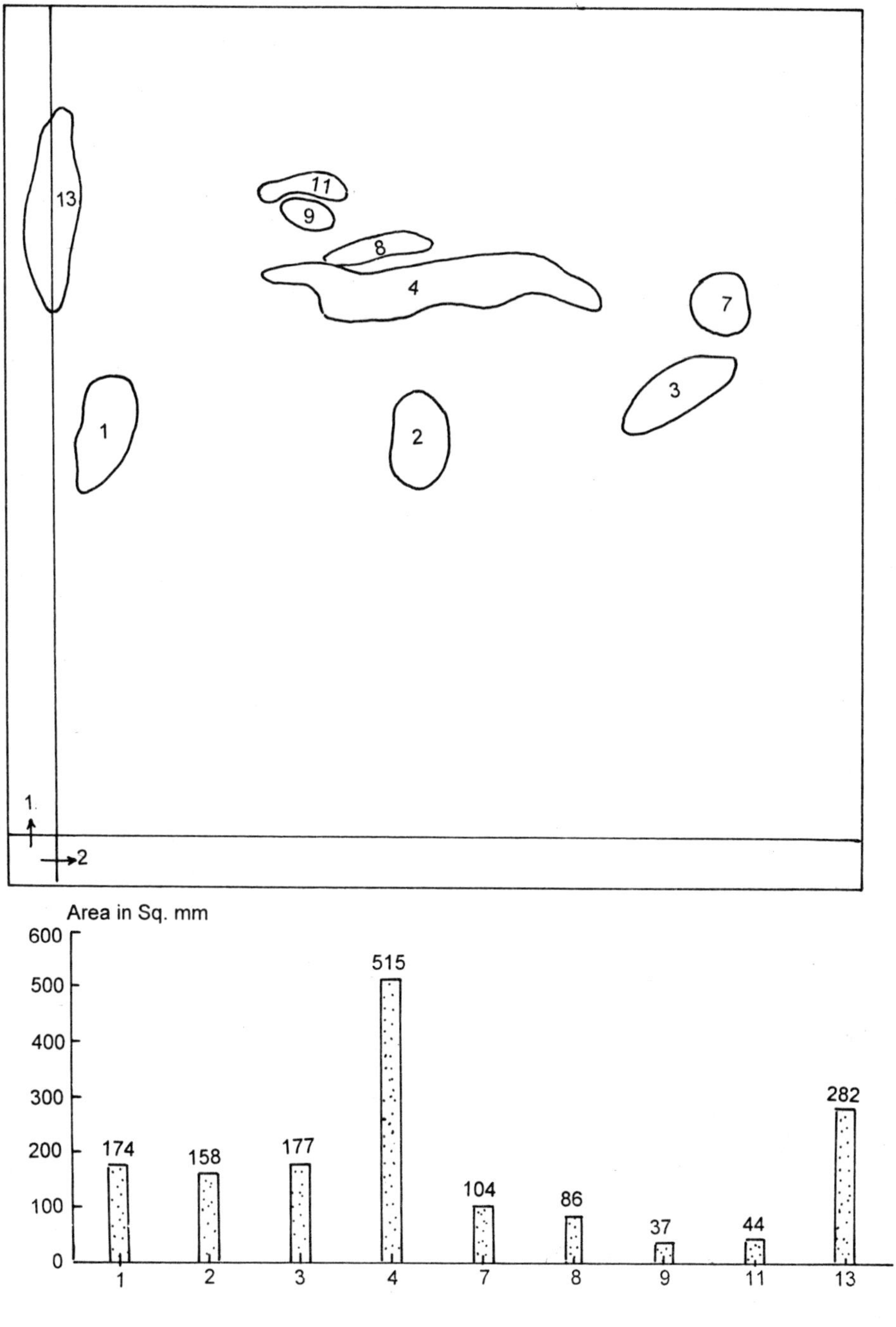

Plate 4.26 Phenolic spot patterns in *Spilanthes calva* (= S.acmella)

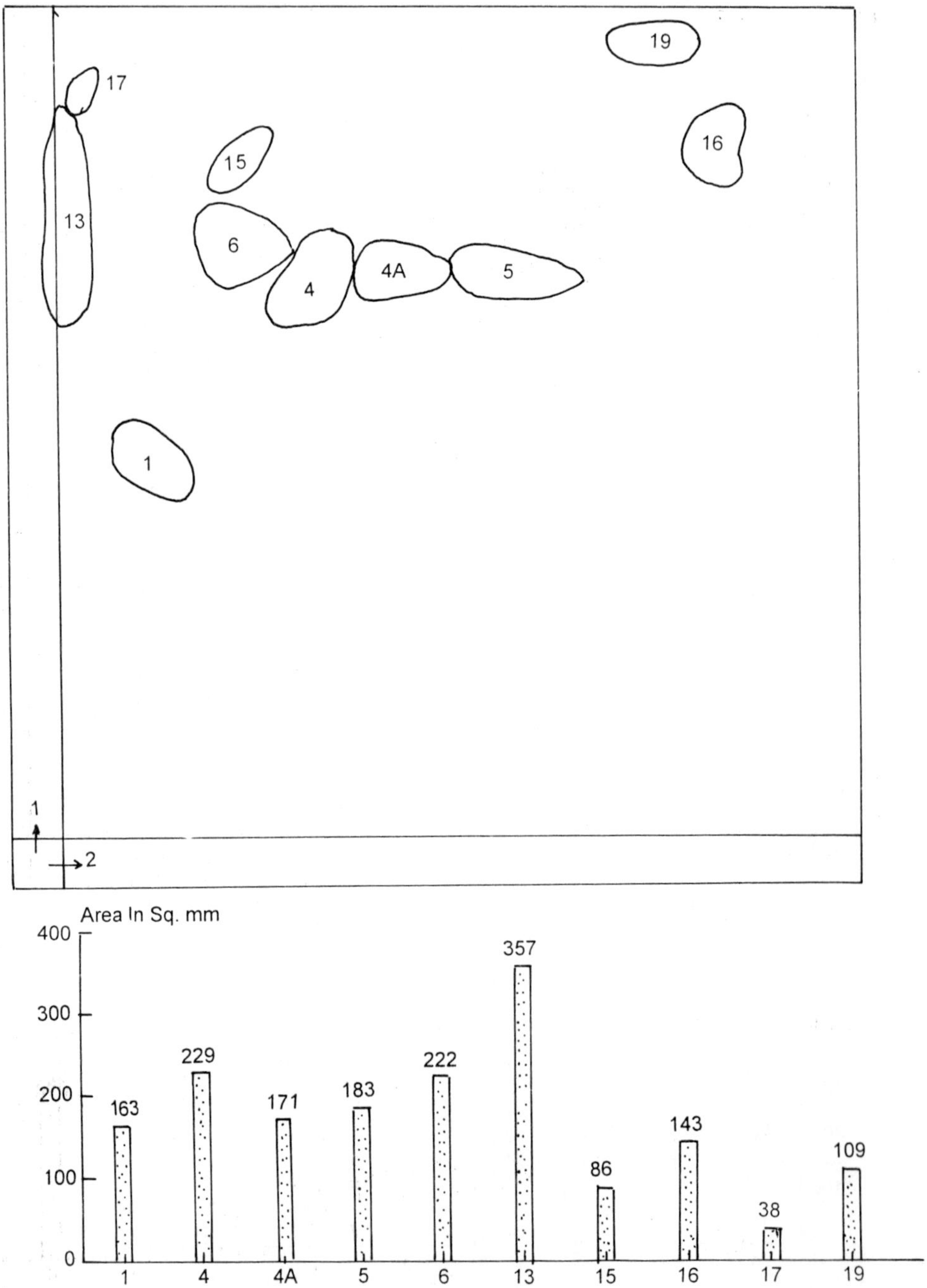

Plate 4.27 Phenolic spot patterns in *Spilanthes calva* (small head)

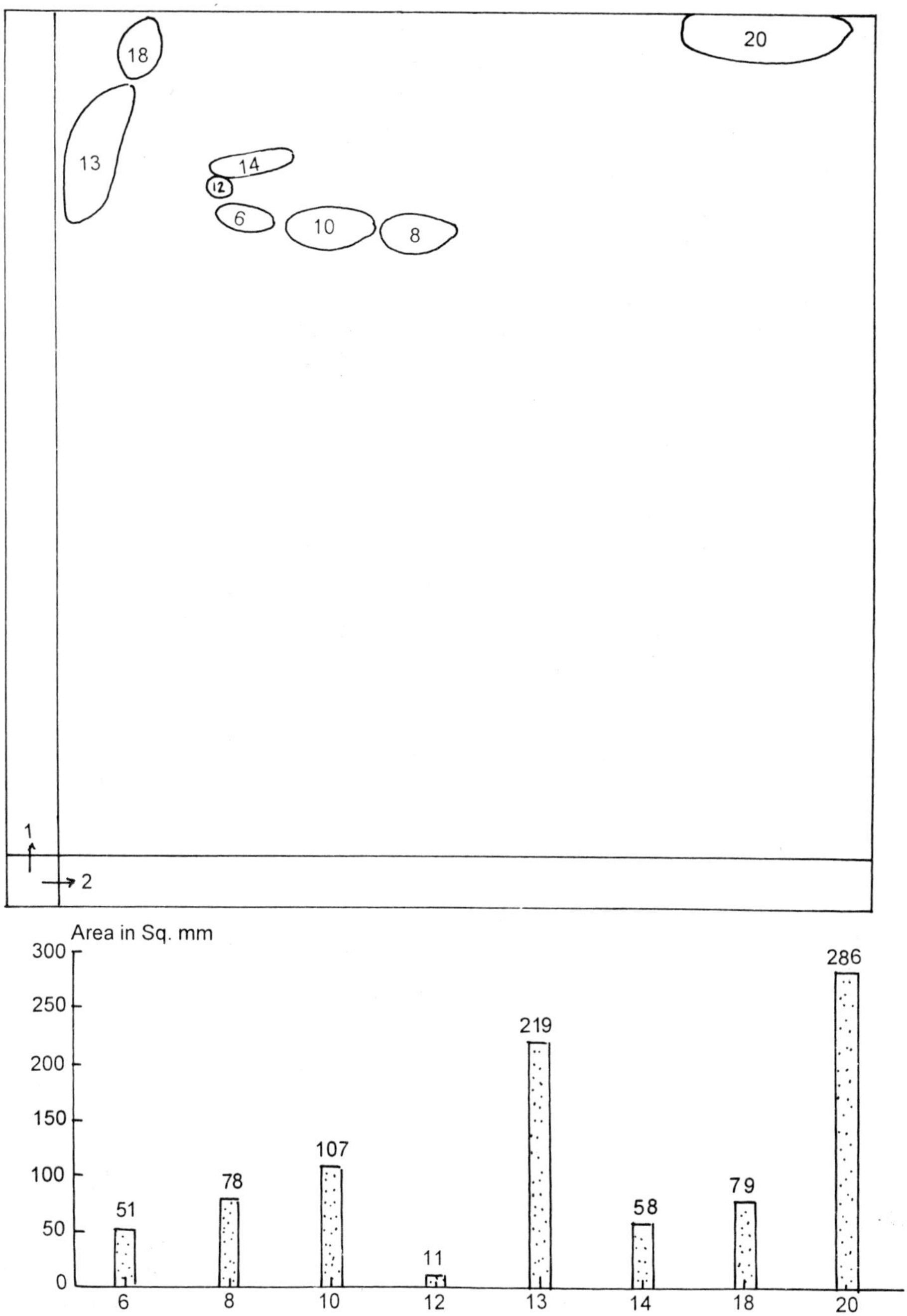

Plate 4.28 Phenolic spot patterns in *Spilanthes paniculata*

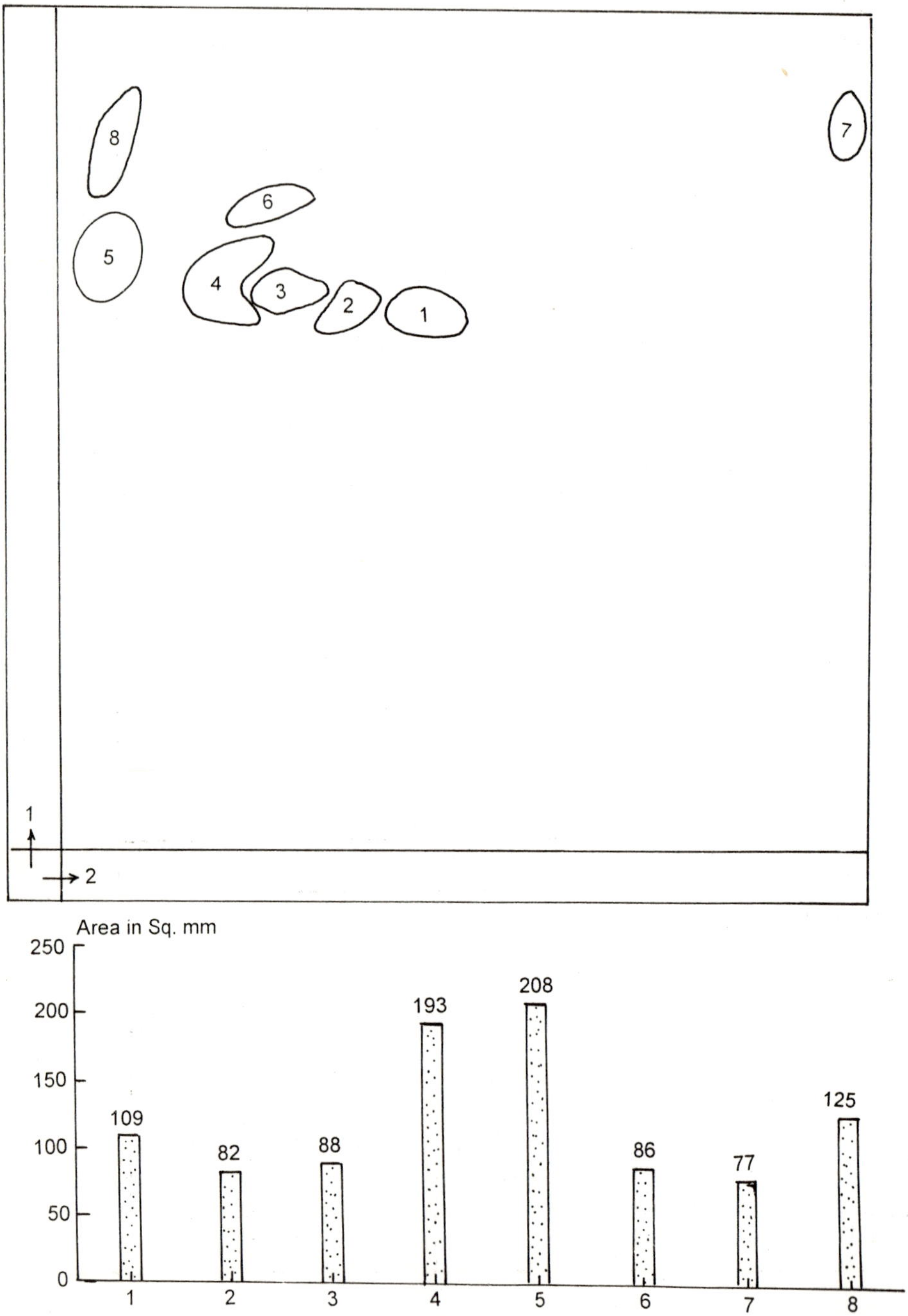

Plate 4.29 Phenolic spot patterns in *Parthenium hysterophorus*

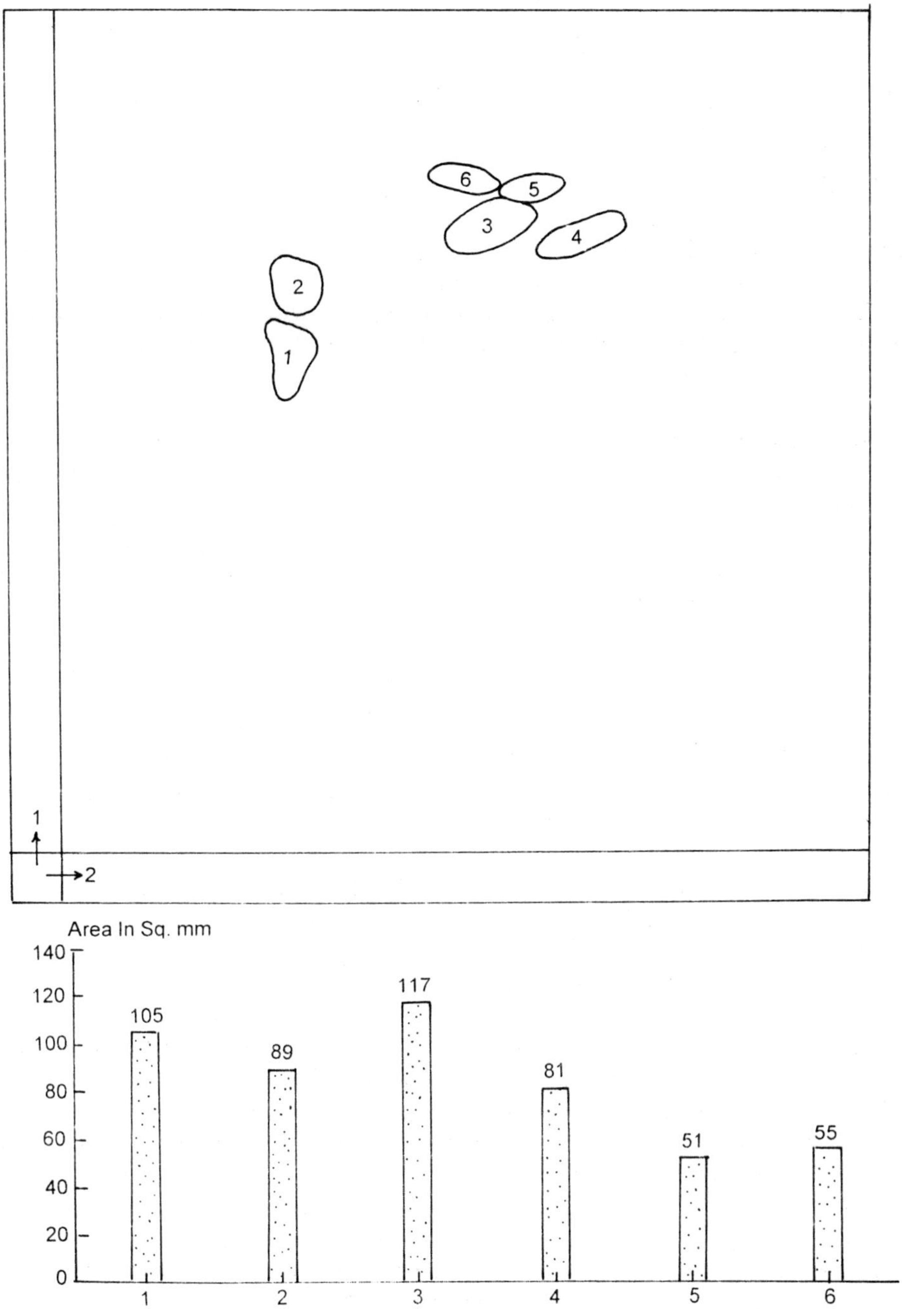

Plate 4.30 Phenolic spot patterns in *Xanthium indicum*

18. Bidens pilosa (normal ray) (Fig. 4.18, Table 4.21). There are 7 spots. Of these, one is greenish yellow, two light grey, one light yellow, two yellow and one grey in colour. Spot no.1 is the largest with an area of 466 sq.mm. This yellow coloured spot can be regarded as generic marker as it is present in all the taxa of the genus (except for intermediate ray).

19. Bidens pilosa (Intermediate ray) (Fig. 4.19, Table 4.22). In this plant, the rays of the flower heads are almost half the size when compared with the normal. The chromatogram of the leaves of this taxon reveals 7 spots. Of these 3 are light green, 3 light yellow, and one greenish yellow in colour. Spot no.1 which is present in all others is absent here. Spot no.18 with an area of 100 sq.mm is the largest.

20. Bidens pilosa (Rayless) (Fig. 4.20, Table 4.23). In this plant, the marginal florets which are female have an actinomorphic corolla, the ray being totally absent. The chromatogram of the leaves of this plant exhibits 10 spots, the highest number for all the three forms of *Bidens pilosa.* Of these spots, two are light green, two light yellow, two greenish yellow, three grey, and one yellow in colour. Spot no.1 which is largest in area (418 sq.mm) and yellow in colour is the common generic spot.

21. Bidens humilis (Fig. 4.21, Table 4.24). There are 5 spots of which, two are yellow, two are light yellow and one greenish yellow in colour. Spot no.1 which is the generic spot (It is present in *B. pilosa* also) is the largest with an area of 264 sq.mm.

22. Moonia heterophylla (Fig. 4.22, Table 4.25). There are 9 spots in all. Of these 2 are green, 3 grey, two light yellow and two greenish in colour. Spot no.5 (green in colour) with an area of 425 sq.mm is the largest.

Galinsoga : In this genus, two species and a possible hybrid have been studied for leaf flavonoids. The spot pattern is as follows :

23. Galinsoga parviflora (Fig. 4.23, Table 4.26). There are 6 spots on the chromatogram. Of these, 2 are greenish yellow, 2 grey, one light yellow and one yellow in colour. Spot no. 12 with an area of 405 sq.mm is the largest.

24. Galinsoga intermediate (Hybrid) (Fig. 4.24, Table 4.27). This is believed to be an intermediate between *G.parviflora* and *G.quadriradiata.* This could be a product of across between these two species. Leaf flavonoids are represented on the TLC platein the form of 6 spots. Of these, 3 are greenish yellow and 3 grey in colour. Spot no.12 with an area of 227 sq.mm is the largest.

25. Galinsoga quadriradiata (= *G. ciliata*) (Fig. 4.25, Table 4.28), There are 8 spots. Of these, 4 are grey, 2 greenish yellow and 2 light yellow in colour. Spot no.2 with an area of 197 sq.mm. is the largest.
In the above three taxa, there are a total number of 14 spots. Spot no.12 present in all the three may be considered as the generic spot. There are 2 common spots (6 and 12) between *G.parviflora* and *G.quadriradiata*, one common spot (12) between *G.parviflora* and *Galinsoga* intermediate and 3 common spots (5, 12 and 14) between *G.quadriradiata* and *Galinsoga* intermediate.

Spilanthes : Three taxa are studied in this genus for leaf flavonoid profiles. Two variants of *Spilanthes calva* (large head and small head) and *S.paniculata* have been investigated.

26. Spilanthes calva (small head) (Fig. 4.26, Table 4.29). There are 9 spots. Of these, 3 are grey, 4 light yellow, one yellow and one greenish yellow in colour. Spot no.4 with an area of 515 sq.mm is the largest.

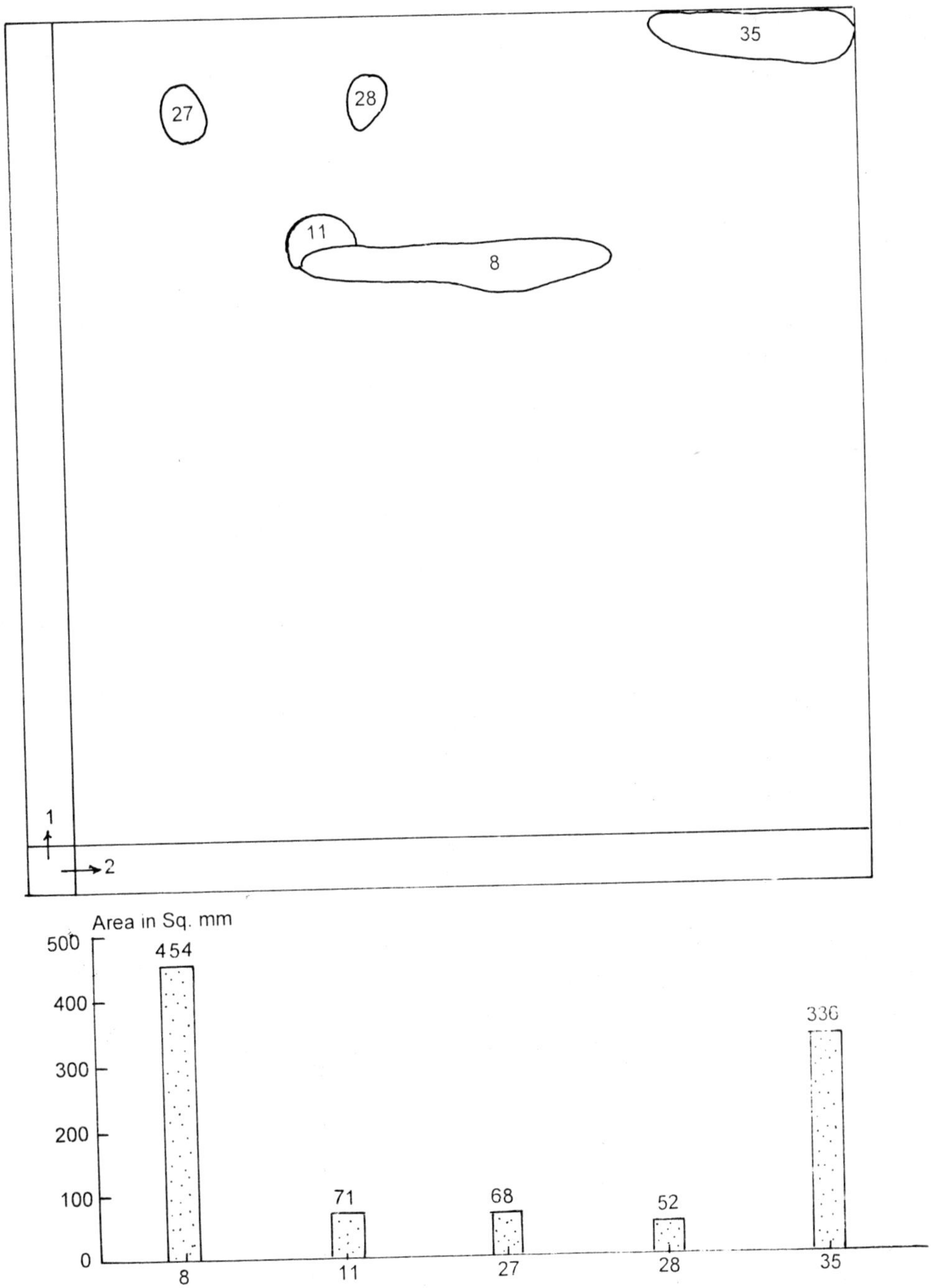

Plate 4.31 Phenolic spot patterns in *Senecio wightis*

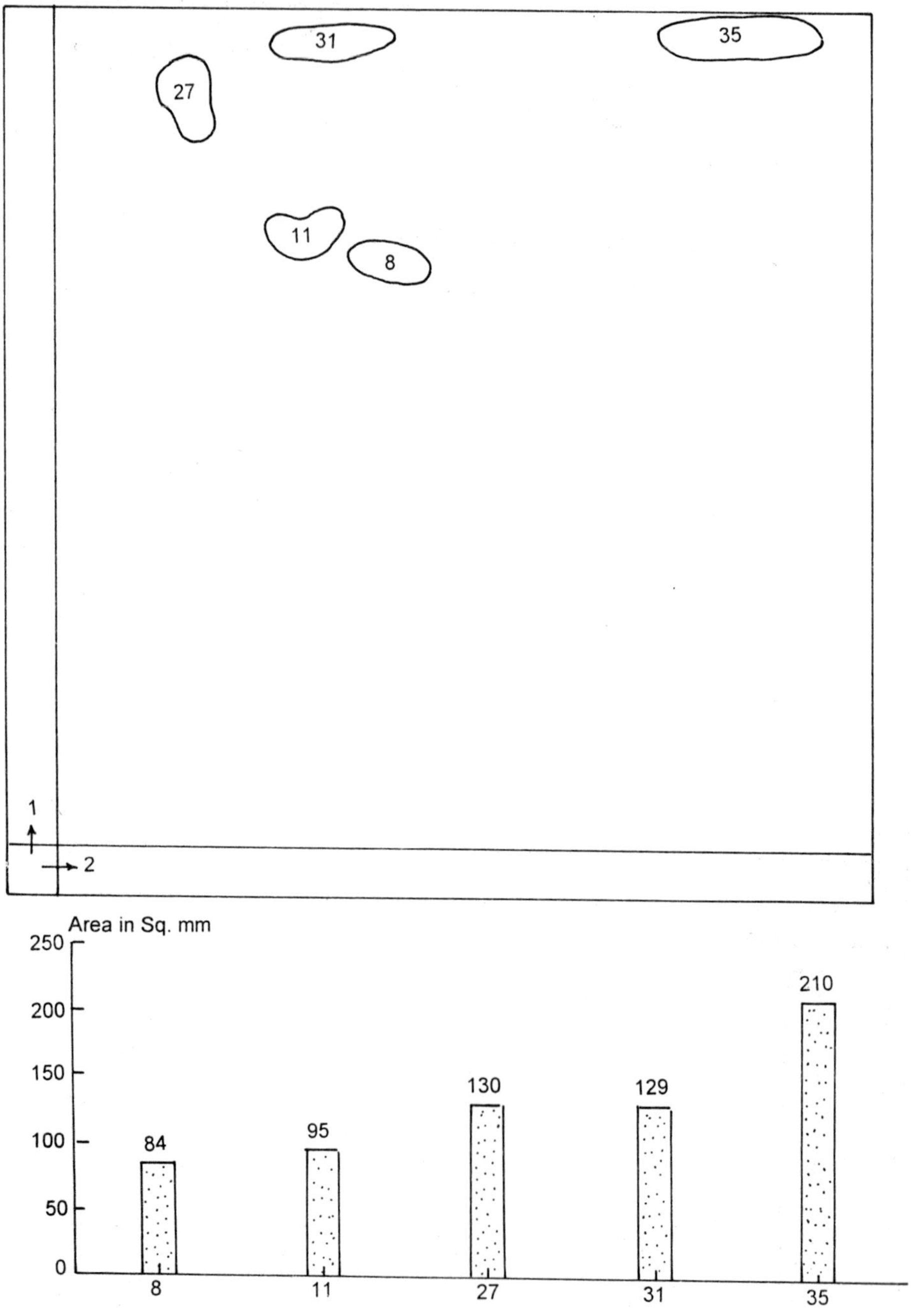

Plate 4.32 Phenolic spot patterns in *Senecio lawsonii*

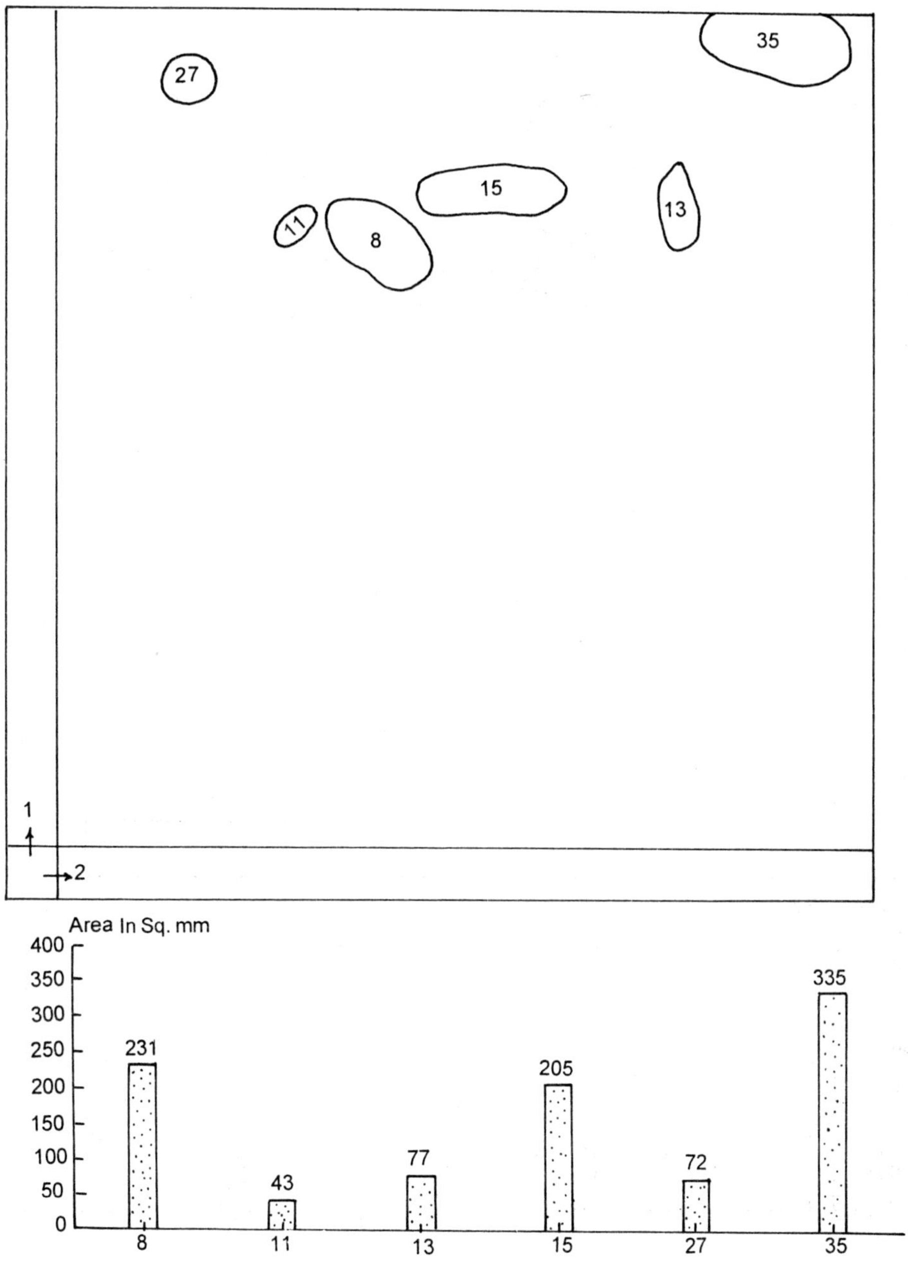

Plate 4.33 Phenolic spot patterns in *Senecio belgaumensis*

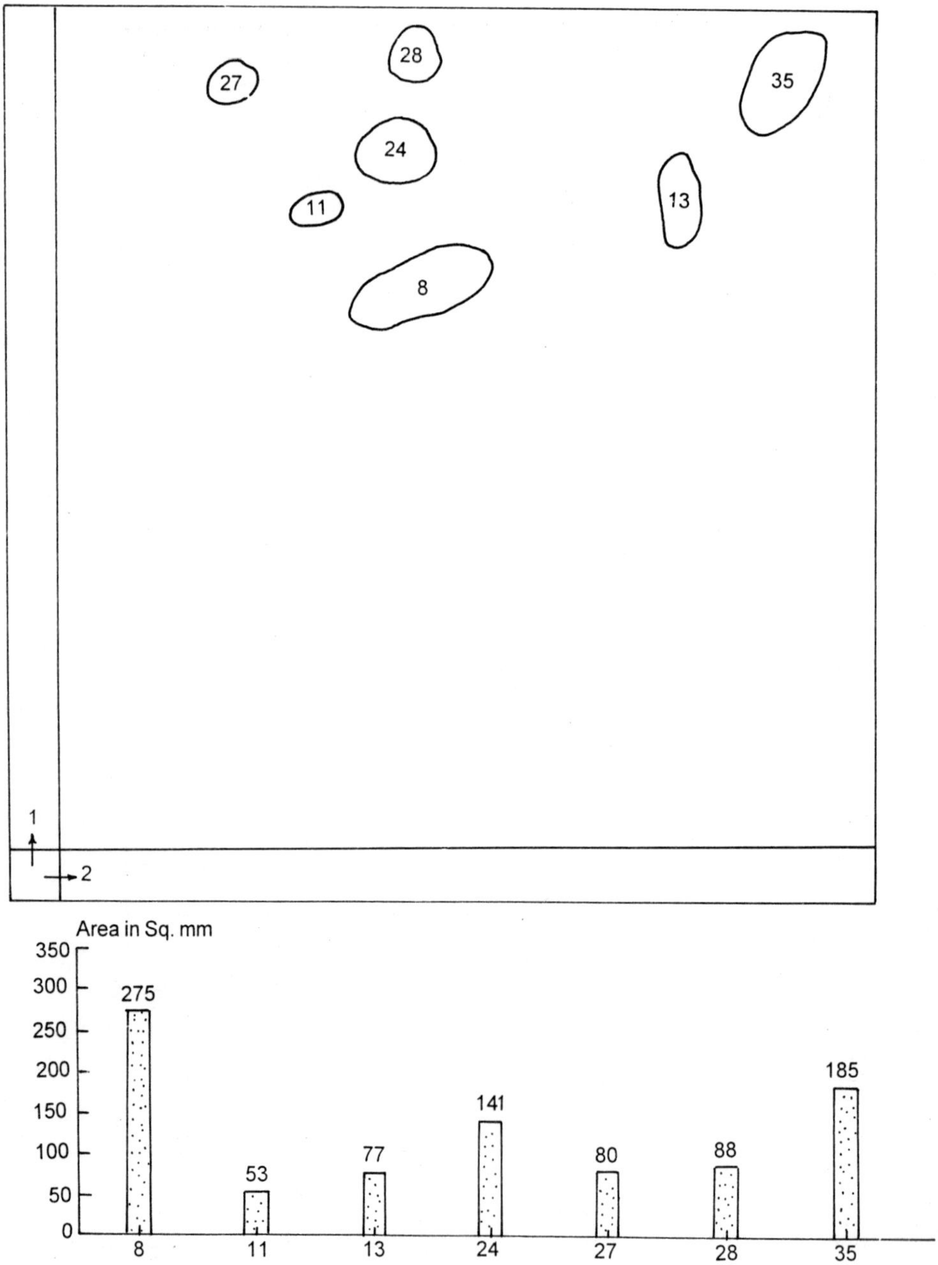

Plate 4.34 Phenolic spot patterns in *Senecio neelgherryanus*

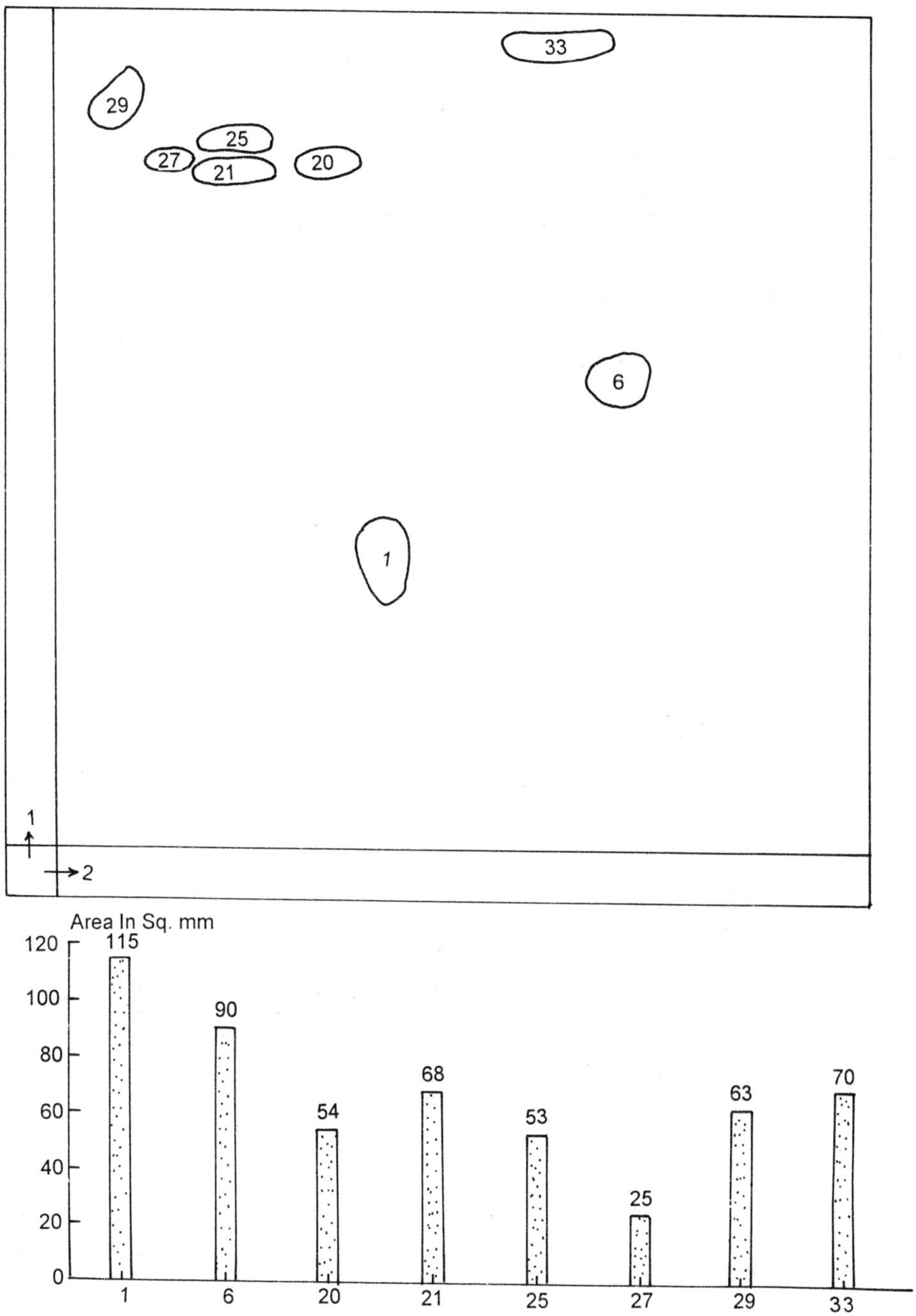

Plate 4.35 Phenolic spot patterns in *Senecio tenuifolius*

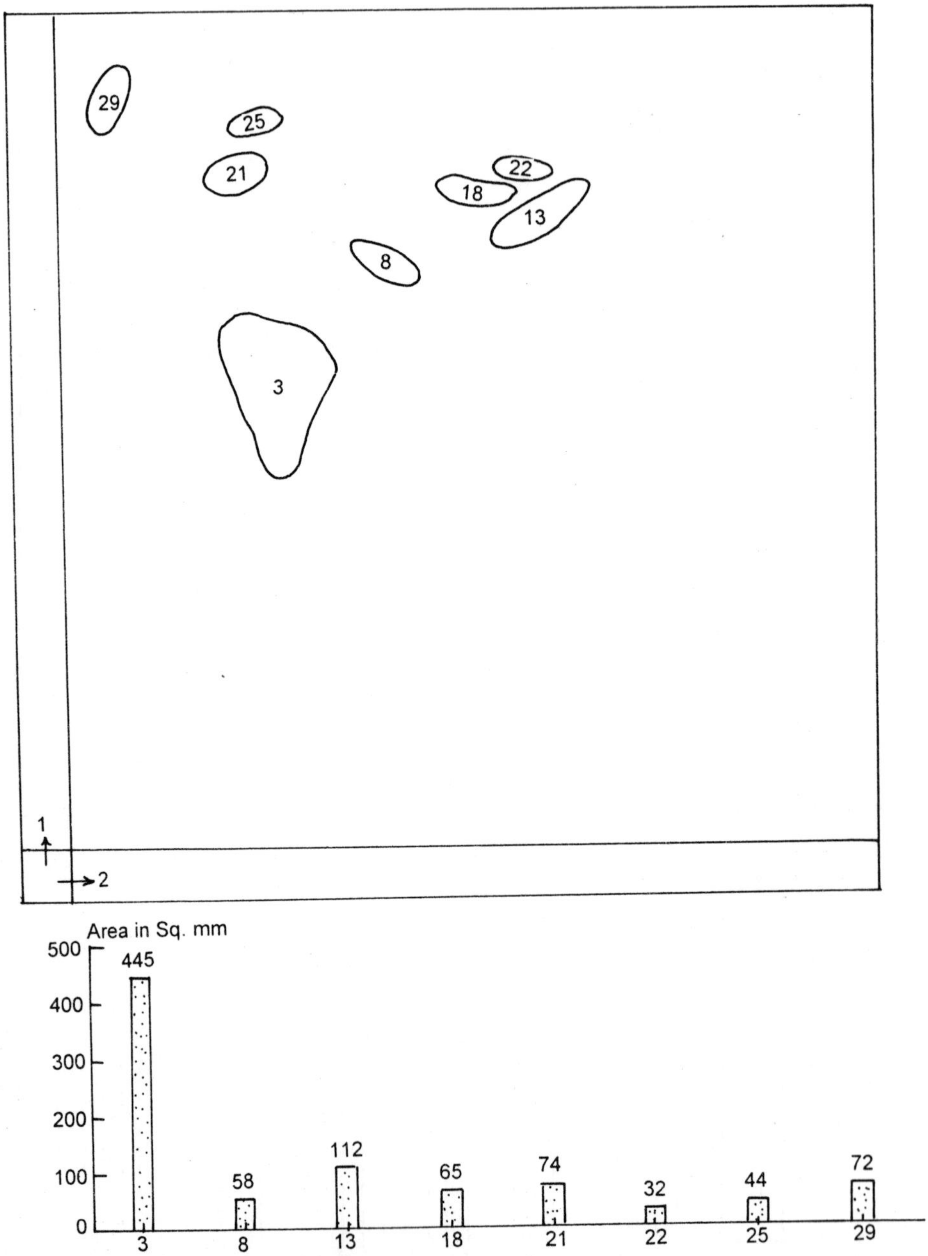

Plate 4.36 Phenolic spot patterns in *Senecio hohenackeri*

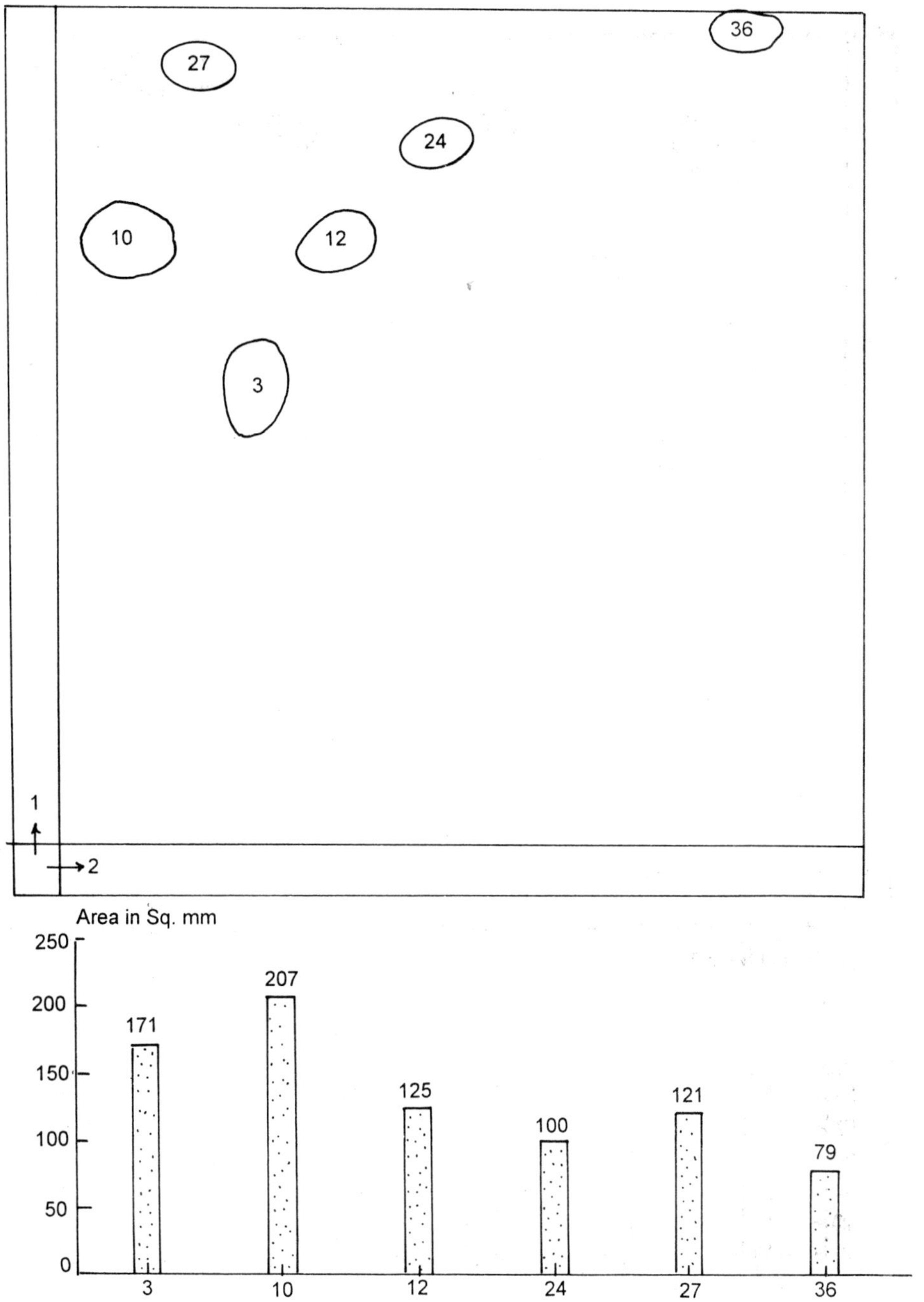

Plate 4.37 Phenolic spot patterns in *Senecio grahami*

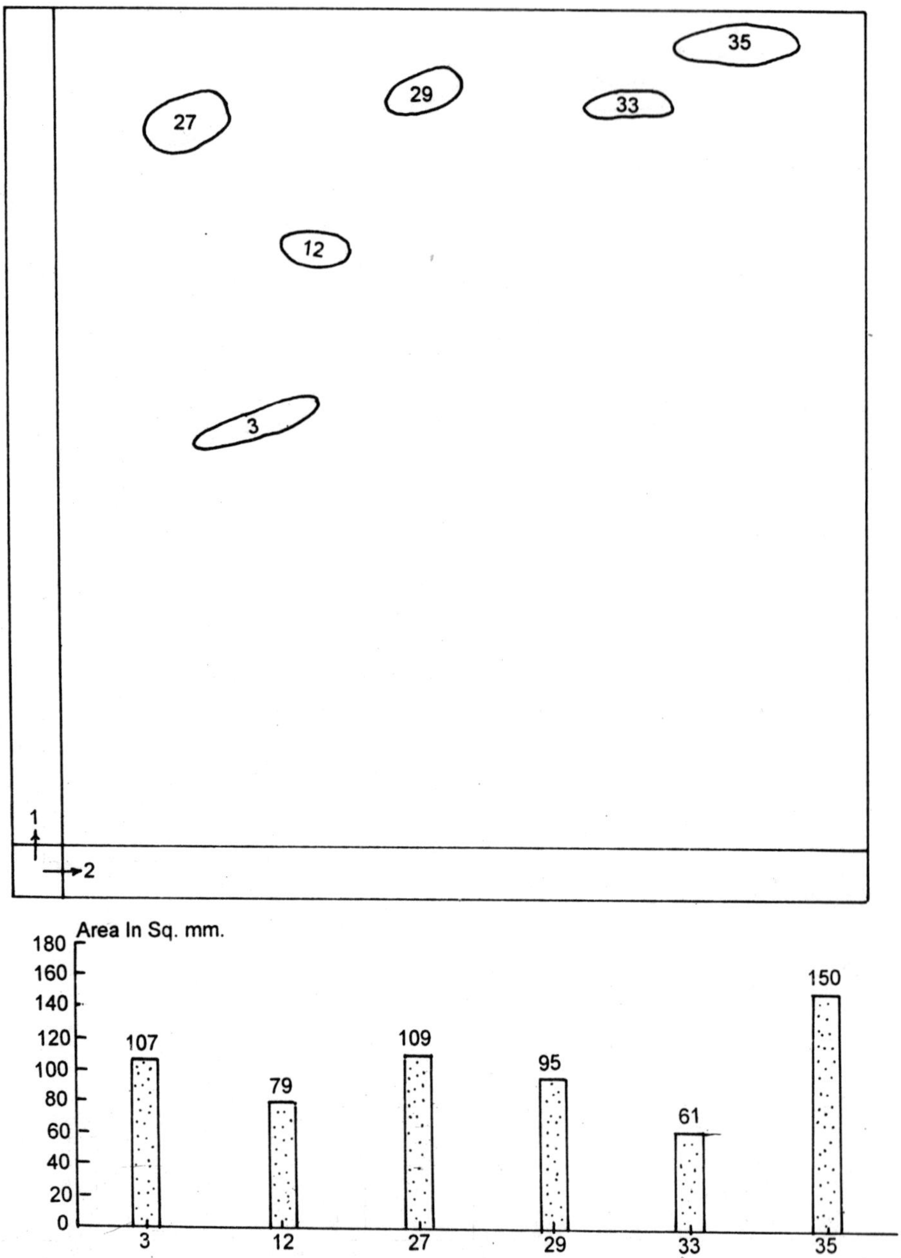

Plate 4.38 Phenolic spot patterns in *Senecio edgworthii*

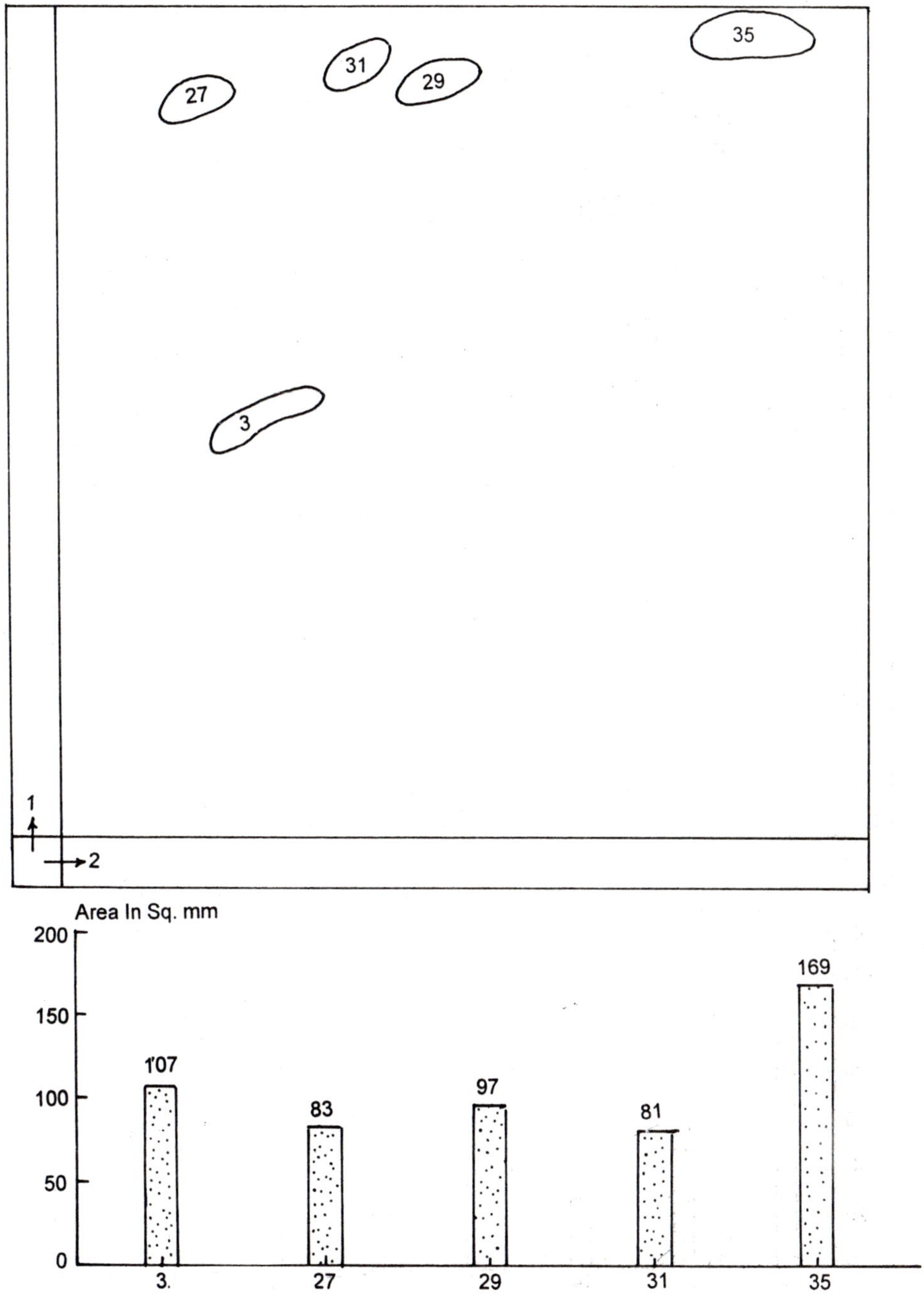

Plate 4.39 Phenolic spot patterns in *Senecio edgworthii* (Small leaf)

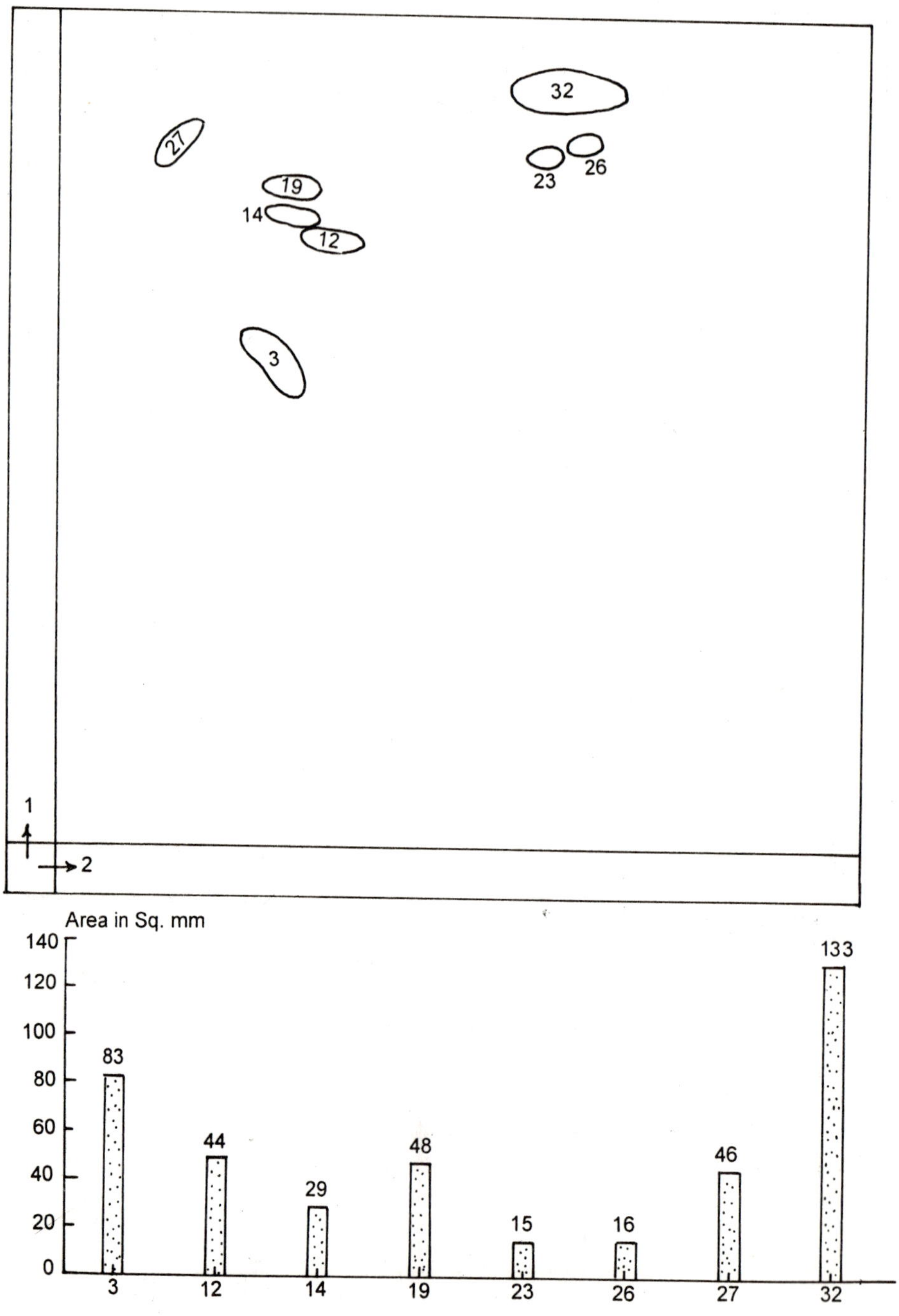

Plate 4.40 Phenolic spot patterns in *Senecio ludens*

27. Spilanthes calva (large head) (Fig. 4.27, Table 4.30). There are 10 spots on the chromatogram. Of these, 4 are greenish yellow, 3 grey, 2 light yellow and one yellow in colour. Spot no.13 (light yellow) with an area of 357 sq.mm is the largest.

28. Spilanthes paniculata (Fig. 4.28, Table 4.31). There are 8 spots of which 3 are green, 3 grey, one greenish yellow and one light yellow in colour. Spot no.20 with an area of 286 sq.mm is the largest.
In the above three taxa, a total of 20 spots have been encountered on the TLC plates. Of these, spot no.13 which is light yellow in colour may be regarded as the generic spot as it is present in all the three taxa. Of the remaining spots, *Spilanthes calva* (small head) and *S. calva* (large mead) share spot nos. 1, 4 and 13; *S. calva* (small head) and *S. paniculata* share nos. 8 and 13, while *S.calva* (large head) and *S. paniculata* share spot nos. 6 and 13.

29. Parthenium hysterophorous (Fig. 4.29, Table 4.32). There are a total of 8 spots, of which 3 are light yellow, 2 grey, one green, one violet and one light one light violet in colour. Spot no.5 with an area of 208 sq.mm is the largest.

30. Xanthium indicum (Fig. 4.30, Table 4.33). The chromatogram shows as total of 6 spots of which 3 are grey, and 3 are light yellow in colour. Spot no.3 with an area of 117 sq.mm is the largest.

Senecio : This genus belonging to the tribe Senecioneae has been subjected to a detailed study of numerical systematics based on flavonoid profiles. A total no. of 12 species with diverse morphological characters have been studied for leaf flavonoids. The details are described below for each taxon separately.

31. S. wightii (Fig. 4.31, Table 4.34). There are 5 spots in all. Of these 2 are grey, on light yellow, one light green, and one violet in colour. Spot no.8 with an area of 454 sq.mm is the largest. There are no species specific spots.

32. S. lawsoni (Fig. 4.32, Table 4.35). There are 5 spots on the chromatogram. Of these 2 are light yellow, 2 light green and one greyish yellow in colour. Spot no. 35 with an area of 210 sq.mm is the largest. Spot on. 31 is the species specific spot being present only in this taxon.

33. S. belgaumensis (Fig. 4.33, Table 4.36). There are a total of 6 spots. Of these 2 are grey, 2 light yellow, one dark green and one green in colour. Spot no. 35 with an area of 385 sq.mm is the largest. There are no species specific spots in this taxon.

34. S. neelgherryanus (Fig. 4.34, Table 4.37). There are 7 spots, of which 2 are light yellow, 2 violet, one light green, one dark green and one grey in colour. Spot no. 8 with an area of 245 sq.mm is the largest. There are no species specific spots.

35. S. tenuifolius (Fig. 4.35, Table 4.38). There are 8 spots on the chromatogram. Of these, 4 are grey, 2 light yellow, one blue and one yellow in colour. Spot no. 1with an area of 115 sq.mm is the largest. This taxon has two species specific spots (1 and 6).

36. S. hohenackeri (Fig. 4.36, Table 4.39). There are 8 spots, of which 4 are grey, 2 light yellow, one light green and one yellow in colour. Spot no. 3 with an area of 445 sq.mm is the largest. There are two spots (18 and 22) which are exclusive to this taxon.

37. S. grahami (Fig. 4.37, Table 4.40). There are 6 spots in all. Of these, 2 are light yellow, one grey, one dark green and one violet in colour. Spot no. 10 with an area of 207 sq.mm is the largest. Spot no.36 is the exclusive spot for this taxon.

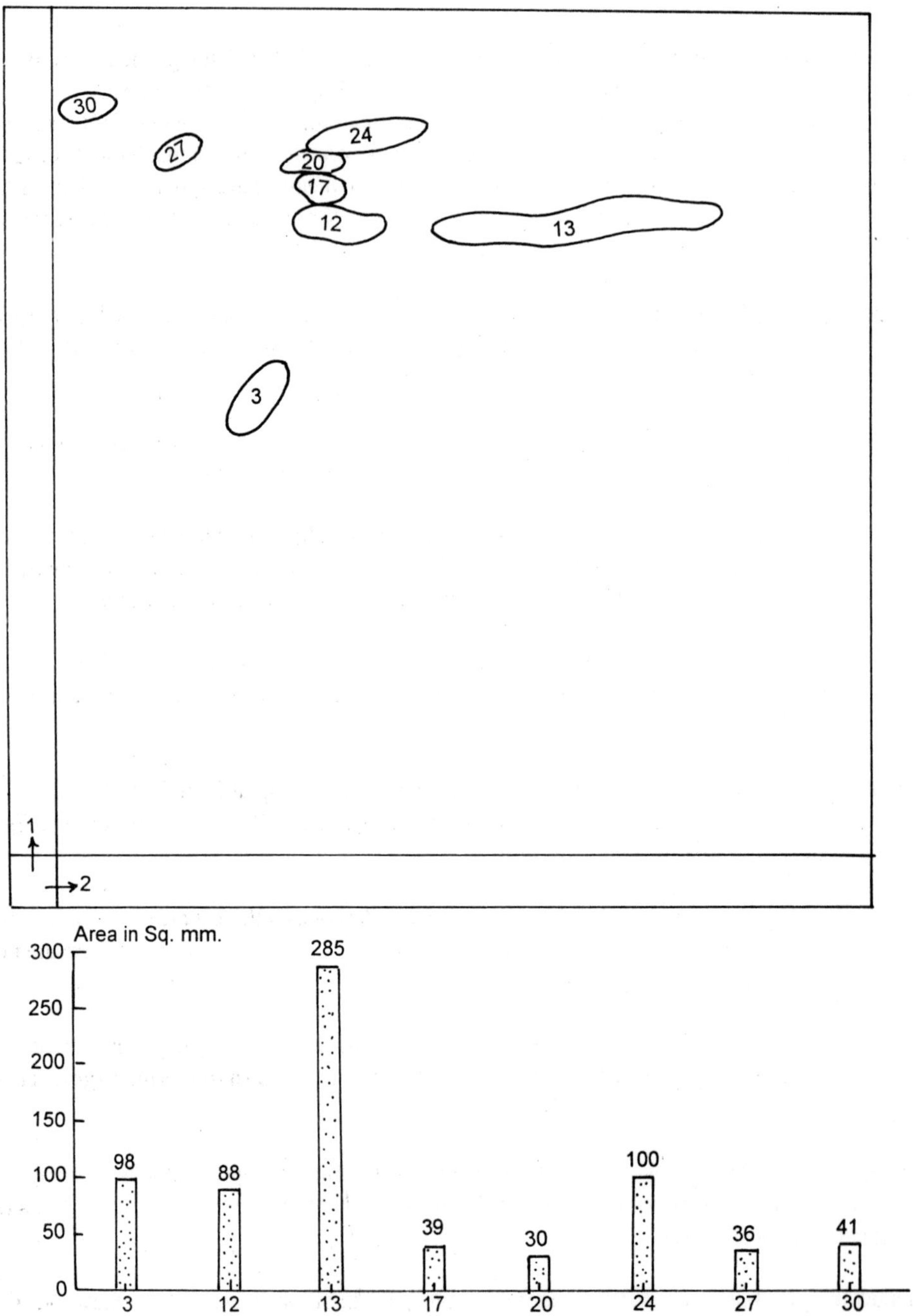

Plate 4.41 Phenolic spot patterns in *Senecio cineraria*

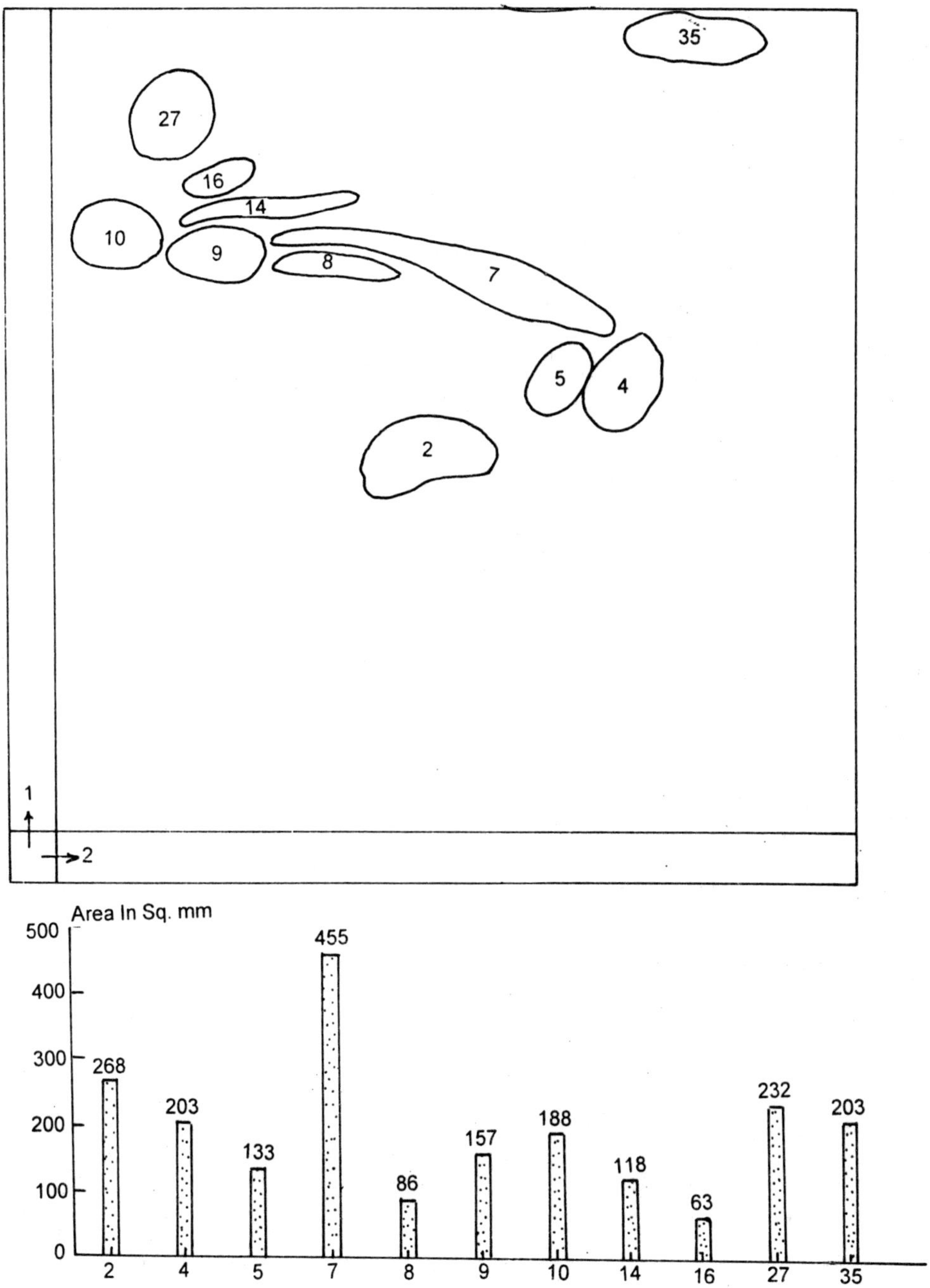

Plate 4.42 Phenolic spot patterns in *Senecio corymbosus*

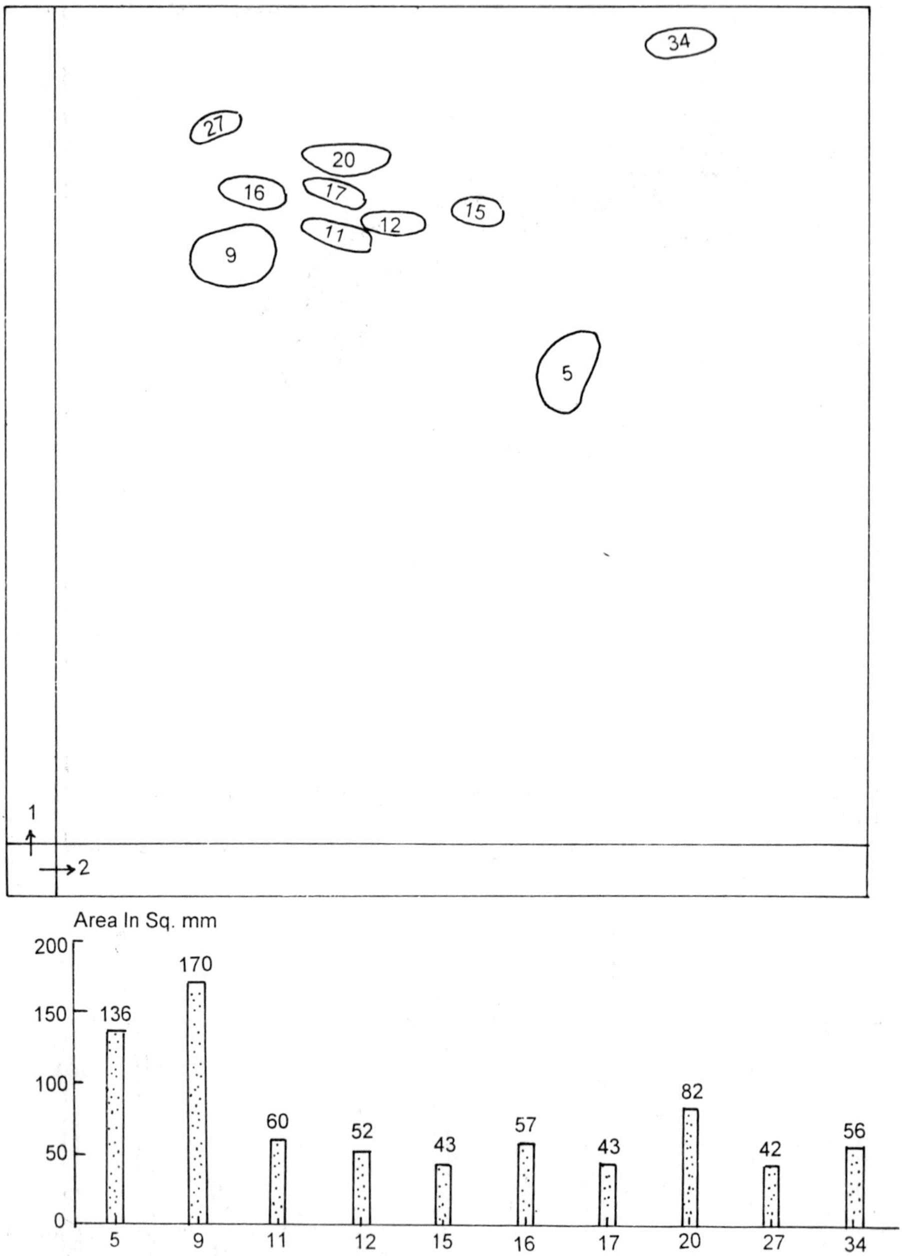

Plate 4.43 Phenolic spot patterns in *Senecio walkeri*

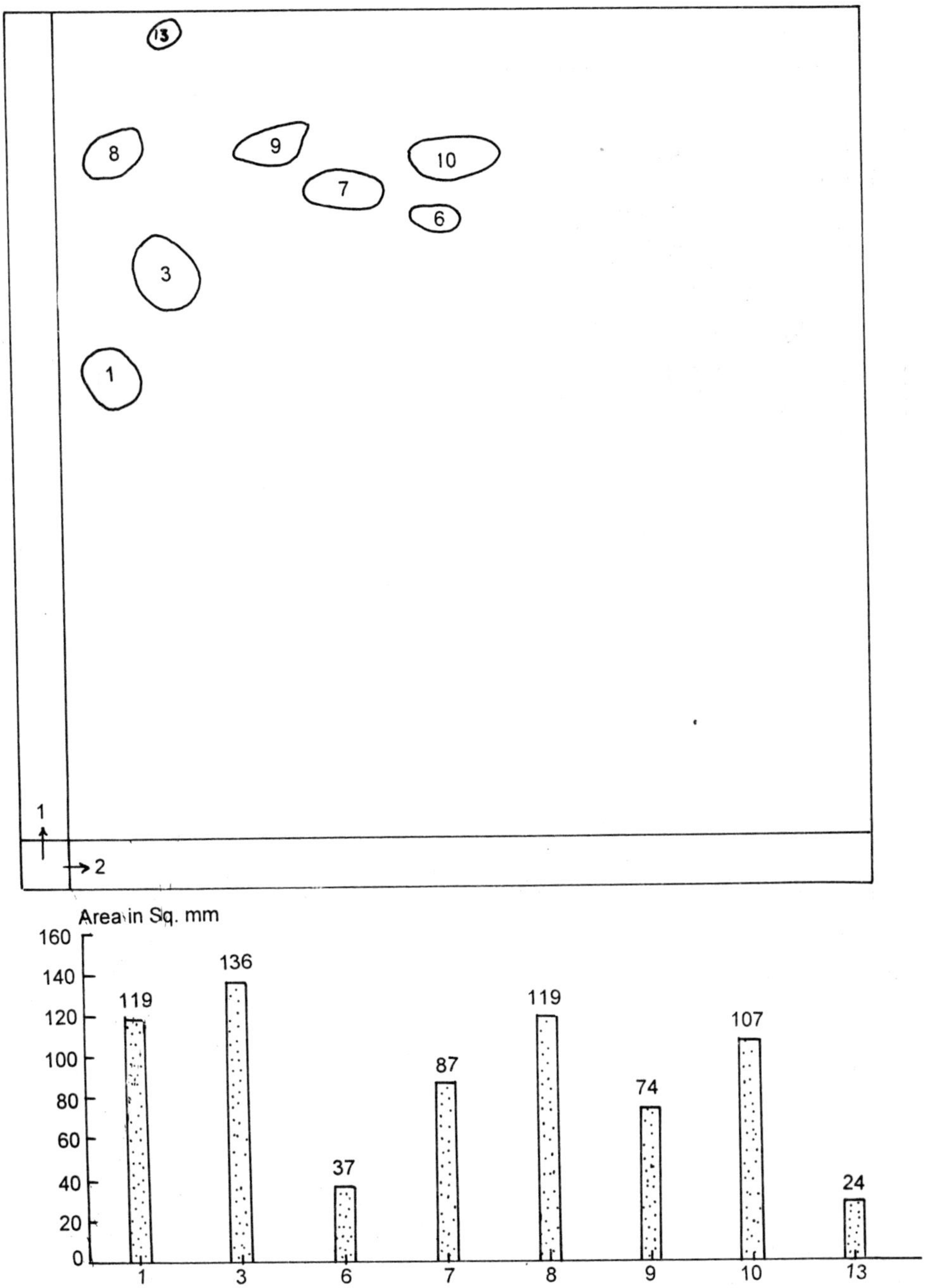

Plate 4.44 Phenolic spot patterns in *Synura nitida*

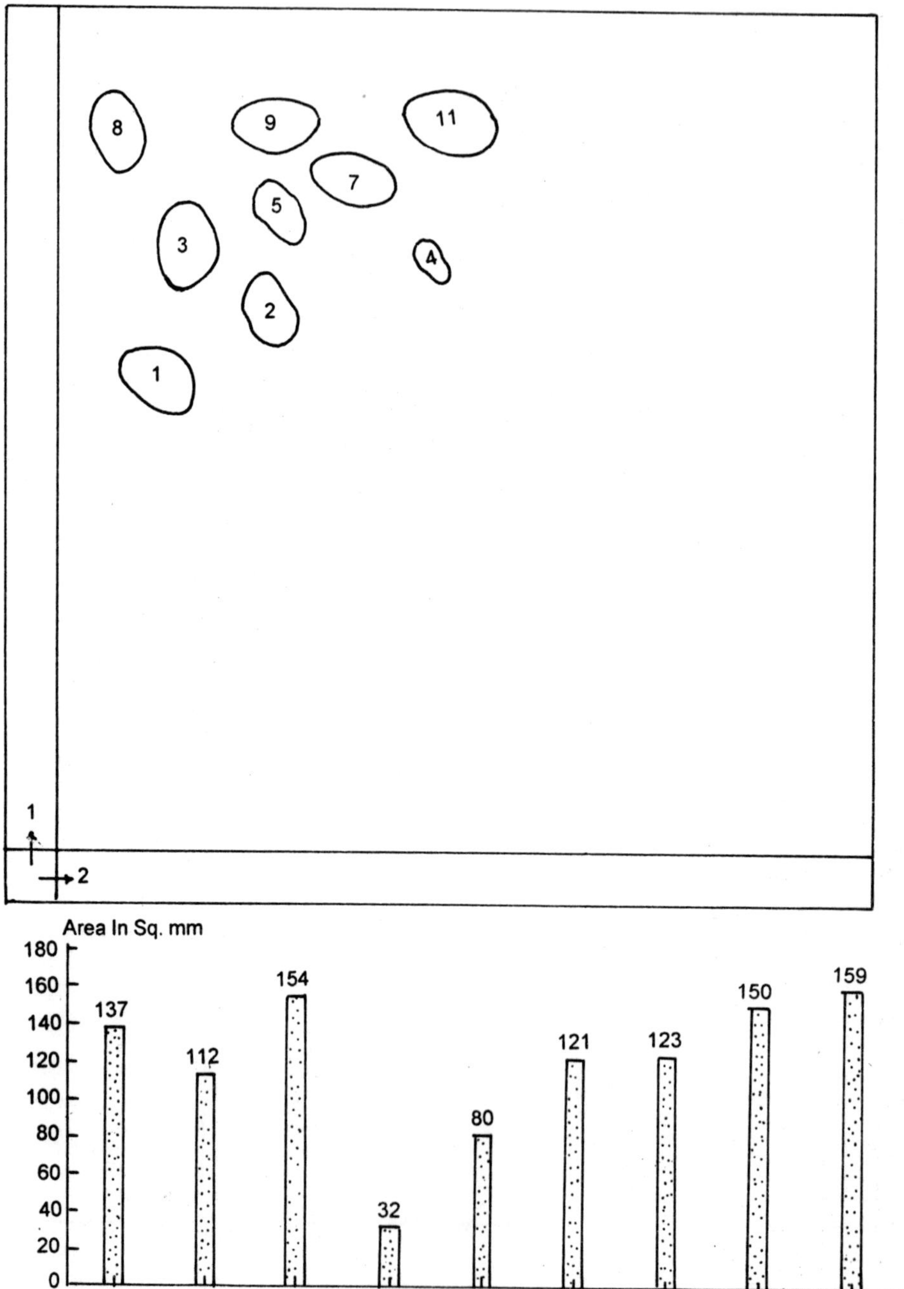

Plate 4.45 Phenolic spot patterns in *Gynura lycopersicifolia*

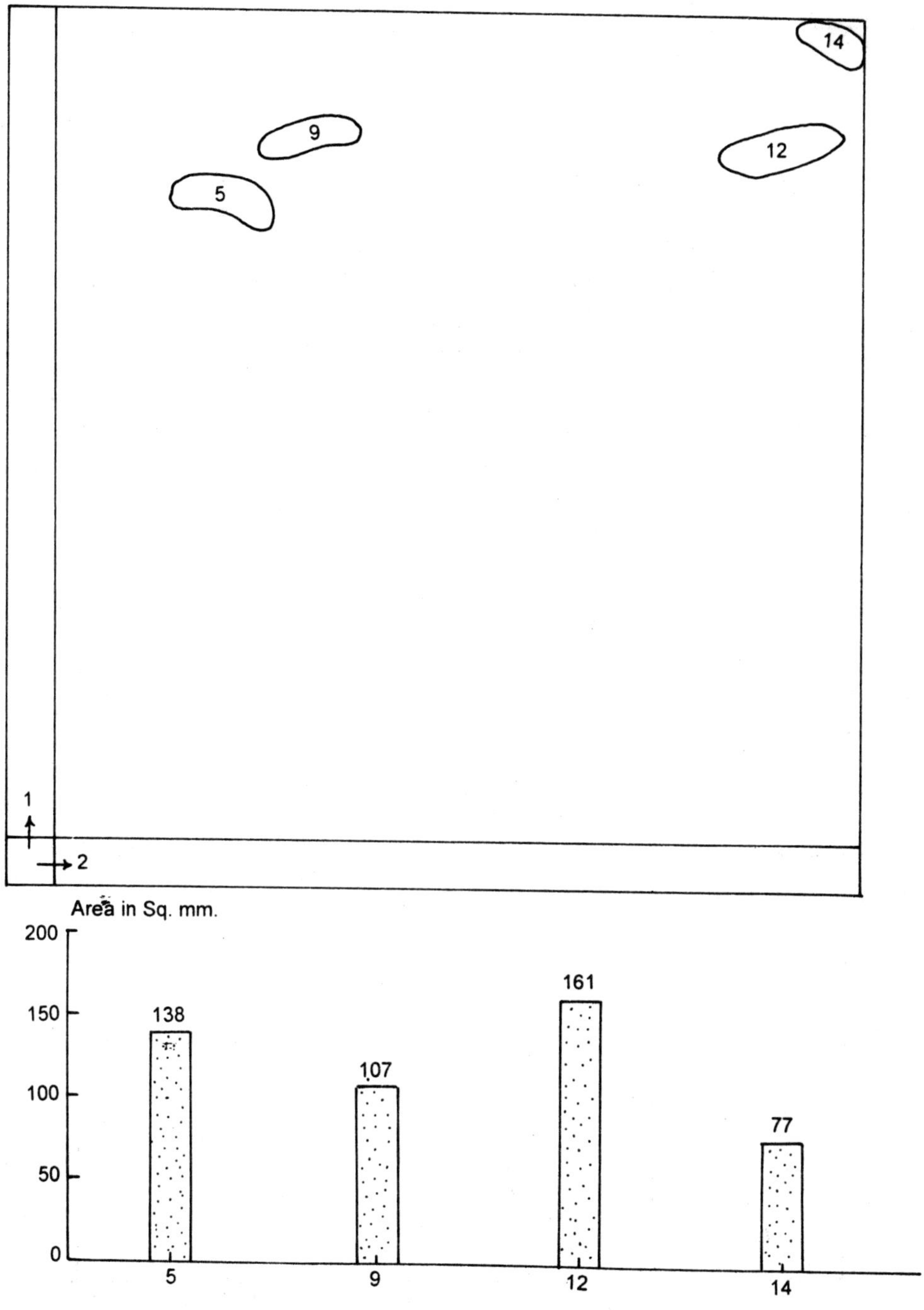

Plate 4.46 Phenolic spot patterns in *Gynura aurantiaca*

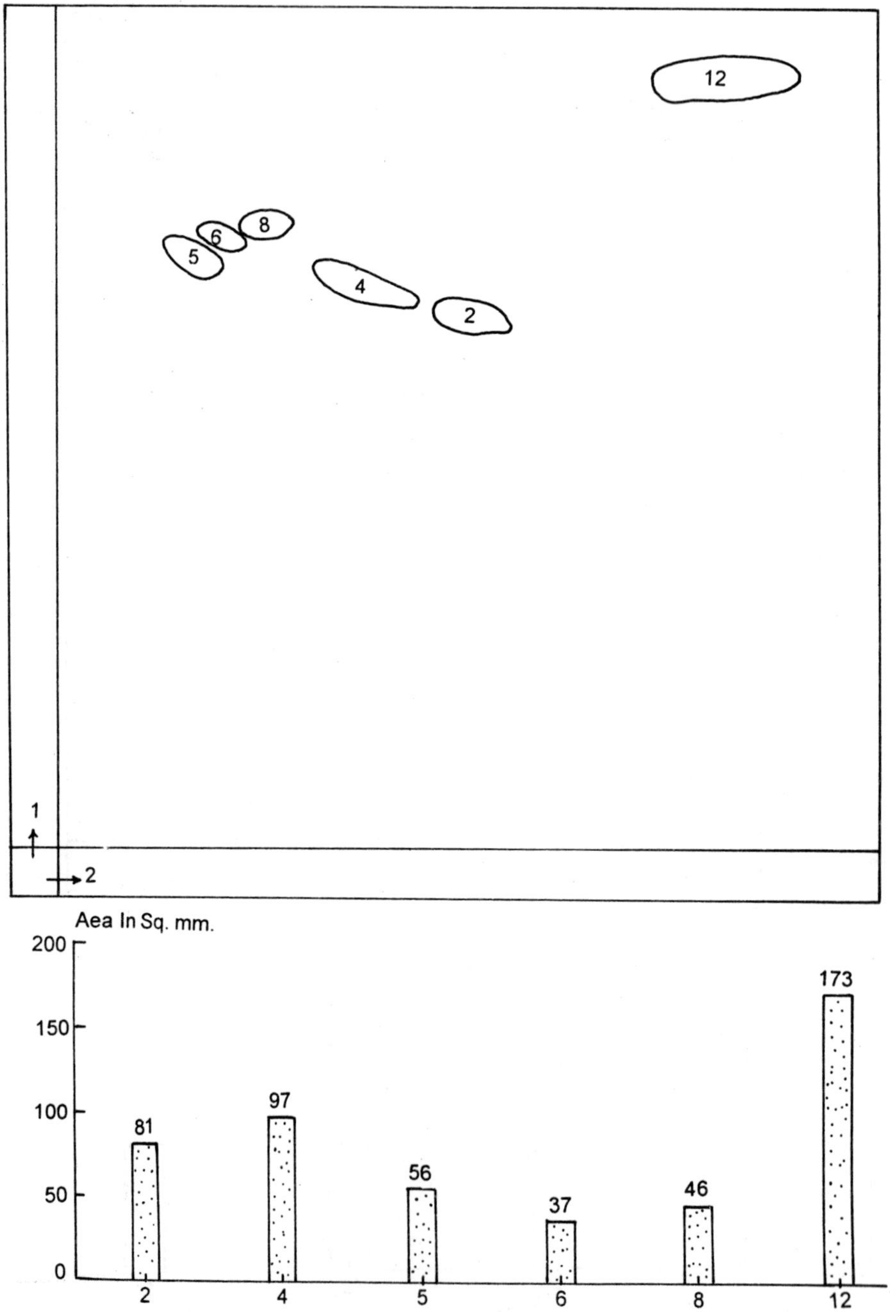

Plate 4.47 Phenolic spot patterns in *Notonia grandiflora*

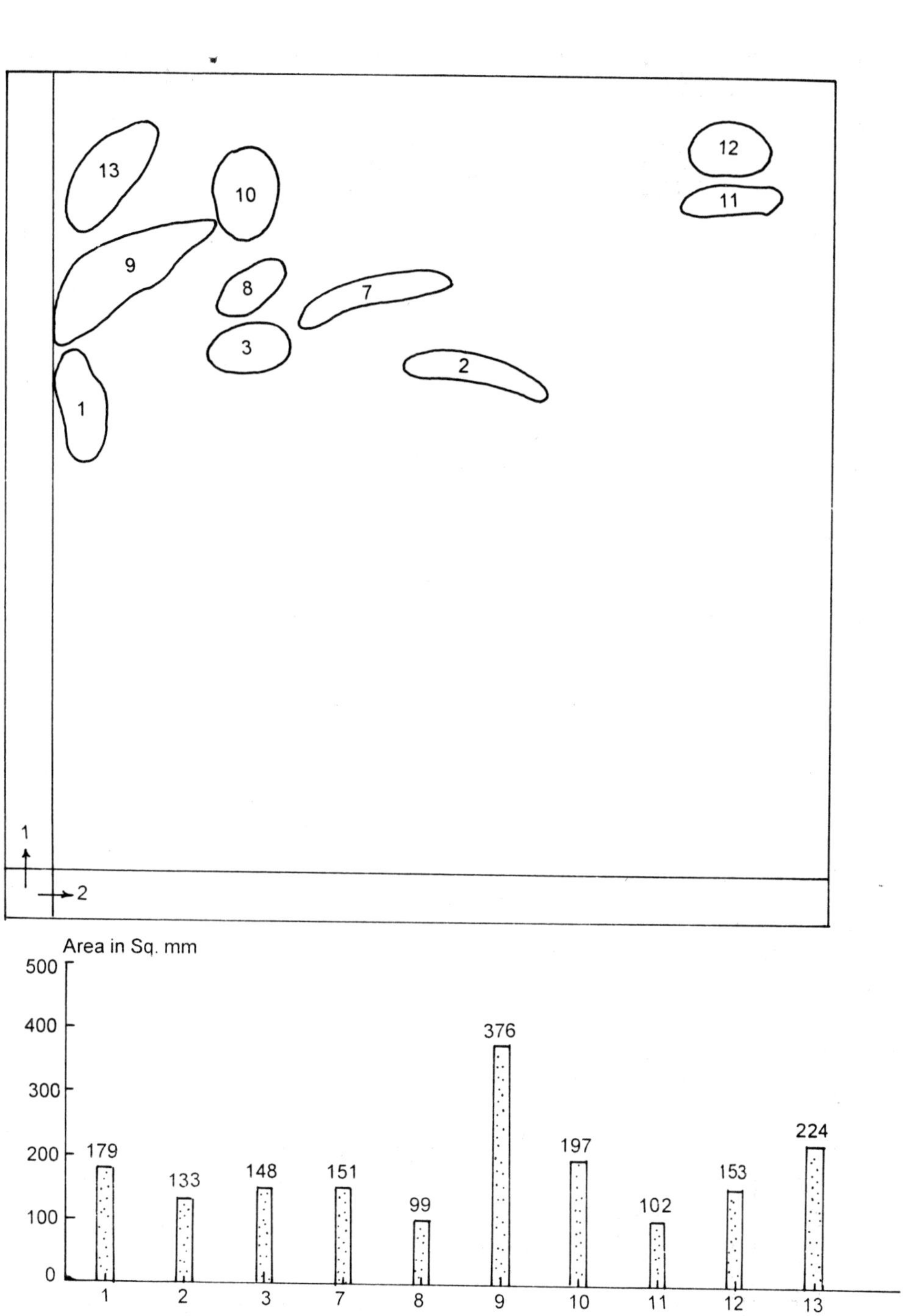

Plate 4.48 Phenolic spot patterns in *Notonia walkeri*

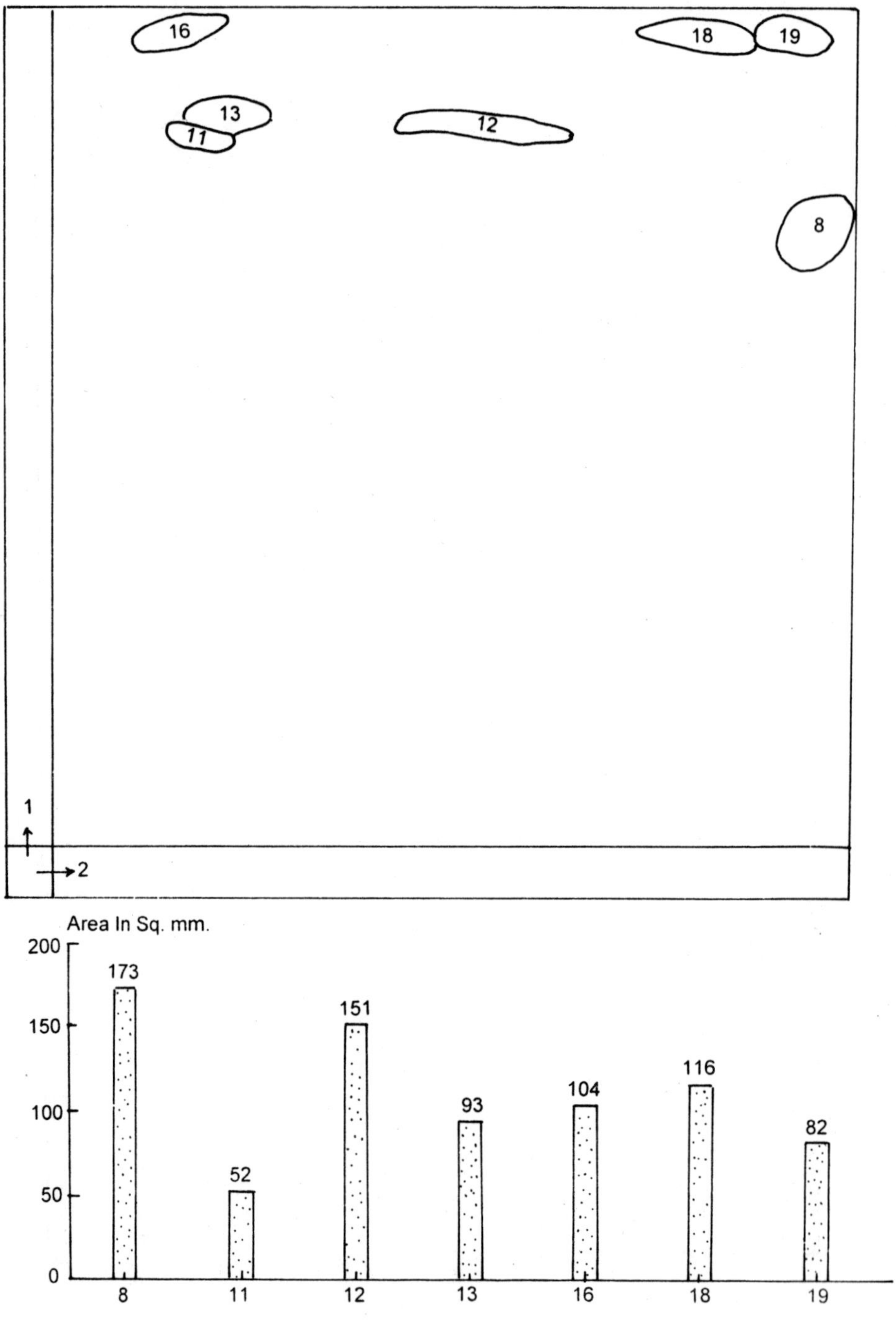

Plate 4.49 Phenolic spot patterns in *Emilia sonchifolia*

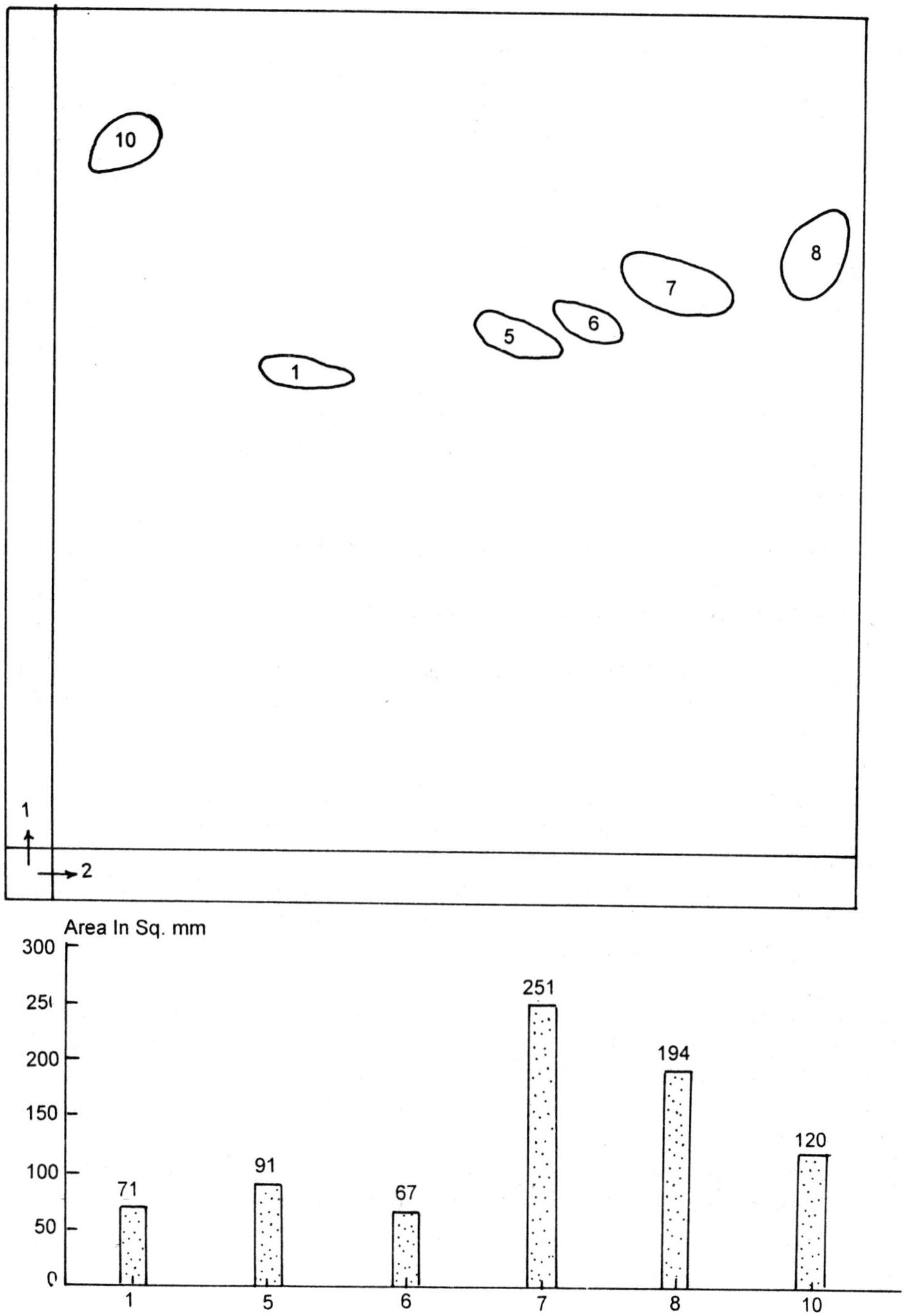

Plate 4.50 Phenolic spot patterns in *Emilia scabra*

38. S.edgeworthii (Fig. 4.38, Table 4.41). There are 6 spots, of which 4 are light yellow and 2 grey in colour. Spot no. 35 with an area of 150 sq.mm is the largest. There are no species specific spots.

39. S. edgeworthii (Small leaf) (Fig. 4.39, Table 4.42). This is small leaf from of *S.edgeworthii* deiffering only in leaf size. Both the taxa however grow in the same area. There are 5 spots in this taxon, of which 4 are light yellow, and one grey in colour. Spot no. 35 with an area of 169 sq.mm is the largest.
There is very little difference in the flavonoid profiles between the small leaf and normal leaf *Senecio edgeworthii*. All spots are common except for spot no. 33 (grey in colour) which is missing in the small leaf form.

40. S. ludens (Fig. 4.40, Table 4.43). There are a total of 8 spots, of which 3 are grey, 3 light yellow and 2 dark green in colour. Spot no. 32 with an area of 133 sq.mm is the largest. This has the largest number of species specific spots among all the taxa of the genus *Senecio* studied. There are four spots (19, 23, 26 and 32) exclusive for this taxon.

41. S. cineraria (Fig. 4.41, Table 4.44). There are 8 spots, of which 3 are grey, 2 light yellow, onw green, one violet and one dark green in colour. Spot no. 13 with an area of 285 sq.mm is the largest. Spot no. 30 is the species specific spot.

42. S. corymbosus (Fig. 4.42, Table 4.45). There are a total of 11 spots in this taxon. Of these, 5 are light yellow, 2 green, 2 light green, one grey and one greenish yellow in colour. Spot no. 7 with an area of 455 sq.mm is the largest. There are 3 species specific spots (2, 4 and 7) in this taxon.

43. S. walkeri (Fig. 4.43, Table 4.46) There are 10 spots on the chromatogram of which 3 are dark green 3 grey, 2 light yellow and 2 green in colour. Spot no. 9 is the largest with an area of 170 sq.mm Spot no. 34 is the species specific spot.

In the genus *Senecio*, a total of 36 spots have been identified as can be seen in the composite key (Fig. 4.66) Table no. 70 gives the distribution of these spots in the 12 species of *Senecio* studied. A perusal of the composite key as well as Table no.65 shows that spot no. 27 which is grey in colour can be regarded as the generic marker as it is present in all the taxa except for *S. hohenackeri*. Highest number of spots (11) are seen in *S.corymbosus*, while the lowest number (5) is found in *S.lawsoni* and *S.wightii*. Out of the 12 species studied, 8 have species specific spots.

Detailed numerical parameters for the 12 species of *Senecio* have been worked out based on the distribution (qualitative and quantitative) of flavonoid spots. Various statistical parameters like paired affinity index (PA) matching coefficient, negative correlation, and biochemical distance have been worked out to assess the relationship between the 12 taxa. These data are computed in the form of similarity matrices and are presented in Tables 66-69. The data and the values in negative correlation are of a negative nature and are used to support the data obtained from PA indices. A higher value in negative correlation indicates that the taxa are distant. Group affinity values (GA) have been calculated based on PA, negative correlation and matching co-efficient to indicate the extent of affinity of the various taxa to the group. These have been presented in tables 70-72.

GA values based on PA, matching coefficient and negative correlation are depicted in the form of comparison histograms in Fig. no. 4.73.

Isolation values also have been calculated for all the taxa based on the number of unique spots in them, in relation to both total number of spots in the taxon (Iv_s) and total number of spots in the group (Iv_n). These have been presented in table no.73.

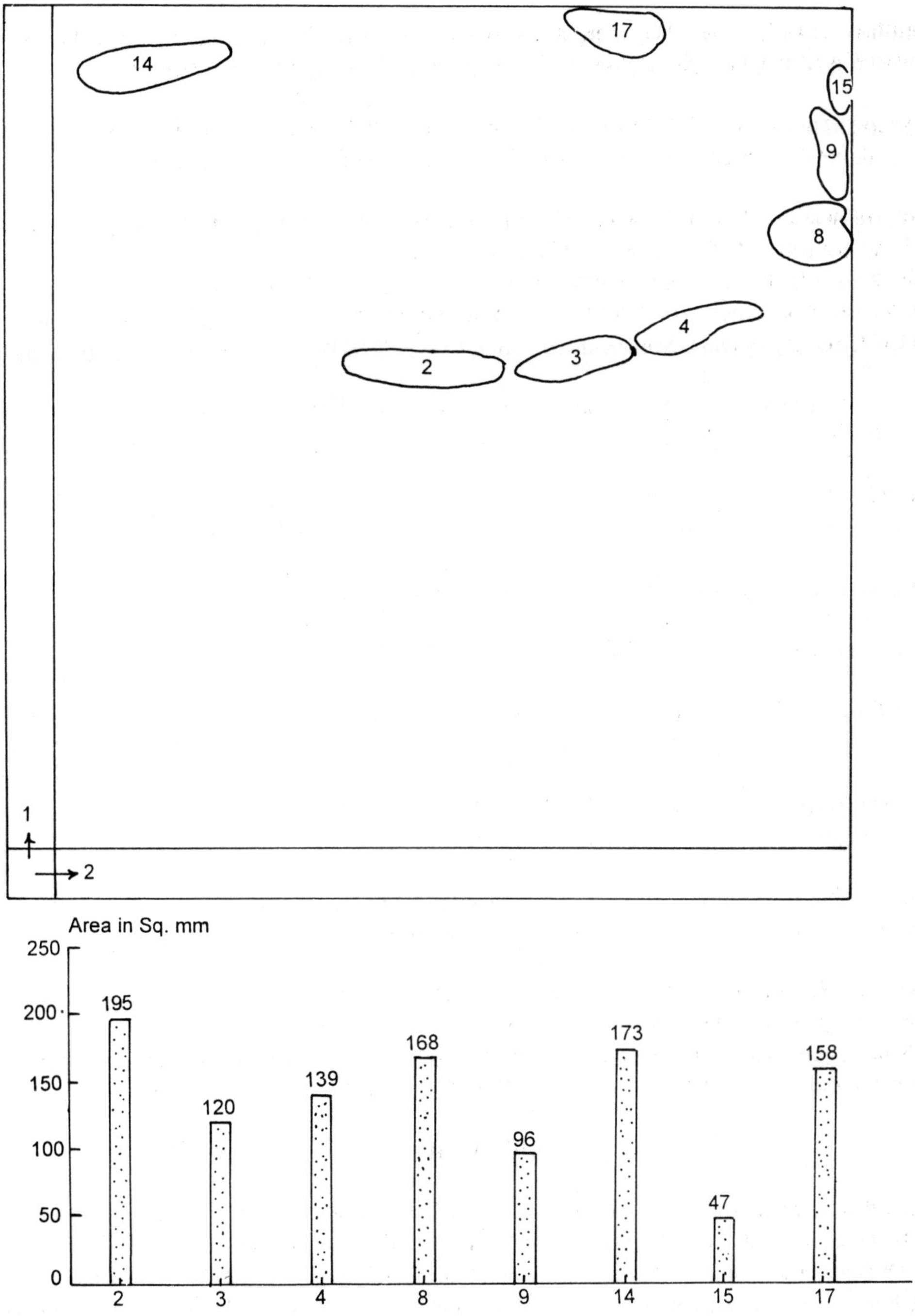

Plate 4.51 Phenolic spot patterns in *Emilia ramulosa*

Gynura : In the genus *Gynura* 3 taxa with diverse morphological appearance have been studied to analyse the flavonoid profiles in the leaves. These are described below.

44. G.nitida (Fig. 4.44, Table 4.47). There are a total no. of 8 spots in this taxon. Of these 5 are light yellow, one green and one light green in colour. Spot no. 3 with an area of 145 sq.mm is the largest.

45. G. lycopersicifolia (Fig. 4.45, Table 4.48). There are a total of 9 spots in this taxon. Of these, 4 are light yellow, 2 yellow and 2 green in colour. Spot no. 11 with an area of 183 sq.mm is the largest.

46. G. aurantiaca (Fig. 4.46, Table 4.49). There are 4 spots, of which 3 are green and one light green in colour. Spot no.12 with an area of 190 sq.mm is the largest.
All the three taxa studied in the genus share spot no.9 indicating that it is the generic marker. *G.aurantiaca* and *G.nitida* share only spot no.9, while *G.aurantiaca* and *G.lycopersicifolia* share spot nos. 5 and 9 *G.nitida* and *G.lycopersicifolia* however, share 3 spots (7, 8 and 9). There are a total of 14 spots in the genus.

Notonia : In the genus, only two taxa have been studied. Morphologically, these are very different. Their flavonoid profiles are described below.

47. N.grandiflora (Fig. 4.47, Table 4.50). There are 6 spots, of which 2 are greenish yellow, 2 green, one light green and one grey in colour. Spot no. 12 with an area of 713 sq.mm is the largest.

48. N. walkeri (Fig.4.48, Table 4.51). There are 10 spots in this taxon, of which 4 are light yellow, 2 green, 2 light green, one grey and one yellow in colour. Spot no. 9 with an area of 376 sq.mm is the largest.
The two taxa share spot nos. 2, 8 and 12 between them. There are a total of 13 spots in the genus.

Emilia : In the genus *Emilia,* three available taxa have been selected for flavonoid analysis. Morphologically the three taxa have only few differences. The flavonoid spot patterns have been described below.

49. E. sonchifolia (Fig. 4.49, Table 4.52). There are 7 spots on the chromatogram, of which 4 are greenish yellow, 2 light yellow and one grey in colour. Spot no.8 with an area of 173 sq.mm is the largest.

50. E.scabra (Fig. 4.50, Table 4.53). There are a total number of 6 spots in this taxon. Of these, 4 are grey and 2 light yellow in colour. Spot no. 7 with an area of 231 sq.mm is the largest.

51. E. ramulosa (Fig. 4.51, Table 4.54). There are 8 spots in this taxon, of which 3 are light yellow, 2 grey, one light green, one greenish yellow and one green in colour.
Spot no. 8 being common for all the three taxa can be regarded as the generic marker. This is the only common spot between the three taxa. There are a total of 19 spots in the genus.

DISCUSSION

It has been estimated that about 2% of all carbon photosynthesized by green plants are converted into flavonoids or closely related compounds (Smith, 1972). The flavonoid pigments - one of the most numerous and widespread groups of natural constituents are of importance and interest both to chemists and biologists in terms of their occurrence, structure and biological function. To the botanist, their widespread occurrence and immense diversity constitutes a very effective systematic tool. The flavonoids are of great taxonomic value below the genus level and of possible use in the classification of higher categories (Radford *et al.*, 1974).

How effective and conclusive are flavonoid data in interpreting phyletic and systematic aspects? To a large extent this depends on the manner in which the flavonoid data are sought to be used and the level of taxonomic hierarchy at which they are applied. Phylogenetic and systematic problems require the user to adopt different approaches in the utilization of the flavonoid data.

In solving phylogenetic problems using flavonoid data, extreme care is required on the part of the taxonomist as the taxa in question may be phylogenetically distant and morphologically distinct. In such circumstances use of flavonoid data should be to a large extent corroborate evidences from other sources and not confront them. When the flavonoid data are used merely to support the evidences from other sources in sorting out a phylogenetic aspect, there does not seem to be much problem. But when the flavonoid data are of crucial importance to provide solution to a phylogenetic problem, enumeration of mere presence of similar flavonoids in the taxa taken for comparison will not do, eventhough flavonoids are direct gene products, reflecting the genome action in the individual. For, it is known, that similar flavonoids can accumulate at the culmination of entirely different biosynthetic pathways, controlled by unrelated genes. In such a situation, phyletic relationship between the two taxa can be established only when they have identical biosynthetic pathways resulting in the production of similar flavonoid compounds.

In the solution of systematic problems, however, flavonoid data have been extremely useful. Here, the taxa taken for comparison are already known to be phyletically related and flavonoid data help us in assigning a position to them within the framework of a classification. This is the reason why, flavonoid data seem to be highly useful in sorting out systematic problems at the lower taxonomic categories like genus or species. At the inter or intraspecific level, even the mere presence of identical flavonoids may serve to indicate relationship, for it is unlikely that two taxa at that level have an altogether different biosynthetic pathway (in the production of flavonoids). In taxa which exhibit a minor gene modification like intraspecific variation, presence of similar flavonoids are indeed of assistance in understanding the adaptive potentiality of the individual in relation to the coenogenic forces of the environment.

Hence it can be safely assumed, that at the generic and more particularly at the species level, presence of similar flavonoids are indicative of relationship. A systematist can use them even without knowing their individual identity to relate the two taxa. It is with this belief, that the present investigation aims to study inter and intraspecific variations in taxa belonging to the tribes Heliantheae and Senecioneae. The tribe Eupatorieae which has only a few local representatives is studied mainly to understand the flavonoid profiles and to find out whether these are of any use at the tribal level. Interpretation of flavonoid profiles is also attempted at the sub tribal level in Heliantheae.

In the tribe Senecioneae, flavonoid profiles are used to assess the intrageneric relationship in *Senecio*, *Notonia*, *Gynura* and *Emilia*.

Tribe : EUPATORIEAE

While the tribe Eupatorieae has been very well delimited, the concepts of subtribal and generic delimitation have been regarded as artificial (Robinson and King, 1977). Prior to the publication of the paper of King and Robinson which appeared in 1977 (The Biology and Chemistry of the Compositae, Eds. Heywood, *et al.*,), the tribe Eupatorieae was divided into 4 subtribes - Ageratinae, Kuhniinae, Adenostemmatinae and Piqueriinae (Robinson, 1913a). Robinson and King (1977) have discarded the four subtribes as unnatural and have proposed a complex subtribal treatment of the Eupatorieae. According to them, the tribe Eupatorieae consists of 19 groups, many of which are potential subtribes. These groups are Adenostemmatinae, Eupatoriinae, *Disynaphia* group, *Gyptis* group, *Acritopappus* group, *Piqueria* group, *Trichocornis* group, *Ayapana* group, *Praxelis* group, *Mikania* group, *Ageratina* group, *Hofmeisteria* group and *Oaxacania* group.

Plate 4.52 Master key for species of *Ageratum*

Plate 4.53 Master key for species of *Eclipta*

Plate 4.54 Master key for species of *Synedrella and Calyptocarpus*

The present work involves four taxa-*Adenostemma lavenia, Chromolaena odorata. Ageratum conyzoides* and *A. houstonianum* which belong to the groups Adenostemmatinae, *Praxelis* group and Piqueriinae respectively. As many genera have not been studied, no attempt has been made to relate the flavonoid profiles to subtribal groupings.

The chromatographic patterns are distinct in the four taxa justifying their inclusion in three separate groups. In *Chromolaena odorata* and *Adenostemma lavenia* there are 13 and 8 flavonoid spots respectively on the chromatogram (Figs. 4.4 and 4.3). A perusal of the chromatogram drawings and tables (6 and 7) will indicate that there are at least 3 spots of light yellow colour common between the two genera. Spot nos. 1,2 and 5 of *Adenostemma lavenia* correspond to 3,5 and 12 of *Chromolaena odorata.* These compounds in all probability might be flavonol glycosides so widely reported for the tribe Eupatorieae and might be indicative of tribal affinity.

In the genus *Ageratum* of which two taxa have been investigated, *A. conyzoides* and *A.houstonianum*, there is distinct chromatographic identity. As can be seen by the master chromatogram (Fig. 4.52), there are two common spots 4 and 5 of greenish yellow colour which may be generic markers. *A.houstonianum* with violet flower heads has 14 spots while *A.conyzoides* has only 7 spots. It has been observed that in *A.conyzoides* itself there are two variants, some with white flower heads and some with lilac coloured flower heads. Both of them however have the same distribution of leaf flavonoids. Of the total number of 14 spots found in the two taxa, only two spots are identical and this justifies their taxonomic categorization as distinct species. The two species have a similarity coefficient of 0.25 (Table 55) only. The fact that both the species involved in chromatography are collected from the same locality further proves their genome difference.

Tribe: HELIANTHEAE

The tribe Heliantheae is studied for two purposes

(a) To assess the intra and interspecific variation among some of the genera and

(b) To analyse how far flavonoid profiles are of assistance in the subtribal delimination. A total number of 26 taxa (see Table 1.1) are investigated for these purposes.

Intra and Interspecific Variation

5 genera, *Eclipta, Synedrella* (including *Calyptocarpu*), *Bidens, Galinsoga* and *Spilanthes* involving 15 taxa are studied to account for the variation found either at the intra or interspecific level. These are discussed below separately.

Eclipta

Eclipta prostrata is the commonly distributed weed found generally in moist places all over South India. Two types of this taxon have been collected.

(a) From Bangalore (South interior Karnataka, South India) and

(b) From castle rock (Northern extreme of Karnataka abuttng Goa, South India).

The Bangalore type has the leaves normal in size in the sense, it is the one which is generally found all over. The castle rock type has smaller leaves (at least 1/3 in size in comparison with the normal). Further, the small leaf type generally grows in dry area while the normal leaf type grows in moist area. The normal leaf type is generally erect, while the small leaf type is generally prostrate. The flower heads are also smaller in size in this plant. In order to ascertain whether the small leaf type is only an ecological variation, the achenes collected from castle rock area were sown in Bangalore and plants raised. The plants appeared to be true breeding in that they all had small leaves, indicating the small leaf nature is genetically controlled.

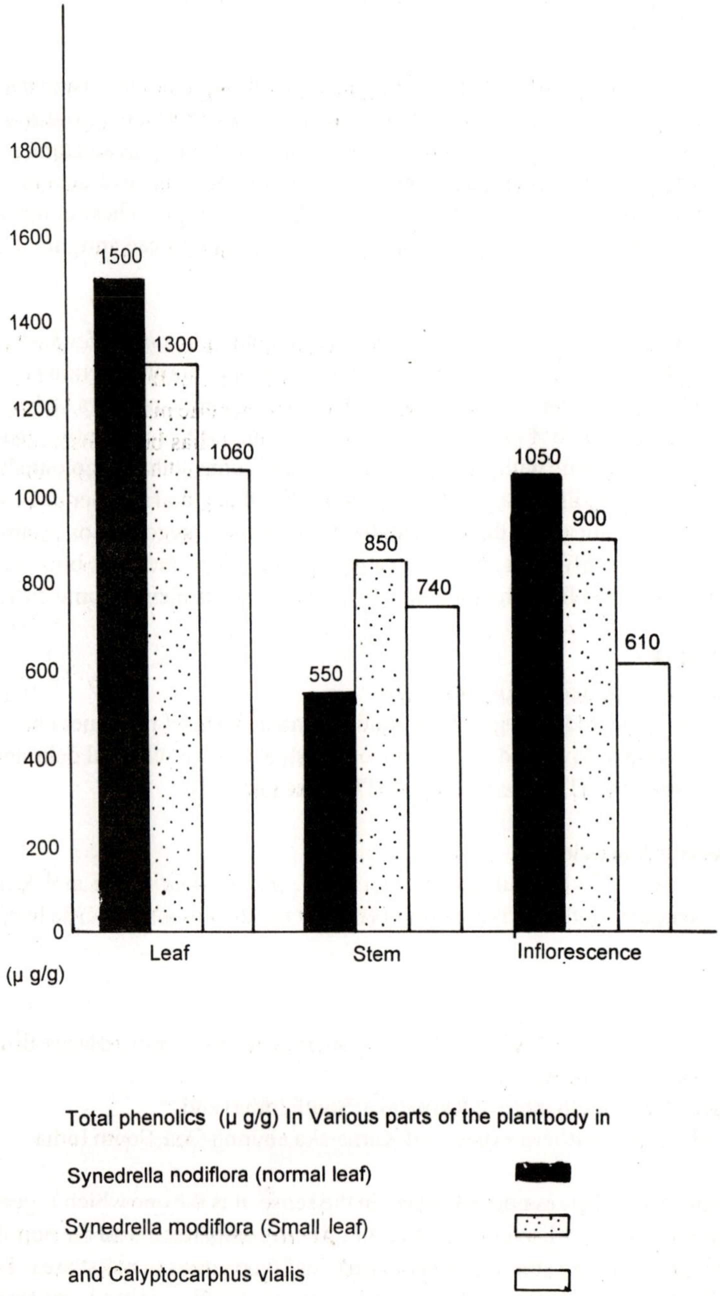

Plate 4.55 Total phenolics in *Synedrella and Calyptocarpus*

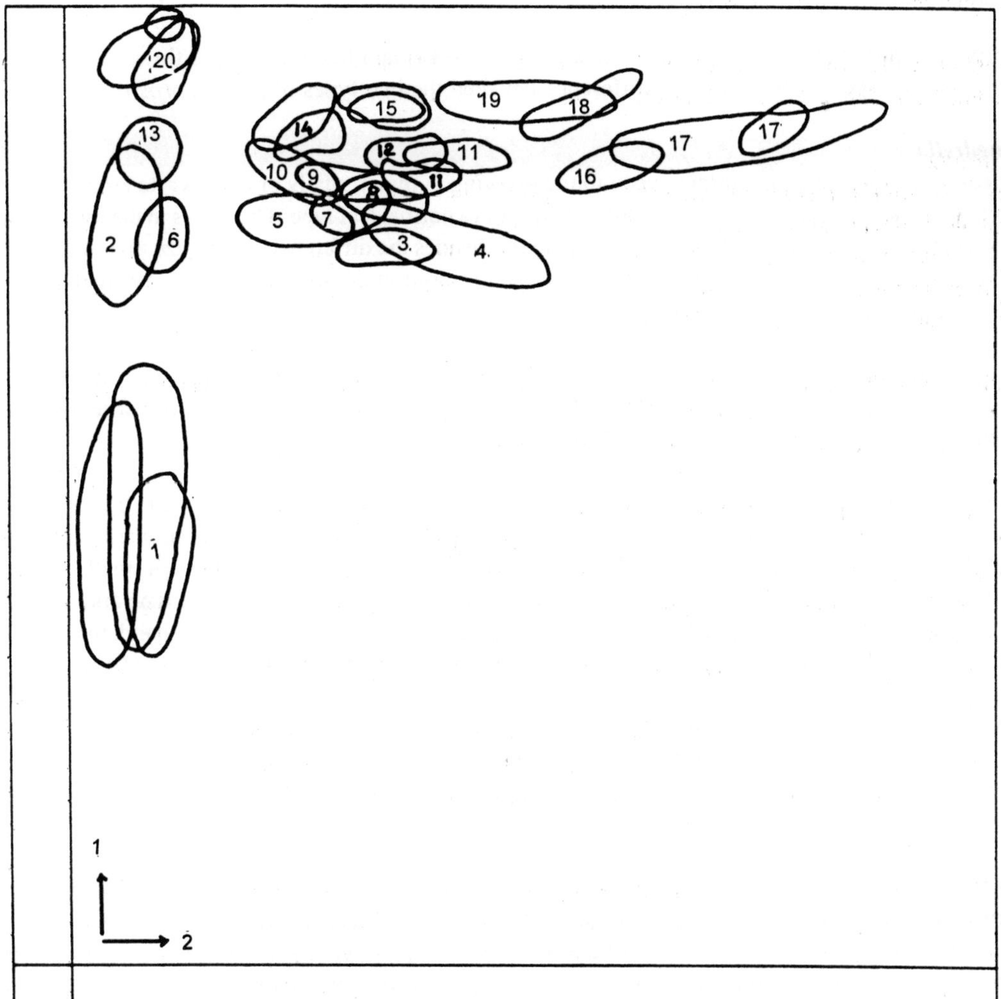

Plate 4.56 Master key for species of *Bidens*

Flavonoid spot patterns have been analysed for both the small leaf and normal leaf types. The small leaf types collected from castle rock area and the plants raised in Bangalore from the seeds of the castle rock plant have the same flavonoid spot pattern providing additional proof that the small leaf type is genetically regulated. A total number of spots 10 are encountered (Fig. 4.55) in both the small and normal leaf types. The normal leaf types has 5 spots (Fig. 4.9, Table 4.12), while the small leaf type (Fig. 4.10, Table 18) has 6 spots. Both the types are chromatographically distinct as can be seen by the fact that they share only spot no.1. In both the types, grey, greenish yellow and light yellow coloured spots are found, but the small leaf type has an additional spot, No.10 which is violet coloured.

Morphologically however except for the variation in size (plant height, leaf and flower head) and posture there does not seem to be any other difference between the two types of *Eclipta prostrata.*

Synedrella

Synedrella nodiflora, is a common weed found growing in moist, shady or open waste lands. There are two morphological types in this taxon - a small leaf type and a normal leaf type. The leaves in the former range from 1 to 2.4 cms long and in the latter 2.8 to 6.4 cm long. Besides the difference in leaf size, the stems are green in colour in the normal leaf type while they are purple in colour in the small leaf type. The small leaf type is of recent occurrence in and around Bangalore.

Another taxon that has been recently introduced to Bangalore and its surroundings is *Calyptocarpus vialis* (= *Synedrella vialis*). This taxon is reported for the first-time here for South India. The earlier two records for North India are those of Ahuja (1968) and Khanna (1980).

An analysis of the flavonoid spot pattern in the three plants clearly reveals (Fig. 4.54) that they are closely related. Spot no.19 is the generic spot. Matching coefficient (MC) (Table no.57) and similarity coefficient (SC) (Table No. 58) show that there is only 17% (0.17) and 23% (0.23) relationship between small leaved type and the normal leaf type (*S.nodiflora*), while the relationship between the normal leaf type and *Calyptocarpus vialis* (based on MC and SC) is 18% (0.18) and 21% (0.21) respectively. On the other hand, the small leaf type and *C.vialis* show greater affinity in terms of both MC and SC (0.33 and 0.44). There is a light violet coloured spot (spot no.4) which is found only in smal leaf *S.nodiflora* and *C.vialis.* This type of homogenity based on flavonoid spot pattern between small leaf *S.nodiflora* and *C.vialis* is also reflected in their morphology in that the habit and leaf size are almost similar in the two, besides similarity in floral structure. From this, it can be safely pointed out that the small leaf *S. nodiflora* has closer affinity with *C.vialis* than with the normal leaf type.

An analysis of the total phenolics in all the three plants (*Synedrella nodiflora* - 2 forms and *Calyptocarpus vialis*) shows that in the quantity of leaf and capitular phenolics, the small leaf *Synedrella nodiflora* is in between normal leaf *Synedrella nodiflora* and *Calyptocarpus vialis* (Fig. 4.55). The stem of small leaf *S.nodiflora* has a higher content of phenolics (850mg/g) when compared with the normal leaf *S. nodiflora* (550mg/g) and *C. vialis* (740mg/g). This is due to the presence of anthocyanins in the stem.

How did the small leaf *Synedrella nodiflora* originate? There are two possible ways in which this would have happened.

1. The small leaf type is true breeding. Hence it is possible that it might have arisen as a mutant from the normal leaf type.
2. As has been observed, there is coincidence of time in the appearance of small leaf type *S.nodiflora* and *C.vialis* in Bangalore. This provides a possibility to speculate whether the small leaf *S.nodiflora* has

Plate 4.57 Master key for species of *Galinsoga*

Plate 4.58 Master key for species of *Spilanthes*

arisen as a hybrid between normal leaf *S.nodiflora* and *C.vialis*. But based on the flavonoid spot pattern alone it cannot be ascertained whether the small leaf type is a hybrid, for the spots in this plant are not exactly additive of normal leaf *S.nodiflora* and *C.vialis*.

Bidens

Several species of the genus *Bidens* belonging to subtribe Coreopsidinae, tribe Heliantheae have been subjected to extensive investigations on their flavonoid contents (Ballard, 1975); floral UV patterns (Scogin and Zakar, 1976) and floral variation and pollination ecology; ray reduction and evolution and intraspecific variation (Sundara Rajan 1974, 1976, 1989).

Bidens pilosa is a cosmopolitan herb growing in waste places and so also is *B.humilis*. In several populations of *B.pilosa* it has been noticed that the rays disappear, though the marginal female florets are still present. The corolla of the ray florets in such cases becomes actinomorphic having five equal lobes just like the disc florets. It has been assumed that the rayless form is a mutant form the normal rayed form (Sundara Rajan, 1974). But it is not certain as to how many genes control the development of the ray. Cineery (1983) working on the *Bidens pilosa* populations of West Indies opined that the presence or absence of rays is determined by a single major gene which expresses incomplete dominance. He found an intermediate form between the rayed and the rayless.

In the populations of *Bidens pilosa* which have been investigated in the present work (populations collected in and around Bangalore, including Nandi Hills 60km., west of Bangalore) there appears to be more than one intermediate raising the question whether more than one gene is involved in governing the development of rays. Further, it has also been observed that the rayless population which appeared about 10 years back in Nandi Hills is increasing in number in contrast to the rayed forms.

A flavonoid analysis of the three forms - normal rayed, intermediate and rayless (for the purpose of flavonoid analysis one form in which the ray is divided into three unequal petals is chosen as the intermediate as it represents half way between rayed and raylessness) of *B.pilosa* together with *B.humilis* reveals an interesting spot pattern. Spot no.1 of yellow colour is common to all the plants except the intermediate. Matching coefficient (MC) and coefficient of similarity (SC) calculations (Tables 59 and 60) show that the rayed and rayless show closer affinity to each other than between rayed and intermediate. Curiosly the affinity values between rayed and rayless and rayless and intermediate are similar (0.26 for MC and 0.35 for SC).

Regarding the affinity between rayed form of *B.pilosa* and *B.humilis*, they in fact show greater affinity (MC:0.18 and SC:0.33) than between the former and intermediate rayed *B.pilosa* (0.095 and 0.14). However all the four plants involved are sufficiently distinct chemically because in no case the affinity value exceeds 0.35 (35%).

How did the rayless and intermediate forms arise and how are they getting stabilized in the environment? It is clear from the observations that have been made over the past several years that the original form is the (normal) rayed *B.pilosa* andthe rayless form is mutant having arisen spontaneously. There is no indication of any coenogenic force having contributed to the origin of rayless individuals. The fact that both rayed and rayless grow in the same area and that rayless is a true breeding trait supports the mutagenic origin of the rayless population. But it is not certain however as to how many genes are actually involved in the ray development. It has been suggested by Sundara Rajan (1977) on the basis of many intergrades, that it (ray development) could be a polygenic inheritance.

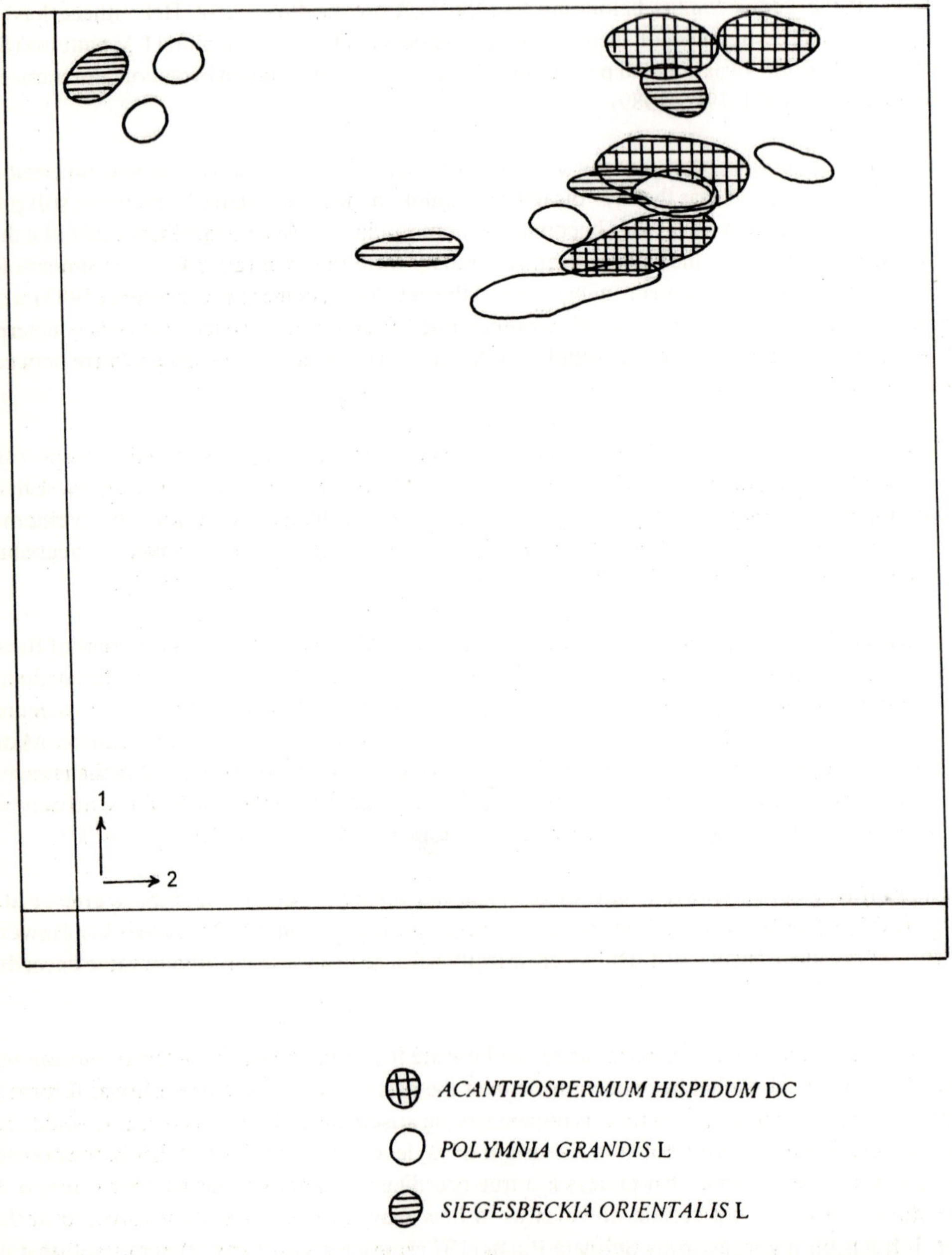

Plate 4.59 Master key for Melan podinae

Regarding the origin of the intermediate types (s), as has been suggested above, it (they) might have arisen as a result of natural hybridization between the rayed and rayless. Another possibility is the origin of the intermediate ray as a mutant from the rayed individuals. But the fact (as has been observed in the past 10 years) that in natural populations there were only the rayed individuals first, raylessness arose later and the intermediates appeared only subsequently goes to show that, the intermediates could not have arisen as mutants from the rayed and the chances are more likely that they (intermediates) are hybrids of the rayed and rayless.

Of what use are flavonoid data in the determination of hybrid nature of the intermediate between the rayed and rayless? Unfortunately the flavonoid data do not seem to be conclusive. As is normally expected of a hybrid (species of *Tragopogon* - Belzer and Ownbey, 1971) the flavonoid spot pattern in intermediate is not an addition of those of rayed and rayless individuals. But as pointed out by Demert and Mabry (1972) for *Thermopsis*, the hybrids need not always be additive. But another peculiarity in the present instance is the absence of generic spot (spot no.1) in the intermediate which is present in all types including *B.humilis*. If indeed the intermediate is a hybrid between the two, it is quite possible that the genic interaction (of the rayed and rayless) might have suppressed the synthesis of the particular flavonoid which is the generic marker. Another interesting aspect that needs a critical analysis is how far the intermediate and rayless individuals are responding to coenogenic forces of the environment in the process of getting stabilized as independent taxa. In this context however the intermediate individuals, it seems can be ignored for they are sporadic in appearance occurring occassionally in places where the rayed and the rayless interact.

The rayless plant however appears to be an independent taxon and in many localities it (rayless) even competes favourably with the rayed individuals. In his papers, Sundara Rajan (1988, 1989) has analysed the selective pressure operating on both the rayed and rayless populatiopns and the response of these individuals. In a study of natural populations of rayed and rayless which is being observed for the past several years at Nandi Hills, it has been noticed that the rayless individuals are gradually increasing in number. It was suggested sometime ago (Sundara Rajan, 1988), that the insect pollinators prefer the yellow colour of the disc florets and the presence of a boundary (it appears red in colour to the instects) in the form of rays distracts their attention. Hence in a competition, rayless individuals seem to be selected as against the rayed individuals. In the populations of rayed *Bidens pilosa* (at Nandi Hills) there is disturbance in fruit setting, while fruit set is normal in the rayless population indicating a favourable selection pressure on the latter.

In studies of anthochlor pigments and floral UV patterns in *Bidens lavies* (Scogin ahd Zakar, 1976), *Coreopsis bigelovii* (Harborne and Smith, 1978), *Lasthemia chrysostoma* (Brehm and Krell, 1975), it has been pointed out that the anthochlor pigments act as honey guides and they are localised in the UV absorptive region of the capitulum. Scogin and Zakar (1976) have pointed out that in *Bidens* species three categories may be recognised based on the localization of anthochlor pigments as evidenced by colour change of corolla when exposed to fumes of ammonia. (It has been shown that the yellow pigments in the corolla of the florets turn orange red on exposure to ammonia if they are anthochlor, but do not show any colour change if they are carotenoid pigments - Leppik, 1977). The three categories recognised by Scogin and Zakar (1976) are;

(a) Uniform distribution of anthochlors throughout the ray,
(b) Anthochlor distribution localized at the basal portion of the ray and
(c) Pigment absent in the rays except for the veins.

In all the three categories however the discs are always rich in anthrochlors.
In the present context, after exposure to ammonia fumes the rayed and rayless capitula of *B.pilosa* showed clearly that the anthochlors are localised only in the discs and the rays are devoid of them except along the veins and that too only on the lower surface.

The above observation perhaps provides a clue as to how the rayless population is reacting favourably to the environment and getting selected in preference to the rayed individuals. The pollinators seem to prefer the rayless heads to the rayed ones resulting in increased fruit seeting in the former. The sporadic appearance of intermediate rayed individuals is perhaps indicative of the gradual appearance of genetic barrier and the islation of the rayless individual as a distinct taxon.

Galinsoga

G.parviflora is a common herbaceous weed found generally in moist places. *G.quadriradiata* (= *G. ciliata*) is a later introduction into India (Babu, 1971). Besides other morphological differences, the two taxa differ in the size of the rays. The ray of the ray florets are larger in *G.quadriradiata* and clearly trifid than in *G.parviflora*. In addition to this, the stem and leaves are hispid in *G.quadriradiata* while they are nearly glabrous in *G.parviflora*.

Populations of individuals of *Galinsoga* clearly intermediate between the two taxa (*G.quadriradiata* and *G.parviflora*) are encountered in two different places (Bangalore and Kemmannugundi in Western Karnataka) at a distance of nearly 250kms from each other. In both these places, all the three populations (*G.quadriradiata*. *G.parviflora* and Intermediate) grow together. The intermediate forms are sparsely hairy, have the ray size intermediate between the two species and tend to appear morphologically more like *G. quadriradiata*.

Flavonoid spot pattern analysis of the three individuals (Figs. 4.23, 4.25 and 4.56; Tables 26, 27.283, 61 and 62) show a clear chemical affinity between them. Spot no. 12 is the generic marker being the only spot common for all the three. Matching co-efficient (MC) (table no.61) and coefficient of similarity (SC) (Table no. 62) show that *G.quadriradiata* and *G. parviflora* have an affinity value of 0.25 (25%) and 0.28 (28%) only. The corresponding values between *G.parviflora* and *Galinsoga* intermediate are only 0.13 (13%) and 0.16 (16%), whereas the affinity values of MC and SC are 0.44 (44%) and 0.57 (57%) for *G. quadriradiata* and *Galinsoga* intermediate. This indicates that *Galinsoga* intermediate is closer to *G.quadriradiata* than *G.parviflora*. This seems to be in conformity with the morphological appearance of the plant.

In view of the above findings, the occurrence of the intermediate form poses an interesting problem as to its mode of origin, either as a hybrid between the two species of *Galinsoga or as a mutant form of G.quadriradiata*. The flavonoid spot pattern observed in the hybrid (intermediate) is not exactly an addition of those of the two species which is a normal expectation for any hybrid. Instead it leans more towards *G.quadriradiata*. But the fact that intermediates have been encountered in two places separated by at least 250 kms or more and that they (intermediate) appear only in those populations where *G. quadriradiata* and *G. parviflora* grow together (no intermediates are found in places where the two species grow as separate populations) lends credence to the hybrid origin of the intermediate. But why the intermediates are more like *G. quadriradiata?* The only possibility perhaps is to assume that the intermediate populations under study are not exactly the F_1 plants but could be F_2 plants produced as a result of back cross between F_1 and *G. quadriradiata*. This explains why the intermediates are more like *G. quadriradiata* both in appearance and in flavonoid spot pattern (on TLC).

Spilanthes

Spilanthes paniculata and *S.calva* (= *S. acmella*) are two small weeds which are often used as medicinal plants for dental problems. The two species are separated by the colour of the capitula (white in *S.paniculata* and yellow in *S.calva*) and a few other morphological traits. These two taxa together with a population of *S.calva* with a larger capitulum (this population has been collected from Kemmannugundi, while the small size capitulum is collected from Bangalore) have been studied in terms of flavonoid profiles to understand intra and interspecific variation.

There are a total of 20 spots in the three populations and they seem to indicate the chemical individuality of the taxa involved. Spot no.13 present in all the three is the generic marker. Statistical calculations based on matching coefficient (MC) (table no. 63) and coefficient of similarity (SC) (Table no.64) show an affinity value of a maximum of 0.31 (31%) only. The small capitulate and large capitulate populations of *S.calva* show greater affinity between themselves (25% MC and 31% CS) than with *S. paniculata*. This shows that the small and large capitulate types are also sufficiently chemically distinct. However the large capitulate type is restricted to higher elevations, while the smaller one is found at lesser heights and plains. It may therefore be said that the large capitulate type with restricted distribution may be an ecotype. The seeds of the large capitulate type however fail to germinate in plains. The variations in the flavonoid spot pattern between the ecotypes is an indication of environmental influence on their (flavonoid) synthesis and accumulation, atleast in this instance. A similar flavonoid variation based on population position along the East West gradient extending from Arizona to Texas (USA) has been reported for species of *Hymenoxys* (Seeligman and Alston, 1967).

SUBTRIBAL CLASSIFICATION OF HELIANTHEAE

The Heliantheae is one of the largest and most morphologically diverse tribes of Asteraceae (Stuessy 1977). It is also regarded as the most primitive tribe in Asteraceae (Cronquist, 1955), though many systematists think otherwise (Small 1919).

The establishment of the tribe Heliantheae was first indicated by Cassini (1819) even though many pre and post Linnaean systematists came close to suggesting a tribe Heliantheae. Cassini's *"Helianthees"* tribe had five sub divisions *Heleniees, Coreopsidees, Prototypes, Rudbeckiees* and *Milleriees*. Cassini (1829) recognised two additional tribes *Ambrosiees* and *Tagetinees*, but he did not include them under "Helianthees". Lessing (1832) and De Candolle (1836) have also provided a classification of Heliantheae. An excellent review of these and earlier classifications are available in "The Origin and development of Compositae" (Small, 1919) and "Biology and Chemistry of the Compositae" (Heywood *et al.*, 1977)

Bentham (1873) has also provided a classification of Heliantheas, which with minor modifications, is in use even today. His Helianthoideae (Heliantheae) consists of 10 subtribes - Lagasceae, Millerieae, Melampodieae, Ambrosieae, Petrobieae, Zinnieae, Verbesineae, Coreopsideae, Galinsogiae and Medieae.

Hoffman (1890) has followed the subtribal classification of Bentham (19873) exactly, though the subtribe names end with *inae* (Lagascinae, Millerinae etc). Since the classification of Hoffmann, several authors (Jepson, 1901, Rydberg, 1914-15; Bessey 1915), have proposed various alterations and modifications in the subtribal treatment of Heliantheae.

Cronquist (1955), suggested that the tribe Helineae should be placed within the Heliantheae. Carlquist (1957) has added a new monotypic subtribe Fitchiinae which has been elevated to tribal level recently (Poljakov, 1967).

The most recent classification of the tribe Heliantheae comes from Stuessy (1977). He has divided the tribe into 15 sub tribes with each subtribe having a number of groups. These subtribes are Melampodiinae, Zinniiae, Ecliptinae, Verbesininae, Helianthinae, Gaillaridiinae, Coreopsidinae, Fitchiinae, Bahiinae, Madiinae, Gaillardiinae, Neurolaeninae, Engelmanniinae, Ambrosiinae and Milleriinae. But as Stuessy (1977) himself points out, "subtribal limits in the Heliantheae should be viewed not as a final statement of relationship but rather as a hypothesis to be tested by future intensive studies on generic groups and whole subtribes". Robinson *et al* (1978) have proposed some new correlations between the subtribes of Heliantheae based on chemical and anatomical characters concentrating more on the placement of genera in the tribes Galinsoginae, Verbesininae and Helianthinae.

Plate 4.60 Master key for *Ecliptinac*

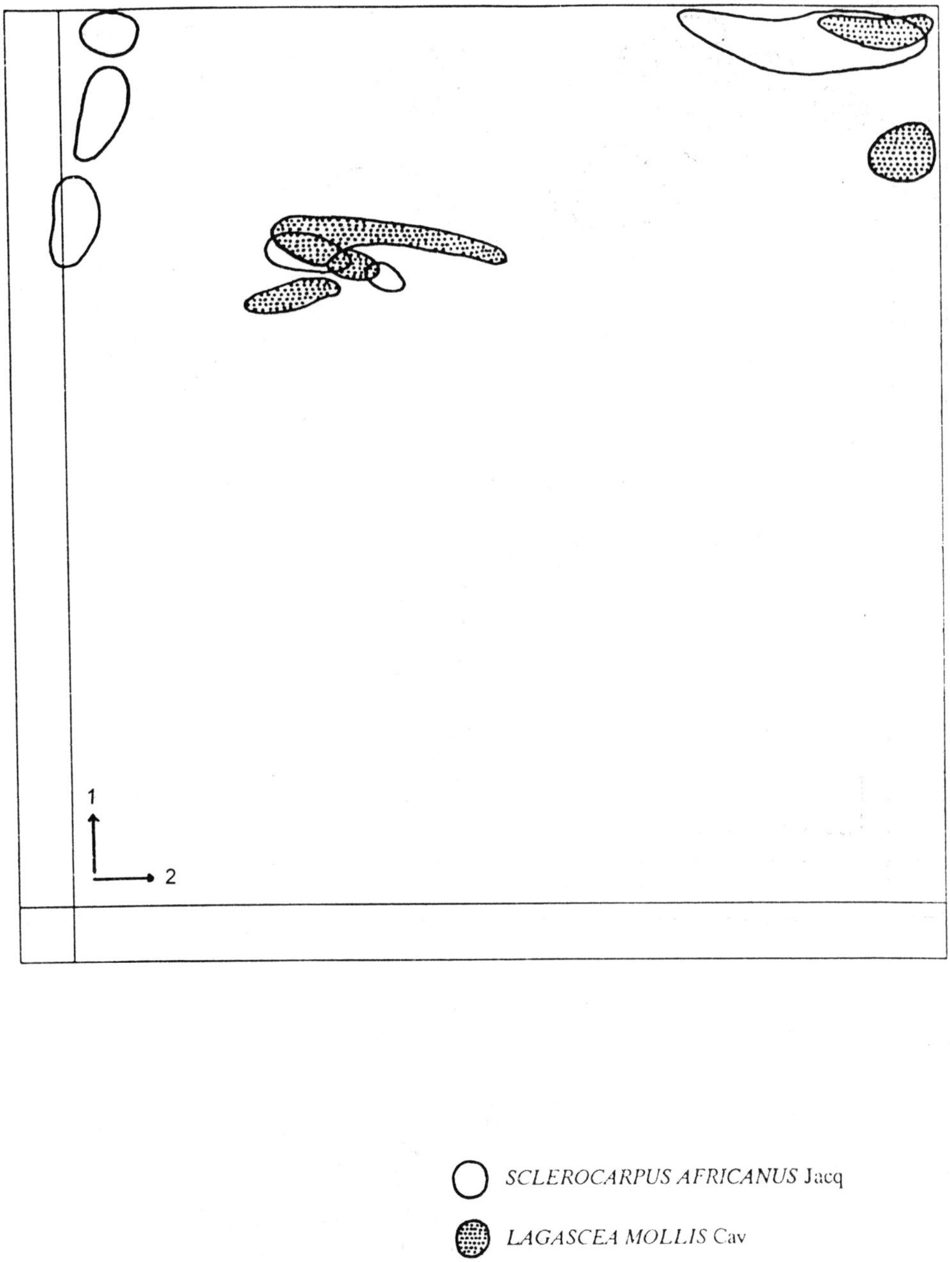

Plate 4.61 Master key for *Verbesininae*

Rudbeckia Nitida Nutt.

Helianthus Rigidus Desf

Plate 4.62 Master key for *Helianthinae*

Plate 4.63 Master key for *Coreopsidinae*

The idea of the present work based on flavonoid spot pattern in neither a detailed analysis of all the subtribes (of Heliantheae) nor any new proposal for a subtribal delimination. It is to find out whether the flavonoid spot pattern of the locally available members of Heliantheae are of any use for the subtribal classification and more particularly for the groups recognised within each subtribe by Stuessy (1977).

The members investigated (26 taxa) fall into subtribes Melanpodiinae, Ecliptinae, Verbesininae, Helianthinae, Coreopsidinae, Galinsoginae and Ambrosiinae. A few observations on these subtribes based on the flavonoid spot pattern of the investigated taxa are given below.

Melampodiinae

Two groups have been recognised in this subtribe. Among the investigated genera, *Acanthospermum hispidum* belongs to group 1, while *Siegesbeckia orientalis* and *Polymnia grandis* belong to group 2. The composite key (Fig. 4.59) will show that spot nos. 2 of *A.hispidum* and *P. grandis* and spot no.3 of *S.orientalis* are light yellow in colour and and occupy almost the same position on the chromatogram (see Rf values in tables 8, 9, and 10). This spot may be a subtribal marker. However, there is nothing to indicate that *S.orientatlis* and *P. grandis* are closer to each other than to *A. hispidum*.

Ecliptinae

This subtribe is also divided into two groups. The taxa *Blainvillea acmella, Calyptocarpus vialis, Eclipta prostrata* and *Syndrella nodiflora* belong to group 2. There is no spot common between all the taxa (Fig. 4.60); spot no. 19 of *S.nodiflora* and *C. vialis* and spot no. 6 of *B. acmella* are common between the three taxa. Similarly spot no. 18 of *C.vialis* and 6 of *E.prostrata* are common between the two. These similarities between the spots show that there is some chemical affinity between the four taxa justifying their inclusion in the same subtribe. The affinity between *Synedrella, Blainvillea* and *Calyptocarpus* seems to be strengthened by the observation - "the floral and vegetative differences among *Blainvillea, Synedrella* and *Calyptocarpus* seem quite minor" (Stuessy 1977).

Verbesininae

This is the largest subtribe in Heliantheae. Stuessy (1977) has included 58 genera in this subtribe. categorized under five groups. The two locally available taxa viz., *Lagascea mollis* and *Sclerocarpus africanus* belong to the first and fifth group respectively. The flavonoid spot pattern (Fig. 4.61) reveals that there are at least two spots common between the two taxa. Spots nos. 3 and 5 of *L. mollis* correspond to spot nos. 2 and 6 respectively of *S.africanus*. While the spot pattern in the two taxa can be cited as an evidence to show affinity between the two and supporting their inclusion in the same subtribe, nothing can be said about their being included under different groups within the subtribe, as only two of them have been investiggated. This can be commented upon only when more taxa are taken for comparison.

Robinson *et al* (1978) however place *Lagascea* in the subtribe Helianthineae based on the presence of a polyacetylene of the dehydrofalcarinone type. Comparison of spot patterns of the four taxa involved in the subtribes Helianthinae and Verbesininae reveals a complex situation. Spot nos 2 and 5 of *Lagascea mollis* and 1 and 7 of *Helianthus rigidus* correspond to one another. So also 7 of *H.rigidus* and *Sclerocarpus africanus*. However there are no common spots between *Rudbeckia nitida* and *L.mollis* and also between *R.nitida* and *S.africanus*. Spot nos. 5, 6 and 7 of *L.mollis, S.africanus* and *H.rigidus* respectively are common. This type of comparison between two subtribes with only the spot pattern being known and not the exact identity of flavonoids is thought be of not much assistance in ascertaining the actual position of disputed genera. There is however no doubt that the taxon *L.mollis* has some common flavonoids found in both the subtribes Verbesininae and Helianthinae. In such a situation data from other sources are to be used in deciding the systematic position of the disputed genus. Robinson *et al* (1978) however are firm in their

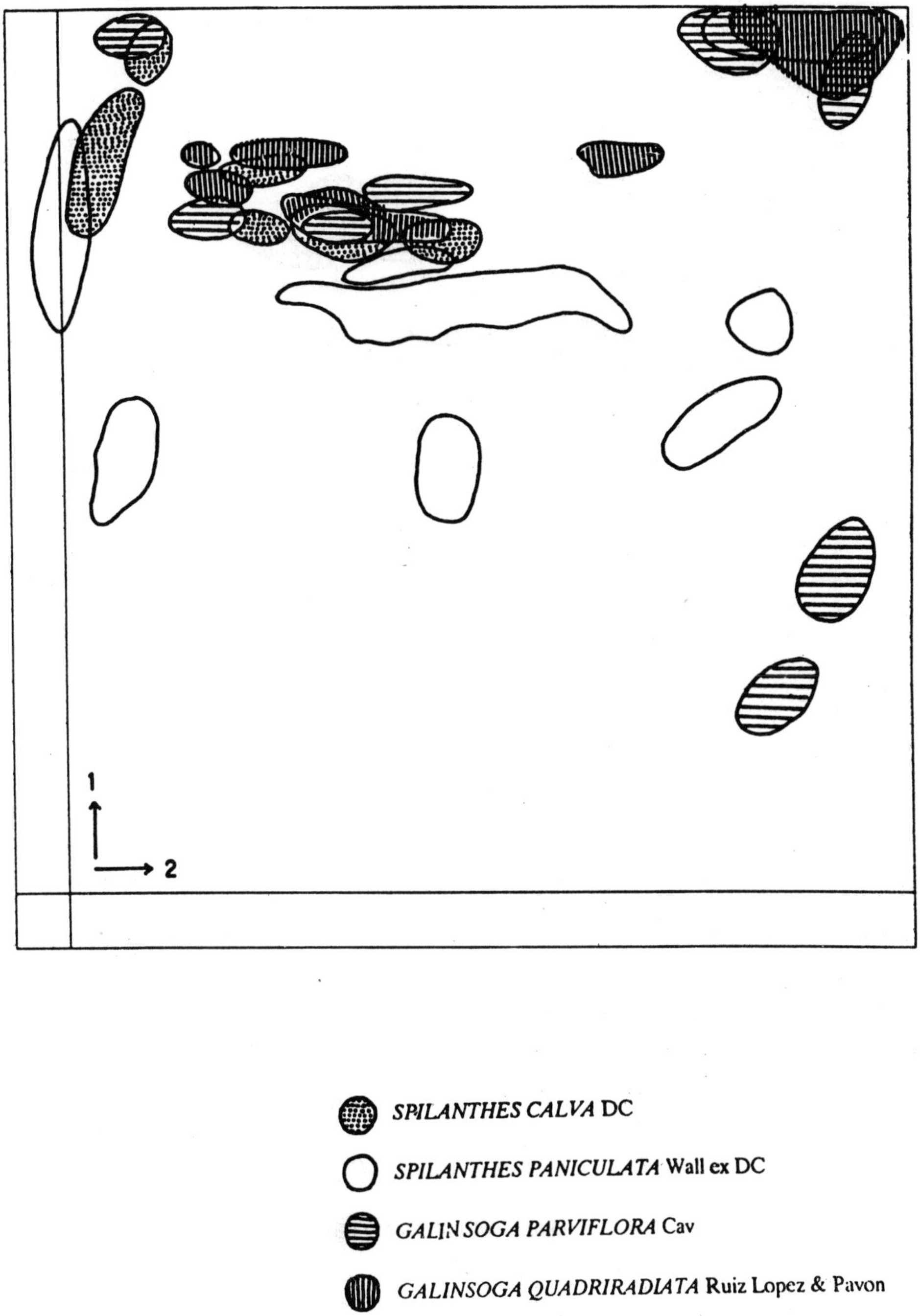

Plate 4.64 Master key for *Galinsoginae*

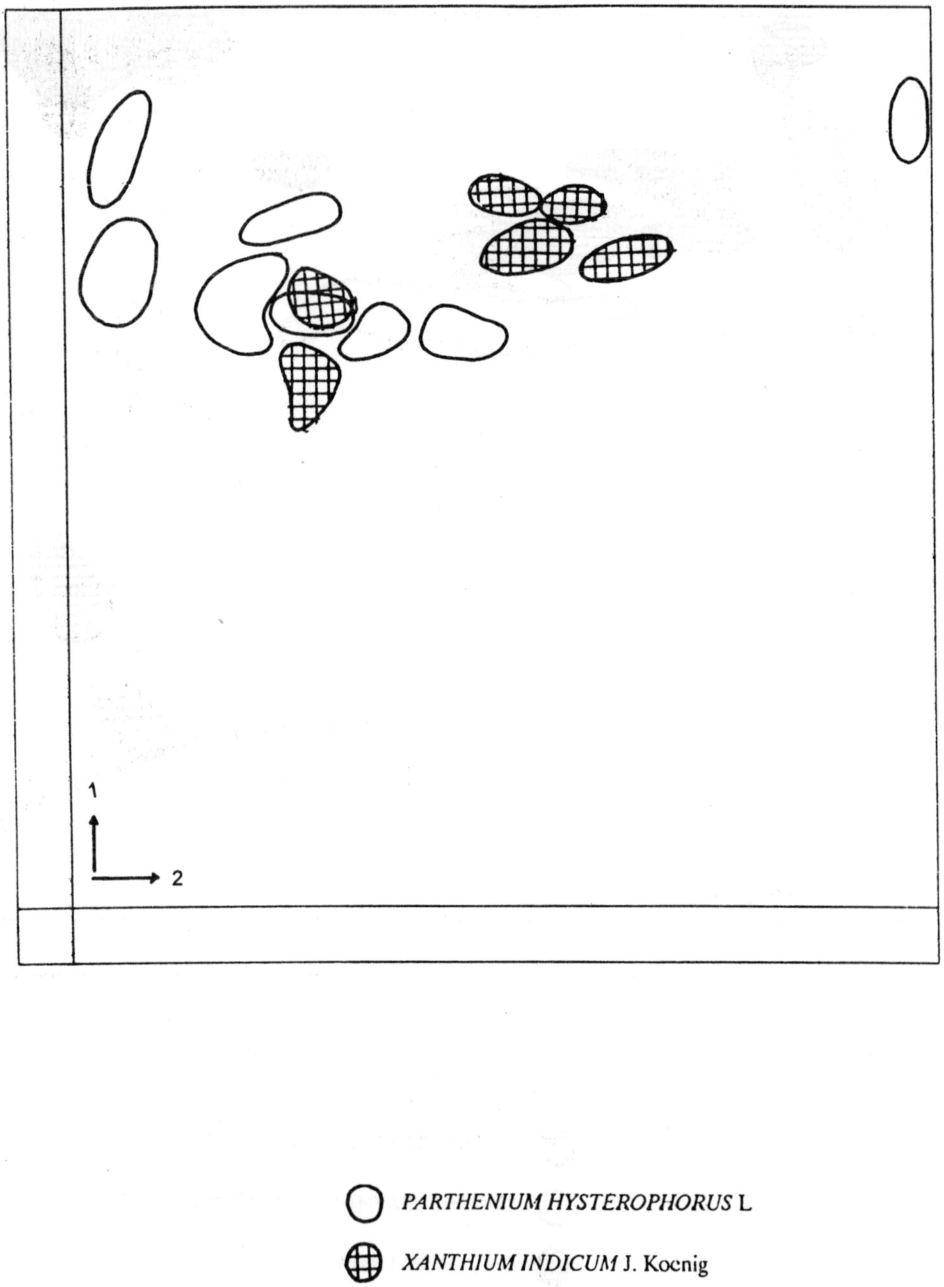

Plate 4.65 Master key for *Ambrosinae*

Plate 4.66 Master key for species of *Senecio*

opinion (relying on chemical characters and more heavily on anatomical characters) that the genus*Lagascea* belongs to the subtribe Helianthinae. They also point out "unfortunately Stuessy's placement of *Lagascea* in the Verbesininae was already superseded by the time of publication by his correct though perhaps inadvertent placement of the genus in the subtribe Helianthinae (Stuessy, 1976)".

Helianthinae

This subtribe includes plants which are perennial herbs. The subtribe is divided into two groups. The two taxa investigated (*Helianthus rigidus* and *Rudbeckia nitida*) fall under different groups. As is shown in the composite key (4.61), there is no common flavonoid spot between the two plants, thus indicating that the spot pattern is of no assistance in the subtrial delimination.

Coreopsidinae

According to Stuessy (1977), this subtribe is well delimited within the tribe by a series of distinctive characters. Chemically also this subtribe has been extensively investigated. The subtribe is divided into seven groups. The three taxa investigated (*Bidens pilosa*, *B. humilis* and *Moonia heterophylla*) belong to the same group within the subtribe. The composite key of flavonoid spot pattern (Fig. 4.62) shows that spot No.1 is common between *B.pilosa* and *B.humilis* and may be regarded as the generic marker. Spot nos. 13 and 7 of *B.pilosa* and 8 of *M.heterophylla* correspond to one another in terms of colour and location on the chromatogram. Similarly spot nos. 3 and 20 of *B.humilis* and 5 and 8 of *M.heterophylla* also correspond to one another. These similarities provide an additional evidence for the inclusion of all these under the same subtribe. However nothing can be said about their being included under the same group as taxa of other groups within the subtribe are not studied.

Galinsoginae

The subtribe includes 15 genera of which two genera and four taxa (*Spilanthes* calva, *S.paniculata*. *Galinsoga quadriradiata* and *G.parviflora*) are studied. Stuessy (1977), has not proposed any groups within the subtribe. Compoite key (Fig. 4.63) shows that spot nos. 6 of *G.quadriradiata* and *G.parviflora* correspond to spot no. 11 of *S.calva*. Except for these two, no spots are common even though they appear to be so on the composite key. (The mere overlapping of the spots on the chromatogram does not indicate similarity unless they have same colour and almost identical Rf. values). There are no spots common between *S.paniculata* and the species of *Galinsoga*. Hence it has to be opined that flavonoid spot pattern of TLC is of no assistance in the subtrial consideration of Galinsoginae, even though there appears to be some similarity between *Galinsoga* species and *S.calva* in as much as they appear to have one flavonoid compound in common.

Abrosiinae

The subtribe as recognised by Stuessy (1977) includes 8 genera. No further categorization of the subtribe into groups is proposed. Two taxa, *Parthenium hysterophorus*, a widely distributed weed and *Xanthium indicum* another common weed are investigated for their flavonoids. The composite key (Fig. 4.64) indicates that there are no spots common. Even though spot no3 of *P.hysterophorus* appears to overlap spot no. 2 of *Xanthium indicum* they are different (see tables 32 and 33).

The discussion on the subtribes of Heliantheae as explained above points out that flavonoid spot pattern on TLC is not of much assistance at the subtribal level and identification of the possible biosynthetic pathways of flavonoids perhaps would provide a better understanding of the situation.

Tribe SENECIONEAE

Bentham (1873) recognised four subtribes in Senecioneae. These are - Senecioneae, Liabinae, Tussilagininae and Othonninac. Hoffmann (1892), followed a similar approach, but recognised only three subtribes. He merged Tussilagininae with Senecioninae. In one of the recent treatments of the tribe, Nordenstam (1977) has proposed the exclusion of the Liabinae from Senecioneae. The Liabinae is raised to the level of a tribe Liabeae. He has included all the genera of Senecioneae under two subtribes. The subtribe Senecioninae includes most of the genera (100 genera), while the subtribe Blennospermatinae includes only four genera with obscure affinities. The subtribe Blennospermatinae is a tentative one (Nordenstam, 1977). All the genera studied in the present investigation belong to the subtribe Senecioninae.

The idea of the present work however is not a subtribal assessment. It is only to analyse and interpret the flavonoid profiles in terms of intrageneric relationship of the species in the locally available genera viz. *Senecio, Gynura, Notonia* and *Emilia.*

Senecio

Senecio is one of the most cosmopolitan and diverse genera of Asteraceae. It includes over 1500 spp (Nordenstam, 1977) and consists of annual or perennial herbs, half shrubs, shrubs, vines and trees. According to Small (1979), the global distribution of *Senecio* indicates it primitive nature and ancient origin. He believed that *Senecio* is in the basal position in Asteraceae with Senecioneae being the most primitive tribe.

In the present work 13 taxa of *Senecio* (including one small leaf form of *S.edgeworthii*) are studied and a detailed statistical appraisal is made of the interrelationships among them based on flavonoid spot pattern. The composite key (Fig. 4.66) reveals that there are a total number of 36 spots distributed between the 12 taxa. Spot no.27 (Table no.65) is common for all the taxa except for *S.hohenackeri.* This spot may be regarded as the generic spot. Lowest number of spots (5) is found in *S.lawsoni* and *S.wightii,* while the highest number (11) is found in *S.corymbosus.*

Apart from spot no.27 which is found in all but one of the species, the next spot which is of frequent occurrence is spot no.35 which is found in 5 species. Spot nos. 1, 2, 3, 6, 7, 18, 19, 22, 23, 26, 30, 31, 32 and 36 are exclusively found in individual species (Tables 65 and 73).

On morphological grounds, the 12 taxa studied roughly fall into four categories (The small leaf form of *S.edgeworthii* is not taken into consideration for statistical analysis of flavonoid spot pattern as it resembles the normal leaf *S.edgeworthii* closely).

Group 1 : *S.grahami, S.edgeworthii, S.cineraria* and *S.ludens.*
Group 2 : *S.lawsoni, S.wigthii, S.belagaumensis* and *S.neelgherryanus.*
Group 3 : *S.corymbosus* and *S.walkeri*
Group 4 : *S.hohenackeri* and *S.tenuifolius*

The distinguishing vegetative characters (apart from floral characters) that serve as a basis of this artificial categorization are the nature of the leaves and habit of the plant. In the first group, the leaves are generally thick and have white wooly outgrowth on them (In *S.ludens* leaves are somewhat hispid). In the second group the leaves are thin and do not have any white wooly outgrowth (in *S.neelgherryanus,* some white tomentum is found). The third group has a climbing habit, with wooly leaves particularly in *S.corymbosus.* In the fourth group the plants are herbaceous with highly dissected leaves.

The morphological groupings of the species have been analysed in terms of various parameters of numerical systematics based on flavonoid data. The parameters selected are :

1. Paired affinity index (PA),
2. Matching coefficient (MC)
3. Negative correlation (NC) and

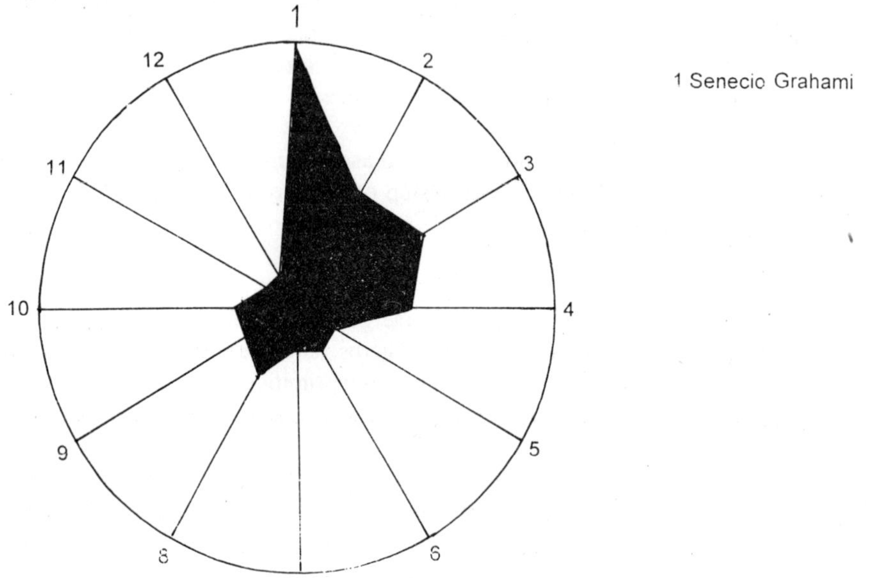

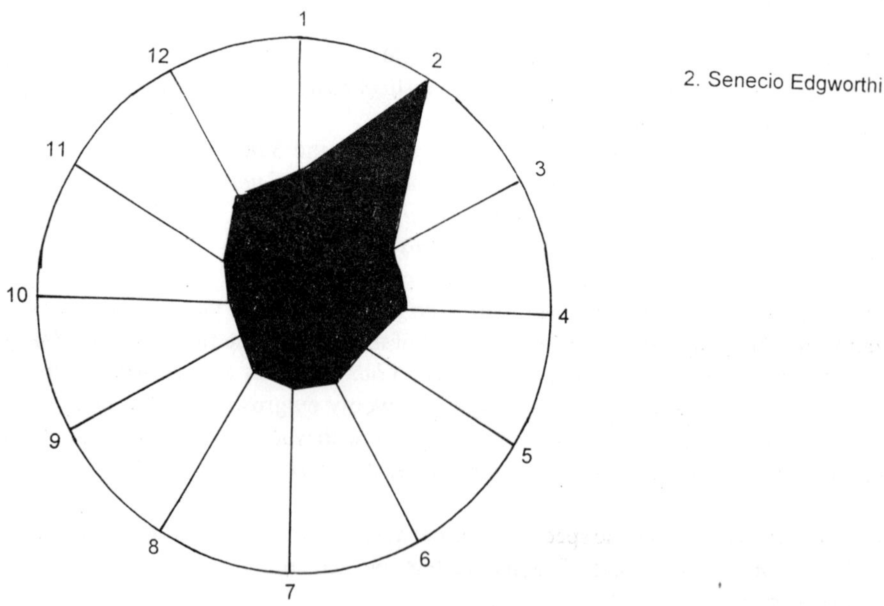

Plate 4.67 Polygonal graphs for species of *Senecio*

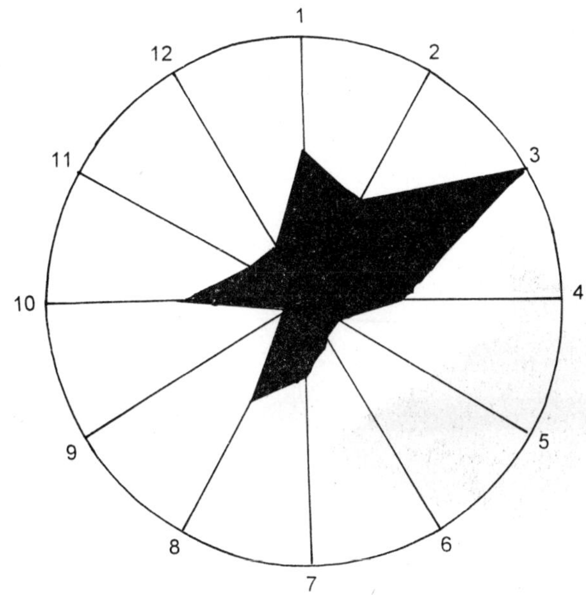

3. Senecio Cineraria

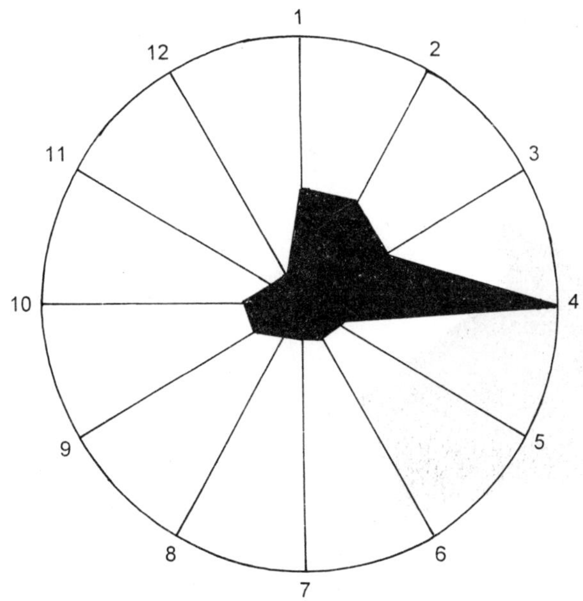

4. Senecio Leudens

Plate 4.68 Polygonal graphs for species of *Senecio*

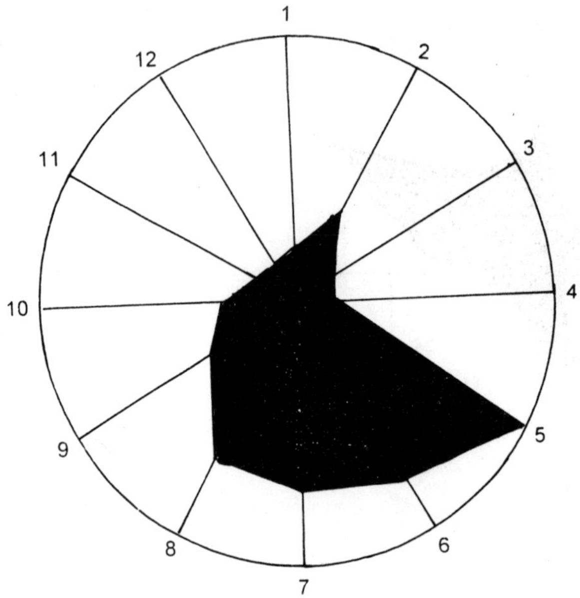

5. Senecio Lawsoni

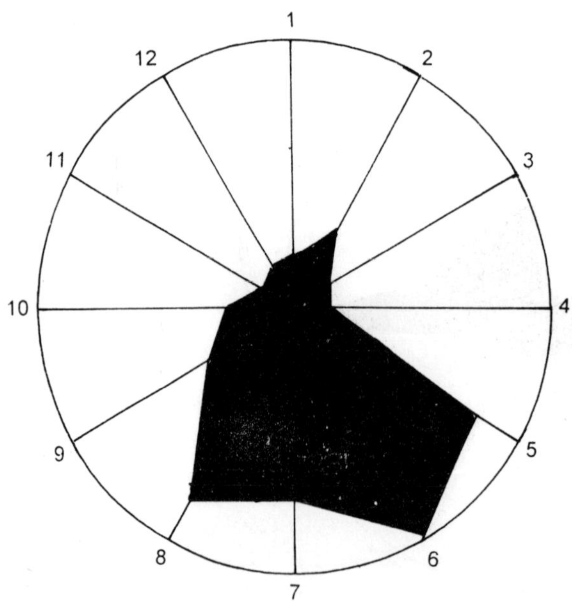

6. Senecio Whith

Plate 4.69 Polygonal graphs for species of *Senecio*

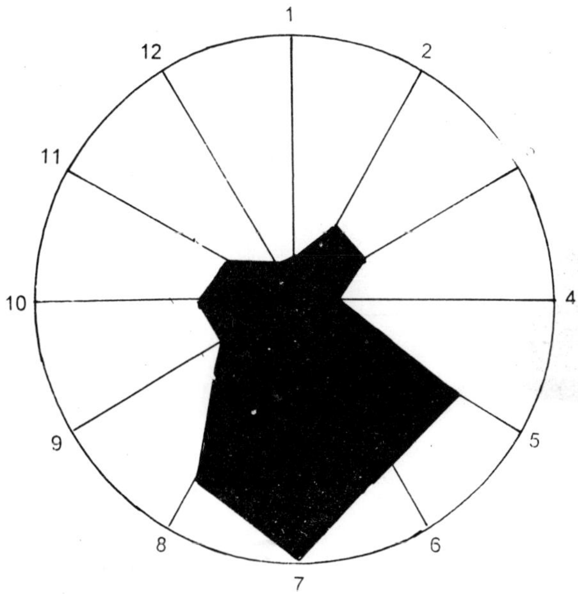

7. Senecio Belgaumensis

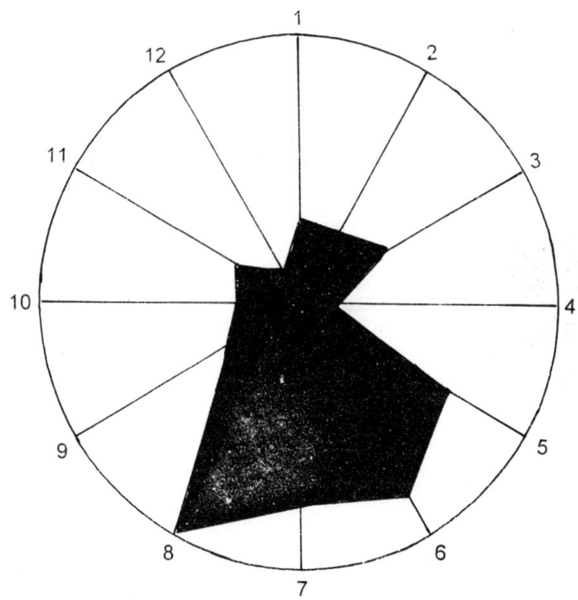

8. Senecio Neelghirryanus

Plate 4.70 Polygonal graphs for species of *Senecio*

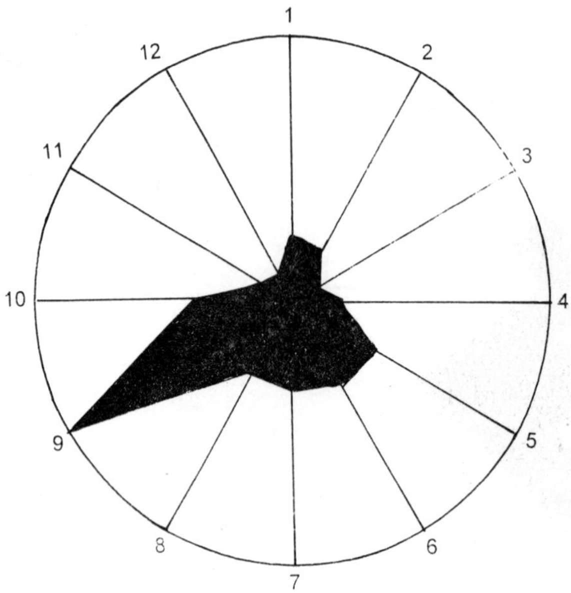

9. Senecio Corymbosus

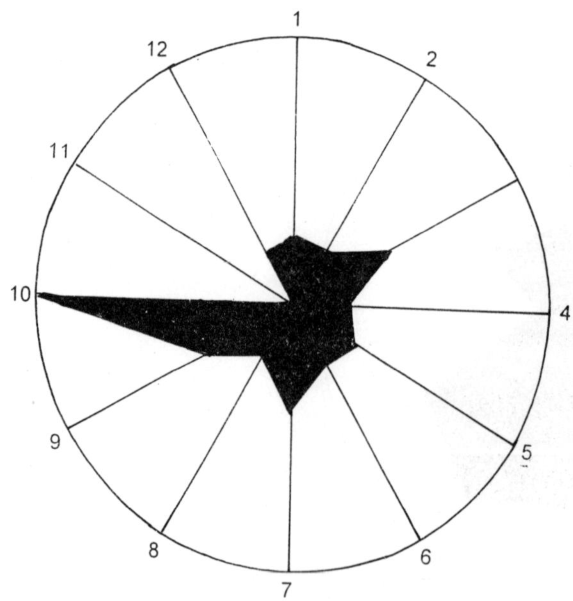

10. Senecio Walkeri

Plate 4.71 Polygonal graphs for species of *Senecio*

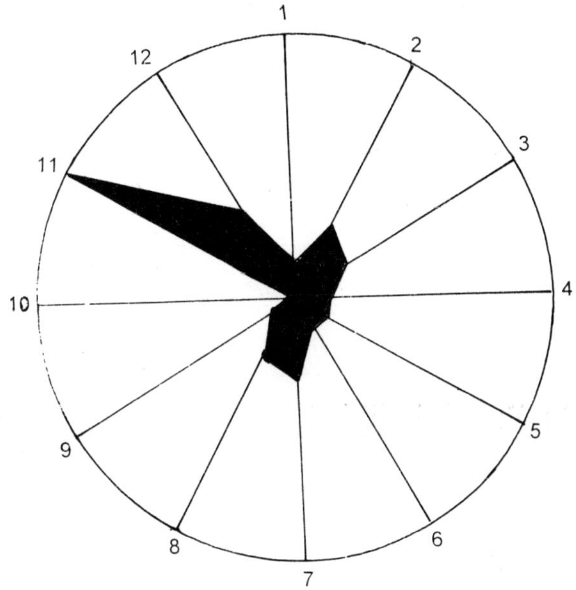

11. Senecio Hohenchari

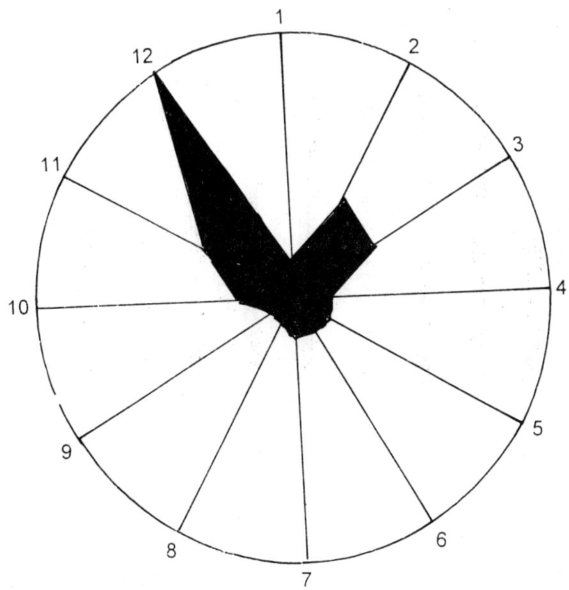

12. Senecio Tenuifolius

Plate 4.72 Polygonal graphs for species of *Senecio*

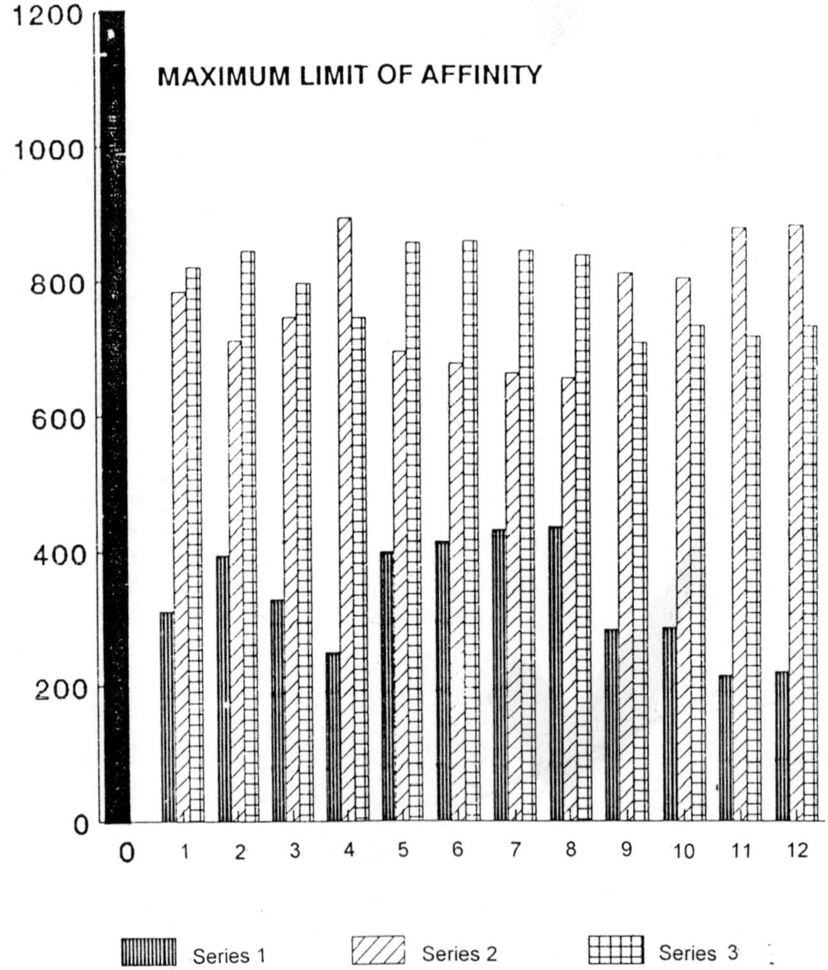

Plate 4.73 Group affinity values for 12 species of *Senecio* based on paired affinity index (Series 1); Negative correlation (Series 2) and matching coefficient (Series 3) (For names of species see Table 4.71)

4. Biochemical distance

The similarity matrices based on the these are given in tables 66, 67, 68 and 69 respectively.
PA indices (Table 66) show that in group 1 the values range between 43 to 57% indicating a close affinity among them. For group two, the values range from 67 to 83%. In group 3 and 4 the value is 38%. This indicates that there is a higher degree of affinity between the taxa within each group than between other groups. There is however a closer similarity between *S.edgeworthii* and *S.tenuifolius* belonging to groups 1 and 4.
The PA indices of individual taxon in comparison with others is totalled up to give group (genus) affinity values (group affinity value refers to the affinity of the taxon with the entire genus and not with the 4 groups identified within the genus) for the genus (Table no.70). This shows that no taxon has a value greater than 439 out of a possible 1200 (100 for each taxon).

Similarly, calculations based on matching coefficient (MC) support the findings in PA and show affinity amongst themselves in the interspecific groupings (Table 67). But here the percentage of affinity between the members of each group seem to be much higher than what is obtained in PA indices. In group one the percentage of affinity varies from 79 to 95%; in group two it varies between 92 to 94%; in group 3 it is 67% and group 4 it is 70%. Group (Genus) affinity values based on matching coefficient similarly show a value as high as 860 for *S.wightii* (Table 72).

Negative correlation (NC) values given in table 68 actually complement the data given for PA indices (Table 66). Here, the values are based on differences between the taxa. Hence a higher value indicates non affinity and a lesser value indicates closer affinity. Group (non) affinity values (Table 71) also similarly complement the GA values based on PA indices.

Another numerical parameter - biochemical distance (Sokal and Sneath, 1962), has also been computed for the purpose of interspecific comparison. In biochemical distance a lesser value indicates a closer relationship and *vice versa*. But the values obtained here seem to be in total disagreement with those obtained for PA, MC and NC in as much as they do not support any closer affinity between the taxa accommodated in the groups mentioned above. This raises a question as to which of the statistical parameters is to be relied upon, particularly when data from one seem to conflict those from the other.

In view of the observations made above, the following questions regarding interspecific relationship, group affinity and most useful statistical parameter for assessing these relationship may be raised.

1. How useful are the various statistical parameters in assessing the interspecific relationships?
2. Why the affinity values in PA indices (and GA based on PA indices) are far less than those obtained for MC (and GA based on MC)?

1. Of the various statistical parameters used here to assess the interpspecific affinity in the genus *Senecio*, PA, MC and NC are qualitative in concept. While PA and NC are mutually complentary in as much as they are concerned with similarities and differences between the two taxa, MC assesses relationship based on the absence of certain characters in both taxa but present in other taxa of the genus. The quantity of the compounds in question based son the area of spots (in mm^2), on the chromatogram is the yard stick of affinity in biochemical distance. It is open to question as to how far a variable trait like quanitity can be relied upon particularly when qualitative data do not support them. In a hypothetical situation, if in three taxa taken for comparison A,B and C, only one spot is comm between A and B, but there are 3 spots common between A and C, it is possible that the single spot common between A and B may be equal in quantity giving a lower value of biochemical distance. Between A and C, because more number of similar compounds are involved there may be difference in quantity giving a higher value of biochemical distance.

In such a situation biochemical distance assumes a closer relationship between A and B than between A and C. But this leads to an anomalous situation as more compounds are identical between A and C. Indeed such a situation does exist between *S.grahami* and *S.edgeworthii* on the one hand and *S.tenuifolius* on the other. There are three spots common between *S.grahami* and *S.edgeworthii*, but only one spot common between the former and *S.tenuifolius*. The biochemical distance value between *S.grahami* and *S.edgeworthii* is 3.38, while between the former and *S.tenuifolius* it is 2.66.
The statistical aspects discussed above clearly point out that biochemical distance based on quantity may not be a suitable parameter in the assessment of affinity between taxa, particularly when it is not in conformity with the data obtained from qualitative parameters.

2. As has been mentioned above, PA and NC depend on the similarities and differences only between the two taxa taken for comparison in a genus, while the MC is based on the characters (in this case flavonoid spots) absent in both taxa but present in others of the groups. If the genus taken for study is a large and diversified one (like *Senecio*) it is but natural that the various members are held together by few common characters but differences are more. When we apply this situation to *Senecio* which is a highly diversified genus, there are only very few spots common between all the species. In fact there is only one common spot (no.27). In such a circumstance in the two taxa taken for comparison, very few spots are common and many are different between them, but a large number of spots will be absent from both. This is the situation that exists in *Senecio* and hence the values of PA indices are generally less and those of MC are high. In a genus or a group which is more homogenous and less diverse, perhaps MC values may tally with the PA values, for a large number of characters (spots) are likely to be common for all the taxa within that genus.

Group (genus) affinity in species of *Senecio*

Group affinity as discussed here refers to the affinity of various species of *Senecio* to the genus (This doesnot refer to the groups 1 to 4 identified within the genus). GA values based on PA, NC and MC are given in tables 70, 71 and 72. In NC, the values are actually non affinity values as a higher value indicates lack of affinity. The GA values are also depicted in the form of histograms (Fig. 72). The PA indices are visualised in the form of polygraphs (Fig. 4.66 to 4.71) following Ellison *et al* (1962), in order to ascertain the intrageneric affinity.

The GA values based on PA (Table 70) show values which are all less than 500 in a possible total of 1200, while in MC (table 72) the values exceed 850. A perusal of the values in Table 70 shows that the species *S. lawsnoi* (399), *S.wightii* (415), *S.belgaumensis* (434) and *S.neelgherryanus* (438) show maximum affinity to the genus, while *S.tenuifolius* (222) and *S.hohenackeri* (216) show least group affinity. The least GA value for the entire group is in *S.hohenackeri* (216) and it is significant that it is in this taxon, the generic marker (spot no.27) is absent.

The GA values based on MC (Table 72) again show that the affinity of four species *S.lawsoni* (857), *S.neelgherryanus* (838), *S.wightii* (860) and *S.belgaumensis* (845), is highest to the genus while *S.hohenackeri* (719) and *S.tenuifolius* (734) are the least affiliated to the genus.

The GA values based on NC (Table 71) also support the assumptions made above in giving higher negative values to *S.hohenackeri* (880) and *S.tenuifolius* (884) indicating very little genus affinity.

The polygonal graphs based on PA indices though generally support the above assumptions of group affinity, eaxct matching of the polygons of taxa are not possible as the values of PA indices are always less in all the species of *Senecio* investigated.

Plate 4.74 Master key for species of *Gynura*

Plate 4.75 Master key for species of *Notonia*

Plate 4.76 Master key for species of *Emilia*

Isolation values

Isolation values indicate how far a taxon is separated in a group (genus) when compared with others (Ellison *et al* 1962). In *Senecio,* these values also (Table 73) show that the taxa with high group affinity value show least isolation (*S.lawsoni, S.wightii, S.belgaumensis* and *S.neelgherryanus*), while *S.hohenackeri* and *S.tenuifolius* show on Iv_n value of 5.5 each. The Iv_s value for the same taxa is 25. Surprisingly the taxon *S.ludens,* which has a slightly higher GA value (Table 70) than *S.tenuifolius* and *S.hohenackeri* has the highest isolation value for the entire genus (Iv_n, 11.11 and Iv_s, 50) This is because it has four exclusive spots (Table 65) even though it has many spots in common with others.

The various numerical parameters discussed above generally support that the 12 taxa investigated roughtly fall into four groups based on flavonoid spot pattern, which seem to be in conformity with their morphological traits.

The very low GA values seen in general in all the taxa and significant isolation values observed may be because of any one or both of the following.

(a) The genus is highly plastic in its genome make up and under selection pressure there is maximum diversity within a very short period. (This is possible even though the genus is very recent in origin).
(b) The genus may be ancient in origin and in course of evolution, the genome has undergone numerous changes leading to diversity in characters both morphological and chemical.

In the case of *Senecio,* both assumptions seem to be true. Small (1919) holds that *Senecio* is a very ancient genus and has the widest geographical distribution among Asteraceae. As for as the plasticity in gemome organisation is concerned there are instances of rapid genome alteration and introgression resulting in complexes as in *S.streptanthifolius* complex (Bovin and Denford, 1985) and *S.radicans* complex (*Glennie et al* 1971).

There can be no doubt however that the genus *Senecio* exhibits a highly diverse and dynamic biological system. The great range of habit and habitat of the genus together with flavonoid variety supports this view.

Gynura

In the genus *Gynura,* three taxa *G.lycopersicifolia, G.nitida* and *G.aurantiaca* which are studied for flavonoid profiles indicate that spot no.9 is the generic marker. *G.aurantiaca* with fleshy violet coloured leaves (violet plant) seems to be distinct from the other two taxa. (Fig. 4.73 table 74 and 75). It has a matching coefficient (MC) (0.13 and 0.25) and similarity coefficient (SC) (0.16 and 0.30) with *G.nitida* and *G.lycoperscifolia* respectively where as, *G.nitida* and *G.lycopersicifolia* show on MC of 0.42 (42%) and SC of 0.47 (47%) indicating they are closer to each other than to *G.aurantiaca.* This seems to be in conformity with the observed morphological characters.

Notonia

Two species of *Notonia* (which are morphologically diverse), *N.grandiflora* and *N.walkeri* are studied. The former is a fleshy under shrub while the latter is a shrubby plant. The leaf has anisocytic stomata in *N.grandiflora,* while the stomata are anomocytic in *N.walkeri.* Besides, the viens are hyphodromous in *N.grandiflora* while they are camptodromous in *N.walkeri.*

These morphological differences are symbolized in their flavonoid profiles also. As seen in the composite key (Fig.4.74) only 3 spots are common between the two taxa. *N.walkeri* has more number of flavonoid spots than *N.grandiflora.* They have a similarity coefficient of only 0.375 (Table 81) indicating their distinct their distinct identity.

Emilia

The classification on the genus *Emilia* is based on very few and sometimes overlapping morphological characters. Gamble (1948) categorizes the species based chiefly on radical or cauline nature of the leaves and the length of the head. Saldanha and Nicolson (1976) have delimited the species of *Emilia* of Hassan disrict (South India) based on the length of corolla lobes which range from 0.7mm to 2mm. They further point out that the genus needs further study.

The present work involving 3 taxa viz. *E.sonchifolia*, *E.ramulosa* and *E.scabra* is undertaken to study whether flavonoid spot pattern is of any assistance in the categorization of the species. In *E.scabra*, the leaves are chiefly radical; cauline in *E.ramulosa* while in *E.sonchifolia* they are both radical and cauline. As the composite key (4.75) shows, only one spot (no.8) is common between the three taxa indicating their flavonoid individuality. Matching coefficient (Table no.82) and similarity coefficient (Table no.83) between the three taxa show that they are very distinct having affinity values ranging from 0.1 (10%) to 0.15 (15%) only. This clearly shows that the morphological delimitation of the species of *Emilia* can be justified chemically also.

LITERATURE CITED

ABBOT, H.C. de.S. 1886. Certain Chemical constituents of plants considered in relation to their morphology and evolution. *Bot. Gaz.* **11** : 270-272.

AHUJA, K.K. and PATASKAR, R.D. 1969. *Synedrella vialis* Less. A. Grey. A New record for India. *Indian Forester.* **95**, 267.

ALSTON, R.E. 1966. Chemotaxonomy or Biochemical systematics. In : *Comparative Phytochemistry* (Swain, T., ed) pp. 33-56 Academic Press. London, N.Y.

______________ and IRWIN, H.S. 1961. The comparative extent of variation of free amino acids and certain secondary substances among *Cassia* species. *Am.J.Bot.* **48** : 35-39.

______________ and TURNER, B.L. 1963. *Biochemical systematics* Englewood Cliffs. N.J. Prentice Hall. Inc.

______________ and TURNER, B.L. 1963b. Natural hybridization among four species of *Baptisia* (Leguminosae) *Am. J. Bot.* **50** : 159-173.

ARDITTI. J and FISCH, 1977. Anthocyanins of the Orchidaceae. In : *Orchid biology,* Reviews and perspectives. I. 119-125.

ASKER, 1973. TLC studies of phenolics in Oata. *Hereditas.* **73**, 17-26.

ASKER, S and FROST, S. 1970. Chromatographic studies of phenolic compounds in apomictic *Potentilla. Hereditas.* **65**, 241-250.

______________ and FROST, S. 1972. Plant age and chromatographic pattern in *Potentilla. Hereditas.* **72**, 149-152.

BABU, C.R. 1971. *Galinsoga ciliata* (Raff) Blake (Asteraceae) A spp. new to India. *Bull. Bot. Survey India* 11 : 184-185.

BAIN, J. and DENFORD, K.R. 1985. Flavonoid variation in *Senecio streptanthifolius* complex. *Can. J. Bot.* **63** (10) 1685-1690.

BALL, E.A. HARBORNE, J.B. and ARDITTI, J. 1972. Anthocyanins of *Dimorphotheca :* pigments in flowers, stems and callus tissues. *Am.J. Bot.* **59** : 924-930.

BALLARD, R.E. 1975. Dissertation abstracts. **36 B**. 1564.

BATE SMITH, E.C. 1962. The phenolic constitutents of plants and their taxonomic significance. 1 Dicotyledons. *J.Linn. Soc. (Bot).* **58** : 95-173.

______________ and SWAIN, T. 1966. The Asperulosides and the aucubins. In *Comparative Phytochemistry* (Swain, T. ed.) pp. 159-174. Academic Press. London.

______________ and RICHENS, R.H. 1973. Flavonoid chemistry and taxonomy in *Ulmus. Biochem. Syst.* **1:1** 41-146.

BEIRNER, M.W. 1973. Chemosystematic aspects of flavonoid distribution in twenty two taxa of *Helerium. Biochem. Syst.* **1** : 55-57.

BELZER, N.F. and OWNBEY, M. 1971. Chromotographic comparison of *Tragopogon* species and hybrids. *Am.J. Bot.* **58** : 791-802.

BENTHAM, G. 1873. Notes on the classification, history and geographical distribution of Compositae. *J.Linn. Soc. (Bot).* **13** : 335-577.

BESSEY, C.A. 1915. The phylogenetic taxonomy of flowering plants. *Ann. Mo. Bot. Gdn.* **2** : 109-164.

BIRCH, A.J. 1974. In : *Chemistry in Botanical classification* (G. Bendz and J. Santesson, eds), pp. 261-270, Academic Press, New York.

BOHM, B.A. and GLENNIE, C.W.M. 1971. A chemosystematic study of the Caprifoliaceae. *Can J. Bot.* **49** : 1799-1807.

______________ SALEH, N.A.M. and ORNDUFF, R. 1974. Flavonoids of *Lasthenia. Am. J. Bot.* **61** : 551-561.

______________ and WILKINS, D. 1978. Chemosystematic studies in the Saxifragaceae. *sensu lato. X.* Flavonoids in *Elmira racemosa. Brittonia.* **30** : 327-333.

BOPIAH, A.K. 1986. Cytological, chemotaxonomical and *in vitro* seed culture studies in some members of Orchidaceae, Ph.D.Thesis, Karnataka University.

BOSE, S. 1972. Comparison of wheat varieties on the basis of their relative phenolic contents. *Hereditas.* **72** : 159-168.

BREHM, B.G. and OWNBEY, M. 1965. Variation in Chromatographic patterns in the *Tragopogan dubius - pratensis - porrifolius* complex (Compositae) *Am.J.Bot.* **52 (8)**, 811-818.

BREHM, B.G. and KRELL, D. 1975. *Science* **190**, 1221.

CARLQUIST, S. 1957. The genus *Fitchia* (Compositae) *Univ. Calif. Publ. Bot.* **29** : 1-143.

* CASSINI, H. 1819. Sixieme memoire sur la familie des Synantherees. Contenant les caracteres des teribus. *J.Phys. Chim. Hist. Nat. Arts.* **88** : 150-163, 189-204.

* ______________ 1829. Tablean Synoptique des Synantherees. *Annls. Sci. Nat. Paris.* **17** : 387-423.

CHENNAVEERIAH, M.S. and RAZDAN, M.K. 1980. Karyomorphological and phytochemical studies in evaluating species relationships in *Garcinia* Linn. and systematic position of *G.xanthochymus* complex. *J.Ind. Bot. Soc.* **59**, 251-262.

CHINNERY, 1983. (Personal Communication to the author).

CHIVALA, F., LIDDLE A.P. and DOGLIA, G. Chemotaxonomy of wormwood *(Artemesis obsinthum L.). Lebsensm unters Forsch.* **176** (5), 363-366.

CRAWFORD, D.J. 1970. Systematic studies on mexican *Coreopsis* (Compositate). *Coreopsis mutica* : flavonoid chemistry, chromosome number, morphology and hybridization. *Brittonia* **22** : 93-111.

______________ 1972. The morphology and flavonoid chemistry of synthetic infraspecific hybrids in *Coreopsis mutica* (Compositae). *Taxon* **21 (1)** : 27-38.

______________ 1974. A morphological and chemical study of *Populus acuminata* Rydberg. *Brittonia* 26, 74-89.

______________ 1978. Flavonoid Chemistry and Angiosperm evolution. *Bot Rev.* **44 (4)**, 431-456.

______________ and MABRY, T.J. 1978. Flavonoid chemistry of *chenopodium fremontii* : infraspecific variations and systematic implications at the interspecific level. *Biochem. Syst* and *Ecol.* **6**, 189-192.

______________ and SMITH, E.B. 1984. Leaf flavonoid chemistry and the taxonomy of Coreopsis sections

Pugiopappus and *Euleptosyne* (Compositae), *Madrano,* **31** : 1, 1-7.

CRONQUIST. A. 1955. Phylogeny and Taxonomy of the Compositae. *Am.Midl. Nat.* **53**, 478-511.

_____________ 1973. Chemical Plant taxonomy, A generalist's view of a promising speciality. In : Noble. 25 (1973), *Chemistry in botanical classification*. Gerd Bendz and John Santesson (eds) : 29-39.

CROWDEN, R.K., HARBORNE, J.B. and HEYWOOD, V.H. 1969. Chemosystematics of the Umbelliferae. A general survey. *Phytochemistry.* **8** : 1963-1984.

DAKSHINI, K.M.M. and PRITHIPAL SINGH, 1977. Chemotaxonomic studies in the genus *Blumea* Dc. (Asteraceae-Inuleae) in India. *Acta Bot. India.* **5 (2)** 143-159.

DASS, H.C. 1972. Analysis of species relationship in *Avenae* by TLC and Disc electrophoresis. *Can J. Genet. Cytol.* **14** : 305-316.

_____________ and NYBOM, N. 1967. The relationships between *Brassica nigra, B.campestris, B.oleracea* and their amphidiploid hybrids studied by means of numerical chemotaxonomy. *Can. J.Genet. Cytol.* 9 : 880-890.

_____________ RANDHWA, G.S. and KAUR, M. 1974. Phylogenetic studies in Cucumerinae by leaf phenolics. *The Nucleus,* **17**, 103-109.

_____________, _____________ and _____________ 1975. Variation in flavonoid patterns and species interrelationships in the genus *Anona. India. Jour. Hort.* **32 (384)**; 111-116.

_____________ RANDHWA, G.S. and PRAKASH, D. 1976. Evaluation of different numerical methods utilizing unidentified phenolic patterns in chemotaxonomy. *Taxon* **25 (5-6)** 617-620.

_____________, _____________ and _____________ 1978. Leaf flavonoid patterns of some important *Citrus* species. *Ind. Jour. Exp. Bio.* **16**, 62-65.

DEDIO, W., KALTSIKES, P.J. and LARTHER, E.N. 1969b. Numerical cheotaxonomy in the genus *Secale. Can. J. Bot.* **47**, 1175-1180.

DOSKOTCH, R.W. MIKHALL, A.A. and CHATERJEE, S.K. 1973. Structure of the water soluble feeding stimulant for *Scolytus multistinatus.* A revision. *Phytochemistry.* 12 1153-1155.

ELLISON, W.L., ALSTON, R.E. and TURNER, B.L. 1962. Methods of presentation of crude biochemical data for systematic purposes with particular reference tot he genus *Bahia* (Compositae). *Am.J. Bot.* **49**, 599-604.

ERDTMANN, H. 1963. Some aspects of chemotaxonomy. In : *Chemical plant taxonomy* (Swain, T.ed) 89-125. Academic Press, London. N.Y.

FAIRBROTHERS, D.E. 1968. Chemosystematics with emphasis on systematic serology. pp. 141-174. In : *Modern methods in plant taxonomy* (Heywood, V.H. ed.) Academic Press, London and N.Y.

_____________, MABRY, T.J., SCOGIN, R.L. and TURNER, B.L. 1975. *Ann.Mo.Bot. Gard.* **62**, 765.800.

FREELAND, W.J. and JANZEN, D.H. 1974. *Am.Nat.* 108 : 269-289.

FROST, S. and HOLM, G. 1971. Studies of phenolic compounds in 20 varieties of barley. *Hereditas.* **69**, 25-34.

FROST, S and HOLM, G. 1971. Studies of phenolic compounds in 20 varieties of barley. *Hereditas.* **69**, 25-34.

FROST, S., ASKER, S., and SALEH, M. 1979. Comparison of flavonoid patterns in wild and cultivated *Hordeum* nspecies. *Hereditas.* **90**, 251-256.

_____________, HOLM, G. and ASKER, S. 1975. Flavonoid patterns and the phylogeny of the barley.

Hereditas. **79**, 133-142.

______________ HARBORNE, J.B., ASKER, S. and SALEH, N. 1979. Comparison of flavonoid patterns in wild and cultivated *Hordeum* species. *Hereditas.* **90**, 251-256.

GALSTON, A.W. 1969. In : Perspetives in Phytochemistry (eds. J.B. Harborne and T.Swain). Academic Press. London. pp. 193-204.

GEISSMANN, T.A. and HEATON, C.D. 1943. Anthochlor pigments IV. The pigments of *Coreopsis grandiflora. J. Amer. Chem. Soc.* **65**, 677-680.

GIBBS, R.D. 1956-57. Biochemistry as an aid to establishing the relationships of some families of Dicotyledonous plants. *Proc. Lonn. Soc. London.* **169 (3)** : 216-230.

______________ 1958. Chemical evolution in plants. *J. Linn. Soc. London.* (Bot). **56**, 49-57.

______________ 1963. History of Chemical Taxonomy. In : *Chemical Plant taxonomy* (Swain, T. ed.) 1963. Academic Press, London. N.Y.

GLENNIE, C.W., HARBORNE, J.B., ROWLEY, G.D. and MERCHANT C.J. 1971. Correlations between flavonoid chemistry and plant geography in the *Senecio radicans* complex. *Phytochemistry.* **10**, 2513-2417.

GRANT, W.F. 1968. Botanical editors and biochemical systematics. *Taxon.* **17** : 259-260.

______________ and WHETTER, J.M. 1966. The 9cytogenetics of *Lotus* XI. The use of TLC in the separation of secondary phenolic compounds in *Lotus* (Leguminosae). *J. chromatag.* **21**, 247-256.

GREGER, H. 1969. Flavonoids and the systematics of Anthemideae. *Naturwissenshaften.* **56**, 467-468.

______________ 1975. Leaf flavonoids and systematics in *Matricaria* and *Tripleurospermum. Plant. Syst. Evol.* 124, 35-55.

GUPTA, P.K. 1978. Evolutionary mechanisms in flowering plants - A multidisciplinary approach. *The Nucleus.* **21(2)** : 117-124.

GUPTA, P.K. and SINGH, R.V. 1977. Variations in chromosomes and flavonoids in *Setaria* Beauv. *The Nucleus.* **20**, 167-171.

HARBORNE, J.B. 1966. The evolution of flavonoid pigments in plants. In : *Comparative Phytochemistry.* Academic Press, London, N.Y. 271-295.

______________ 1967a *Comparative biochemistry of the flavonoids.* Academic Press, London, N.Y.

______________ 1967b. Comparative biochemistry of the flavonoids. IV. Correltions between chemistry, pollen morphology and systematics in the family Plumbaginaceae. *Phytochemistry.* 6 : 1415-1428.

______________ 1967c. Comparative biochemistry of the flavonoids VI. Flavonoid patterns in the Bignoniaceae and Gesneriaceae. *Phytochemistry* **6** : 1643 - 1651.

______________ 1969. Chemosystematics of the Leguminosae. Flavonoid and Isoflavonoid patterns in the tribe Genisteae. *Phytochemistry.* **8** : 1449 - 1456.

______________ 1971a. Distribution of flavonoids in the Leguminosae. In : *Chemotaxonomy of Leguminosae* (eds). Harborne, J. B. *et al.* Academic Press. London. 31-12.

______________ 1971b. In *'The biology and Chemistry of the Umbelliferae* (Heywood, V.H. ed) 293-314.

______________ 1975. Biochemical systematics of flavonoids. *The Biology and Chemistry of the Compositae* (Holywood V.H. *et al.* eds), Academic Press. London. N.Y.

______________ WILLIAMS, C.A. 1973. A chemotaxonomic study of flavonoids and simple phenols in leaves of the *Ericaceae*. *Bot. J. Linn. Soc.* **66**, 37-54.

______________ and SIMMONDS, N.W. 1964. In : *Biochemistry of phenolic compounds.* J.B. Harborne (ed.) 77. Academic Press, London. N.Y.

______________ and SMITH, D.M. 1978. *Biochem. Syst. Ecol.* **6**, 287.

______________ and MABRY, T.J. (EDS.) 1982. The flavonoids-Advances in Research. Chapman and Hall. London. N.Y.

______________ GIRIJA DEVI, H.R. and LAKSHMI, H.M. 1983. Anthochlor pigments from the petals of *Musseanda hirsutissima Phytochemistry.* **22(12)**, 2741-42.

HARNEY, P.M. and GRANT, W.F. 1964. A chromatographic study of the phenolics of species of *Lotus* closely related to *L.corniculatus* and their taxonomic significance. *Am.J.Bot.* 621-627.

______________ 1956. A polygonal presentation of chromatographical investigations on the phenolic content of certain species of *Lotus. Can. J. Genet. Cyt.* **7**, 40-51.

HEGDE, S.N. and JORAPUR, S.M. 1975. Chemosystematic studies in the subtribe Lipardinae, Benth. (Orchidaceae). *Ind. J. Expl. Biol.* **13**, 476-480.

HEGNAUER, R.C. 1962. *Chemotaxonomie der Paflanzen.* Birkhauser Bael Stuttgart. Birkhauser.

______________ 1967, *Pure Appl. Chem.* **14**, 173-197.

HEGNAUER, R. 1977. The Chemistry of the Compositae. In : *The Biology and Chemistry of the Compositae* (Heywood, V.H., Harborne, J.B. and Turner, B.L. eds.) Academic Press. London. N.Y.

______________ 1969. In : *Prespectives in phytochemistry* (J.B. Harborne and T. Swain, eds.) pp. 121-38. Academic Press. New York.

HERZ, W. and BHATT, S.V. 1970. Isolation and structure of two new germacronolides from *Polymnia uvedalia* (L.) L. *J.Org. Chem.* **35**, 2605.2611.

HEYWOOD, V.H. 1966. Phytochemistry and Taxonomy. pp. 1-20. In : *Comparative phytochemistry* (ed. Swain, T.) Academic Press, London and N.Y.

______________, HARBORNE, J.B. and TURNER, B.L. 1977. *The Biology and Chemistry of the Compositae.* Academic Press. London. N.Y.

HOFFMANN. 1890. Tubuliflorae - Heliantheae and Helenieae. In *Die naturlichen pflanzenfamilien* (Engler and Prantl. eds.) **4 (5)**, 210-267.

IKRAMOV, M.T., KALMATOV, K.K., BATIROV, E.K. and MALIKOV, V.M. 1986. Flavonoids from *Anaphalis velutina Khim. Prim. Soedin* (Task) **2**, 245.

ISAMAN, M.B. and DUFFEY, S.S. 1981. *Ent. Exp. and Appl.* **34**, 370.

ITIYAMA, K and GRANT, W.F. 1972. A correlation of nuclear DNA content and TLC patterns resolving genome relationships in *Avena. Can. J. Bot.* **50**, 1529-1545.

JAWORSKA, H. and NYMBOM, N. 1968. A TLC study of *Saxifraga caesis, S.aizoides* and their putative hybrids. *Hereditas.* **57**, 159-177.

JEPSON, W.L. 1901, A flora of western middle California, Encina Publ. Co. Berkeley.

JONES, D.A. 1979. *Am. Nat.* **113**, 445-451.

KALTSIKES, R.J. and DEDIO, N. 1970a. A TLC study of phenolics of the genus *Aegilops.* I Numerical chemoaxonomy of diploid species. *Can. J. Bot.* **48** (10), 1175-1180.

______________ 1970b. A TLC study of the phenolics in the genus *Aegilops* II. Numerical chemotaxonomy of polyploid species. *Can. J. Bot.* **48** (10), 1781-1786.

KAUR, M., DASS, H.C. and RANDHAWA, G.S. 1973. Systematic status of round melon (*Citrullus vulgaris* var. *fistulosus*)as studied by leaf phenolics. *Curr. Sci.* **42**, 730-731.

KHANNA. 1980. (quoted In : Key works to the taxonomy of flowering plants of India, M.P. Nayar, 1. BSI, 1984).

LEELA RAO, K. 1983. Cytotaxonomical and Phytochemical studies in Campanales of Bentham and Hooker. Ph. D. Thesis. Karnataka University.

LEPPIK, E.E. 1978. The evolution of capitulum types of the Compositae in the light of insect flower interaction. In : *The Biology and Chemistry of the Compositae* (Heywood, V.H., Harborne, J.B. and Turner, B.L. eds.) Academic Press. London and N.Y.

LESSING, C.F. 1832. "Synopsis generum Compositarum" Berlin.

LORENZ, H and SCHULZ-SCHAEFFER, J. 1964. Biosystematic investigations in the genus*Agropyron* II. A chromatographic approach. *Z.pflazenzuchtung.* **52**, 12-26.

MABRY, T.J. 1966. The betacyanins and betaxanthins. In : *Comparative phytochemistry* (T. Swain, ed.) pp. 231-244. Academic Press. London and N.Y.

______________ 1974. In *Chemistry and Botanical* classification (G. Bendz J. Santesson, eds.) p. 63. Academic Press. London and N.y.

______________ 1973. *Pure. Appl. Chem.* **34**, 377-400.

______________, BASET, Z.A. PADOLINA, W.G. and JONES, S.B. 1975. Systematic implications of flavonoids and sesquiterpene lactones in species of *Vernonia. Biochem. Syst. Ecol.* **2**, 185-192.

MARKHAM, K.R. 1982. *Techniques of flavonoid identification.* Academic Press. London and N.Y.

MANSOUR, R.M.A., SALEH, N.A.M. and BOULDS, L. 1983. A chemosystematic study of phenolics in *Sonchus. Phytochemistry.* **22 (2)**, 489-492.

MATHEWS, I.F., 1966. A paper chromatographic study of north and south American species of the genus *Tradescantia. Bot. Gaz.* 127, 74-78.

MENTZER, C. 1966. Biogenetic classification of plant constituents (in Swain T. ed.) *Comparative phytochemistry* 21-31. Academic Press. London and N.Y.

MERFORT, I., MARCINEK, C and EGGERT, A. 1986. Flavonoid distribution in *Arnica* subgenus*chamissonis. Phytochemistry* **25(12)**, 2901-2903.

MOORE, D.M. 1972. In *Taxonomy Phytogeography* and *Evolution* Ed. Valentine, D.H.). Academic Press London and N.Y. pp. 115-140.

______________, HARBORNE, J.B. and WILLIMAS, C.A. 1970. *J.Linn. Soc. (Bot.)* 63, 277.

MOTHES, K. 1966. *Lloydia.* **29** : 156-171.

NAGESHWAR, G., RADHAKRISHNIAH, M and NARARYANA, L.L. 1984. Numerical chemotaxonomy of *Bauhinia. Proc. Ind. Acad. Sci.* **B. 93** (6) 621-627.

NAIR, G.G. DANIEL, M and SABNIS, S.P. 1986. Chemosystematics of *Ipomea* and some other related data. *Curr. Sci.* **55 (19)** 961-965.

NAKOKI, T. 1936. Glucoside of the flavone of the white flower IV. Constituents of *Cosmos bipinnatus* Cav. *J.pharm. Soc. Japan.* **55**, 967-968.

NARAYANA. B.L. 1986. A survey of leaf flavonoids in Portulaceae. *Biochem. Syst.* and *Ecol.* **14 (6)**, 633-635.

NATARELLA, M.J. and SINK, K.C. (Jr). 1974. A chromatographic study of phenolics of species ancestral to *Petunia hybrida J. Hered.* **65**, 85-90.

NORTHINGTON, D.K. 1974. Chemosystematic studies of the genus *Pyrrohappus* (Cichorieae). *Spec. Publ. mus. Tex. Tech. Univ.* **6**, 1-38.

NORDENSTAM, B. 1978. Senecioneae and Liabeae - systematic review. In : *The Biology and Chemistry of the Compositae* (Heywood, V.H., Harborne, J.B. and Turner, B.L. eds.) Academic Press. London. N.Y.

OKUDA, J., MORI, K. and HATANO, T. 1980. The distribution of geranin and mallotusinic acid in the order Geraniales *Phytochemistry.* **19**, 547-551.

PEGAY I. 1986. Les flavonoids de *Senecio aureus.* J. Pharamazie. **41 (7)**, 524-525.

PETIVER, J. 1699. Some attempts made to prove that herbs of the same make or class for the generality have the like virtues and tendency to work the same effects. *Phil. Trans. Roy. Soc.* **21 B**, 284-294.

POLJAKOV, P.P. 1967. *Systematika i prnsxozdeni sloznocvetnyx* (Systematics and origin of the Compositae) Acad. Sci. Kazakasthan. Alma Ata.

RADFORD, A.E., DICKISON, W.C., MASSEY, J.R. and BELL, C.R. 1974. *Vascular plant systematics.* Harper and Row Publishers. N.Y. and London.

RAJHATHY, T., SHEAREV D.A. and WARENE, E.M. 1971. A TLC study of some amphidiploids in *Avena. Can. J. Genet. Cytol.* **13**, 749-759.

RAEN, R.H., EVERT, R.F. and CURTIS, H. 1976. *Biology of plants.* Worth, New York.

RAO, C.K. 1980. The methodology of phytochemistry in relation to plant Taxonomy pp. 209-221. In : *Glimpses in plant research.* Vol. V. Modern trends in plant taxonomy (P.K.K. Nair. ed.) Vikas Publications. New Delhi.

REES, S. and HARBORNE, J.B., 1984. Flavonoids and other phenolics of *Cichorium* and related members of the Lactuceae (Compositae) *Bot. J. Linn soc.* **89(4)**, 313-319.

RILEY, H.P. and BRYANT, T.R. 1961. The separation of nine species of the Iridaceae by paper chromatography. *Am. J. Bot.* **48** : 133-137.

ROBINSON, B.L. 1913a. A key to the general of the Compositae - Eupatorieae. *Proc. Am. Acad. Arts. Sci.* 49, 429-437.

ROBINSON, T. 1974. *Science.* **183**, 430-435.

ROBINSON, H. and KING R.M. 1977. Eupatorieae - systematic review. In : *The Biology and Chemistry of the Compositae* (Heywood, V.H., Horborne, J.B. and Turner, B.L. eds). Academic Press, London, N.Y.

ROBINSON, H., BOHLMANN, F. and KING, R.M. 1978. Chemosystematic notes on the Asteraceae : New correlations in the subtribes of the Heliantheae. *Phytologia* **41(1)**, 50-54.

ROCHELDER, F. 1847. In : *Principles of Angiosperm Taxonomy* (Davis and Heywood, 1967), Oliver and Boyd. Edinburgh.

RAMUSSI, G. and PAGANI, F. 1970. Constituents of *Bidens frondosa. Boll. Chim Farm.* **109**, 467-475.

RUBKHOVASKAYA, T.V., MANADILOVA, A.M. and SAPCO, O.A. 1985. The flavonoids of *Artemesia sublessingiana. Khim. Prim. Soedin* (Task). **(3)**, 407.

RYDBERG, P.A. 1914 - 1915. Carduceae, Helineae and Tageteae. In : *North American Flora.* Vol. **34** pp. 1-80. New York Botanic Garden.

SALDANHA, C.J. and NICOLSON, D.H. 1976. *Flora of Hassan District.* Amerind Publishers, New Delhi.

SANTA RAM, A and DEVI, H.M. 1983. Chemotaxonomic study of Umbelliferae. Flavonoid aglycones. *Indian J. Bot.* **6 (1)** 21-24.

SCOGIN, R. and ZAKAR, K. 1976. Anthochlor pigements and floral UV patterns in the genus *Bidens. Biochem. Syst. Ecol.* **493**, 165-167.

SEELIGMAN, P. and ALSTON, R.E. 1967. *Brittonia.* **19**, 205-211.

SEIGLER, D.S. 1974. Chem. Br. **10**, 339-342.

______________ 1981. Secondary metabolites and plant systematics. In : *The Biochemistry of Plants* Vol. **7**. Secondary plant products pp. 139-176. (Stumpf, P.K. and Conn. E.E. eds). Academic Press, London. N.Y.

______________ and PRICE, P.W. 1976. *Am. Nat.* **110**, 101-104.

SINGH, P and DAKSHINI, K.M.M. 1977. Chemotaxonomic studies on the genus *Blumea* Dc. (Asteraceae, Inuleae) in India. *Acta. Bot. Ind.* **5(2)**, 143-159.

SINGH, B.P., VARMA, S.K. and DOGRA, J.V.V. 1986. Chemotaxonomy of Vitaceae of Bihar - A study of flavonoids. *J. Indian. Bot. Soc.* **65**, 335-339.

SMALL, J. 1919. The Origin and Development of Compositae. New Phytol. reprint No.11. William Wesley and Sons, London.

SMITH and LEVIN, D.A. 1963. A chromatographic study of reticulate evolution in the Appalachian *Asplenium* complex *Am J.Bot.* **50**, 952-958.

SOKAL, R.R. and SNEATH, P.H.A. 1963. Principles of numerical Taxonomy, W.H. Freeman, San Fransisco, USA.

SMITH, D.M. and HARBORNE, J.B. 1971. Xanthones in the Appalachian *Asplenium* Complex. *Phytochemistry,* **10**, 2117-2119.

SMITH, D.M. GLENNIE, C.W., HARBORNE, J.B. and WILLIAMS, C.W. 1977. Flavonoid diversification in the Polemoniaceae. *Biochem. Syst. Ecol.* **5**, 107-115.

STEBBINS, G.L., HARVEY, B.L., COX, E.L., RUTGER, J.N., JELLENCOVIC, G., and YAGIL, E. 1963. Identification of ancestryof an amphidiploid *Viola* with the aidof paper chromatography *Am.J.Bot.* **50**, 830-839.

SUNDARA RAJAN, S. 1977. Ray reduction in flowers and evolution in *Bidens pilosa, New Botanist* : **IV : 1-4** : 95-98.

______________ 1979. Introgression in *Bidens pilosa. Geobios,* **6** : 12-14.

______________ 1984. Floral variation in *Bidens pilosa,* (Abstract) International plant population congress, Waginengen, Netherlands.

______________ 1988. Pollination mechanism, palynology and anthecology in *Bidens pilosa* (Abstract). 7th international palynological congress, Brisbane, Australia.

______________ 1989. Intraspecific variation in *Bidens Pilosa* (Abstract). 4th international symposium of plant biosystematists, Kyoto, Japan.

SUNDARI, K.T., RADHAKRISHNIAH, M and NARAYANA, L.L. 1982. Chemotaxonomy of *Ceratophyllum. Acta. Bot. Indica.* 10(2) 304-305.

STUESSY, T.F. 1977. Heliantheae - systematic review. In : *The Biology and Chemistry of the Compositae* (Heywood, V.H. Harborne, J.A. and Turner, B.L. eds.). Academic Press, London. N.Y.

SWAIN, T. (ED) 1963. *Chemical Plant taxonomy.* Academic Press N.Y. and London.

SWAIN, T. 1973. Flavonoids as evolutionary markers in primitive tracheophytes. 81-92. In Nobel **(25)** (1973). *Chemistry in botanical classification.* (eds. Gerd B and Santesson, J.). Academic press. London. N.Y.

SWAIN, T. 1976. In *Secondary metabolism and coevolution* (M.Luckner, K. Mothes and L. Nover, eds.). pp. 411-421. Dtsch. Akad. Naturf. Leopold, Halle. oll.

SZEPCZYNSKA, K. and WOLBIS, M. 1984. Polyphenolic compounds of *Galinsoga ciliata* (Raff) Blake, *Acta. Pol. Pharm.* (Engl. Trans.) **41 (1)** 117-122.

TURNER, B.L. 1969b. Chemosystematics. Recent Developments. *Taxon.* **18**, 134-151.

TURNER, B.L. 1984. *Pure. Appl.* Chem. **14**, 189-213.

VAN EMDEN, H.F. 1973. (ed). *Insect plant relationships.* Wiley. New York.

VYAS, A.V. and MULCHANDANI, N.B. 1986. Polyoxygenated flavones from *Ageratum conyzoides. Phytochemistry.* **25 (110)**, 2625-2627.

VERYKOKIDOUVITSAROPOULOS, E. and BECKER, H. 1983. Flavones of *Spilanthes leracea* Jacq. *Arch. Pharm.* (Weinheim). **316(9)**, 815.

WADDELL, T.G. 1973. Hymenoxin - a flavone from *Helianthus angustifolius. Phytochemistry.* **12**, 2061.

WHITTAKER, R.H. and FEENY, P.P. 1971. *Science.* **171** : 757-770.

WIEMARCH, H. 1963. Skanes flora. Lund.

WILLAMS, A.H. 1966. Dihydrochalcones. In : *Comparative Phytochemistry* (Swain, T. ed). Academic Press. London and N.Y.

WILLIAMS C.A., HARBORNE, J.B. and GIESSMANN, S.F. 1983. Flavonoids as taxonomic markers in some Cocosoid palms Plant Syst. Evol. **142 (3-4)**, 157-159.

YOUNG, D.A. Flavonoid chemistry and the phylogenetic relationships of the Julianaceae. *Systematic Botany.* **1**, 149-162.

ZIELSKE, A.G., SIMONS, J.N. and SILVERSTEIN, R.M. 1072. A flavone feeding stimulant in alligator weed. *Phytochemistry.* **11**, 393-396.

* **Original not seen.**

(references included in tables 4,5 and 6 are not listed separately).

5

SYNOPSIS

The family Asteraceae (Compositae) is the largest among the flowering plants. The innumerable representatives of the family on this planet have made it (the family) the most versatile and most enigmatic among all the flowering plants. The family enjoys almost a universal distribution being present practically in almost all geographical areas of the world. The members of Asteraceae have a wide range of adaptation to suit the different ecological niche and can thrive well even in the most inhospitable terrain. Mesophytes, Xerophytes, marsh plants and even eiphytes have their representatives in the family. With such diverse adaptive radiation no wonder then Asteraceae is the largest family of the flowering plants with over 900 genera and 13000 species (Willis 1960). One of the recent works on Asteraceae (Heywood etal Ed. 1977) lists nearly 1400 genera and 25,000 species. Of the total number of flowering plants on earth, the share of Asteraccae is more than 10%

A family as diverse as Asteraceae has attracted the attention of researches since a long time. Having occupied almost the pinnacle of evolution among sympetalous dicotyledons, the family offers a very favourable material for the inquiry into various processes that have characterized the most successful line of evolution.

Classification of the family

The basic classification of the family recognised today is practically same as established by Bentham in his celebrated essay "Notes on the classification, history and geographical distribution of the Compositae" (1873 b). There are 13 tribes in the family as recognised by Bentham 1873 a). These are :-

VERNONIAEAE
EUPATORIAEAE
ASTERAE
INULEAE
HELIANTHEAE
HELENIEAE
ANTHEMIDEAE
SENECIONEAE
CALENDULEAE
ARCTOTIDEAE
CYNAREAE
MUTISEAE and
CICHORIEAE

Carlquist (1966) has surveyed the distribution of certain characters in the tribes of the Asteraceae and has revealed the existence of two rather distinct sub families -

1. ***Sub family cichorioideae***
   Mutisieae, Vernonieae, Cardueae, Arctotideae, Cichorieae and Eupatorieae.

2. ***Sub family Asteroideae***
   Heliantheae, Astereae, Inuleae, Calenduleae, Senecioneae and Anthemideae. Jeffrey (1978) also splits the family into two sub families in a way similar to Carlquist, but suggests that the "tribes Eupatorieae and Senecioneae might warrant seggregation into a third sub family as they are intermediate between others".

There is considerable controversy regarding the phyletic interrelationships and subtribal classification in the family. Small (1919) recognised Senecioneae to be the basic or primitive tribe while Cronquist has proposed a phylogenetic scheme deriving most of the tribes (except Lactuaceae and Calenduleae) from Heliantheae.

The present work deals with the tribes Eupatorieae, Heliantheae and Senecioneae and takes into consideration (a) Foliar Studies, (b) Embryology and (c) Phytochemistry. A total number of 35-50 species representing all these tribes have been studied with reference to the aspects mentioned above. Heliantheae and Senecioneae have been studied in detail while the tribe Eupatorieae (3 genera) has been studied with the limited purpose of finding out whether phytochemical data could be of any use to test the hypothesis of Jeffrey (1978). In the tribes heliantheae and Senecioneae sub tribal classification, mutational variation and hybridization observed in nature have been assesed from the point of view of phytochemistry.

FOLIAR ANATOMY

Foliar anatomical studies provide one of the most reliable evidences for classification and possibly for phytogenetic studies.

In the members of Asteracene included in the present investigation, the results of anatomical studies are presented from the view point of comparative as well as systematic anatomy. Comparative anatomical studies provide an overview of the situation, while systematic anatomy analyses, the level (tribe, genus, species) at which foliar characters can be of diagnostic value.

In the present work three aspects of anatomy viz, nodal anatomy, foliar venation pattern and epidermal morphology are studied.

Nodal Anatomy

Most of the taxa in the present investigation conform to 3 lacunar 3 trace type except for *Gynura aurantiaca* and *Notonia grandilflora*, which show split lateral condition at the foliar nodes.

Petiole anatomy

The members do not present a uniform picture, with regard to the number of leaf trace bundles at different levels in the petiole. 15 of the 29 taxa show 3 traces throughout the petiole, with only the median extending up to the leaf apex while the laterals diverge into the basal part of the lamina. The remaining 14 taxa (out of 29 investigated) show deviations from this common pattern with an amplication in the number of traces resulting often from a branching of the lateral (5 trace condition in 11 taxa) traces or by a branching of both the lateral and median traces in some taxa. *Gynura aurantiaca* shows branching of the lateral traces only, while in *Xanthium indicum* both the lateral and median traces branch with 5 traces extending up to the tip of the leaf. In *Rudbeckia nitida* also both median and lateral traces branch, resulting in a 10 trace conddition, with all the traces extending through the midrib and reach the leaf apex.

Venation pattern
In most of the members, the venation pattern is pinnate and semicraspidodromous or eucamptodromous. In *Notonia grandiflora,* the venation is hyphodromous. In all the taxa, the divergence of secondaries is at an acute angle, the lowest pair being more acute than the rest. Areoles are well developed in all the members, but compactly arranged only in some taxa. Veinlets may be simple, linear, curved or branched once or twice.

Vein density
Vein density is calculated as the total length of the veins in a unit area. Vein density value is low for species of moist and shady habitats while it is high for taxa growing under xeric conditions.

Stomatal/epidermal studies
The members investigated present a univorm picture in having anomocytic type of stoma except for *Notonia grandiflora* of the tribe senecioneae where the stoma is of the anisocytic type.

In the present work variations are met with in the shape of epidermal cells, stomatal frequency and stomatal index.

Systematic considerations
None of the anatomical features studied can be used to distinguish the two tribes Heliantheae and Senecioneae. Even at thesubtribal level (in tribe Heliantheae) the foliar anatomical characters are of little value.

Diagnostic foliar characters in genera and species
At the generic/specific level, foliar anatomical characters prove to be of great significance as they can be relied upon for (a) diagnosis of species, (b) species inter-relationship and (c) analysis of genetic/ecological variation at the intraspecific level.

Foliar anatomical characters clearly demarcate the species in the following genera - *Ageratum, Emilia, Gynura, Notonia, Senecio* and *Spilanthes.*

In *Senecio* where 10 species are studied for stomatal structure, leaves are both hypostomatic as well as amphistomatic. Stomatal character can be correlated with their habit as well as to delimit the species.

EMBRYOLOGY

There is uniformity in the anther development in the members investigated. All of them follow a dicotyledonous type of anther development.

Anthers are bisporangiate in all the members of the tribe Heliantheae, while in Senecioneae - *Emilia sonchifolia* has biosporangiate anthers, while *Senecio ludens* and *S.tenuifolius* have tetrasporangiate anthers.

Endothecium is fibrillar in all the members investigated except in *Emilia sonchifolia.*

Tapetum is plasmodial in all the members investigated. In *Emilia sonchifolia* the tapetal breakdown occurs at a very late stage. Tapetal cells are uninucleate to start with but become multinucleate at a later stage.

Archesporial cells occur in two rows in *Synedrella nodiflora Spilanthes acmella* and *Blainvillea rhomboidea.*

In *Senecio ludens, S. tenuifolius and Emilia sonchifolia* there is a single row of archesporial cells. In all the members studied microsporogenesis is regular and tetrahedral or decussate microspore tetrads are formed. Cytokinesis is simultaneous and wall formation is by furrowing.

Pollen grains are 3 celled at the time of shedding in all the members studied.

Male gamets may be circular or elliptical in shape.

The female archesporium is usually single celled but multicellular archesporium is found in *Spilanthes acmella, S.paniculata, Senecio ludens* and *S.tenuifolius.*

In *Synedrella nodiflora* more than one megaspore shows signs of functioning. In the two species of *Senecio* investigated unusual features like twin tetrads, may functioning megaspores etc. have been noticed.

In *Senecio ludens,* a megaspore mother cell haustorium is noticed.

In *Emilia sonchifolia* a hypostase has been recorded for the first time in Asteraceae.

An endothelium is recorded for all the members investigated. In *Senecio ludens* and *S.tenuifolius* the endothelium is partly biseriate.

Antipodals are either two or three in number. Antipodal nuclei divide and sometimes fuse resulting in a polyploid condition.

Antipodal haustoria are seen in *Senecio ludens* and *Emilia sonchifolia.*

Endosperm is *ab initio* cellular in all the ____________ the tribe Heliantheae, while it is free nuclear for in the members of the tribe Senecioneae.

In all the members investigated embryogenyn conforms to Senecio variation of the Asterad type.

Somatic apospory is found in *Spilanthes acmella.*

In the tribe Senecioneae gonial and somatic apospory is found in *Senecio ludens* and *S.tenuiflius.*

In *S.tenuifolius* aposporous embryo sacs invade the position of the normal embryosac and replace it.

Carpology and Systematics

Fruit wall structure helps in the delimitation of specis studied in the tribe Heliantheae, as well as Senecioneae. *Species categorization in Emilia.*

The two species, *Emilia flammea* and *E.sonchifalia* which are embryologically investigated can be separated based on their embryological characters.

Phyletic appraisal of the tribes Heliantheae and Senecioneae on embryological grounds cannot lead to any conclusion as the data are ambiguous. But it is not possible to indicate that Heliantheae is a more primitive tribe than Senecioneae on embryological grounds.

PHYTOCHEMISTRY

The plants were collected from various parts of Karnataka except for *Notonia walkeri* which was collected from Nilgiri Hills, Tamilnadu.

Flavonoid spot profiles have been used to study the inter and Intra-specific relationship in taxa belonging to the tribes Heliantheae and Senecioneae. The tribe Eupatorieae which has only a few local representatives is studied mainly to understand whether these are of any use at the tribal level. Interpretation of the flavonoid profiles is also attempted at the subtribal level is Heliantheae.

In the tribe Senecioneae, flavonoid profiles are used to assess the intrageneric relationship of species in *Senecio, Notonia, Gynura* and *Emilia.*

In the tribe Eupatorieae four taxa vix. - *Adenostemma lavenia, Chromolaena odorata, Ageratum conyzoides and A. houstonianum* belong to Adenostemmatinae, Praxelis and piqueriinae groups respectively. The Chromotographic patterns are distinct in the four taxa justifying their inclusion in separate groups.

In the genus Ageratum of which two taxa have been investigated (*A. conyzoides and A.houstonianum*) there is distinct chromatographic identity.

Tribe Heliantheae

The tribe Heliantheae is studied for two purposes.

(i) To assess the intra and interspecific variation among some of the genera and
(ii) To analysi how far flavonoid profiles are of assistance in the subtribal delimination. A total number of 26 taxa are investigated for this purpose.

Intra and interspecific variation

Five genera, *Eclipta, Synedrella* (including *Calytocarpus*), *Bidens Galinsoga and Spilanthes* involving 15 taxa are studied to account for the variation found either at the intra or interspecific level.

In *Eclipta prostrata* two variants have been collected (large leaf and small leaf) and they are distinct from the point of view of flavonoid spot pattern. In *Synedrella nodiflora* also a similar instance (large and small leaf) is found and the two forms have distinct Chromatographic pattern.

Calyptocarpus vialis a recent introduction to Karnataka and Bangalore appears to be closer to the small leaf form of *Synedrella nodiflora.*

Quantitative analysis of total phenolics in the two forms of *Synedrella nodiflora* and *Calyptocarpus vialis* also indicates the affinities of *C.vialis* with small leaf form of *S.nodiflora.*

In the genus *Bidens* two species are investigated - *B.pilosa* and *B.humilis.* The two are distinct from the point of view of flavonoid spots.

In several populations of *Bidens pilosa* several rayless mutants have been detected. Here the marginal female florets are present but the corolla becomes actionomorphic just as in disc florets. The rayed and rayless are undergoing natural hybridization producing many intermediate ray sizes pointing to the possibility of a polygenic inheritence.

Flavonoid analysis of rayed, rayless, and intermediate rayed plants shows that the rayed and rayless are closer to each other than between the rayed and intermediate.

An attempt has been made to assess the adaptove potentiality of the rayed and rayless population with reference to the promotion of pollination by insect visitors. The accumulation of anthochlor pigments in the disc florets and their relative absence in the rays renders therayless population being selectively pollinated by insects. Anthochlor pigments are known to act as honey guides and a non anthochlor boundary in the form of rays acts as a hinderance though originally the rays are meant to attract the pollinating agents.

In one of the natural populations of *Bidens pilosa* under observation for the past 10 years. (in Nandi hills) fruit setting is more in the case of rayless than the rayed.

Galinsoga

Introgressive hybridization has been observed between *G.parviflora* and *G.quadriradiata* and it is supported by the flavonoid spot pattern.

Spilanthes

The two species (*S. parviflora and S.calva*) investigated exhibit distinctive flavonoid sport pattern on the Chromatogram. Further in *S.calva,* two forms - a large capitulate type and a small one have been analysed from the point of view of spot pattern. It has been concluded that the two forms are ecotypes.

Sub tribal classification of Heliantheae

The classification of the tribe by Stussey (1977) into 15 sub tribes has been analysed from the members investigated. These belong to the sub tribes Melampodinae, Ecliptinae, Verbesininae, Helianthinae, Coreopsidinae, Galinsogineae and Ambrosineae.

It has been concluded that flavonoid spot pattern is not of much assistance at the sub tribal level either to establish sub tribal affinity or intra tribal relationships.

Tribe Senecioneae

The present work alalyses and interprets flavonoid profiles in terms of intrageneric relationship of the species in *Senecio, Gynura, Notonia* and *Emilia.*

Senecio

The 12 taxa studied in the genus roughly fall into four categories -

Group 1. *S. grahami, S. edgeworthii, S. cineraria* and *S. ludens*
Group 2. *S. lawsoni, S. wightii, S. belgaumensis* and *S. neelgherryanus*
Group 3. *S. corymbosus* and *S.walkesi*
Group 4. *S. hohenackeri* and *S.tenuifolius*

The morphological grouping mentioned above have been analysed in terms of various parameters of numerical systematics based on flavonoid data. These are, paired affinity index (PA), matching coefficient (MC) and biochemical distance. The data obtained largely support the morphological grouping given above. Some other conclusions arrived at on the basis of numerical analysis are -

(i) The values of paired affinity index show very less value and when plotted on a polygraph the shapes of graph between the species show very little similarity.

(ii) The values of matching coefficient based on negative values (i.e., comparison between species based on the common absence of certain characters) show a higher value.

(iii) In a diversified genus such as *Senecio* biochemical distance - a quantitative parameter is not of much assistance in under-standing the interspecific relationship.

Group affinity (GA) in genus Senecio

GA values are based upon the total value for each species (based on PA, MC etc) when compared with all other species of the genus.

Isolation values calculated for all the species show that *S. tenuifolius* and *S. hohenackeri* have the highest isolation values, while *S. wightii, S. belgaumensis* and *S. neelghyrranus* have least isolation values indicating their higher generic affinity. Based on these it has been concluded that *Senecio* is a highly diversified and ancient genus and that the species have only few characters in common among them. This is supported by the great variety in the habit and habitat of the genus.

Gynura

The three species - *G.nitida, G.aurantiaca* and *G.lycopersicifolia* have distinct chromatographic identity as been by the distributian of flavonoid spots.

Notonia

N. Walkeri and *N. grandiflora* which are distinct morphologically exhibit the same at the biochemical level also as seen by flavonoid spot pattern.

Emilia

In the genus *Emilia,* where species are categorized based on very few and overlapping characters, flavonoid distribution serves as an important tool in species delimitation.

The present work involves three species viz., *E.sonchifolia, E.scabra and E.ramulosa.* The flavonoid spot pattern supports their delimitation into distinct species based on morphological data.

6

APPENDIX I

(Tables - General Introduction)

Table 1.1 : Chemical pattern of the Asteraceae*

	Class of compound	Location and biological activity
PRESENT IN ALL TRIBES		
1.	Inulin in-type fructans	in storage organs
2.	Characteristic fatty acids	in seed oils
3.	Sesquiterpene lactonse	mainly in leaves; bitter tasting
4.	Pentacyclic triterpene alcohols	asters in fruit pericarps, and in lipids generally
5.	Caffeic acid esters	in leaves; cynarin is diuretic
6.	Methylated flavonoids	in leaves and flowers (as yellow pigments)
PRESENT IN MOST TRIBES		
7.	Acetylenic compounds	in roots and leaves; antimicrobial activity
8.	Essential oils, including phenolic monoterpenes	in leaves and fruits
9.	Cyclitols	in leaves
10.	Coumarins	in leaves and flowers
PRESENT IN FEW TRIBES		
11.	Rubber (polyisoprene)	in roots and stems
12.	Pyrrolizidine alkaloids	in leaves; highly toxic
13.	Triterpene acids	free in flowers; combined with sugar (as saponin) in leaves
14.	Diterpenes	in all tissues
15.	Cyanogenic glycosides	in leaves and fruits; toxic
16.	Anthochlor pigments	in yellow flowers
17.	Chromenes	in leaves and roots; insecticidal
18.	Fatty acid amides	in roots; insecticidal

* (Adopted from " The Biology and Chemistry of the Compositae. Eds. Heywood *et al.* 1977)

Table No.1.2 : List of Taxa studied

Sl. No.	Name of the species	Petiole anatomy	Stomatal studies	Venation pattern	Embryology	Phyto-Chemistry
1	2	3	4	5	6	7
1.	*Ageratum conyzoides*	+	+	+	–	+
2.	*Ageratum houstonianum*	+	+	+	–	+
3.	*Chromolaena odorata*	–	+	–	–	+
4.	*Adenostemma lavenia*	–	+	+	–	+
5.	*Acanthospermum hispidum*	+	+	+	–	+
6.	*Polymnia grandis*	–	–	–	–	+
7.	*Siegesbeckia orientalis*	–	+	+	–	+
8.	*Blainvillea acmella*	–	–	–	+	+
9.	*Eclipta prostrata* (normal leaf)	+	+	+	+	+
10.	*Eclipta prostrata* (small leaf)	–	–	–	–	+
11.	*Synedrella nodiflora* (normal leaf)	+	+	+	+	+
12.	*Synedrella nodiflora* (small leaf)	–	–	–	–	+
13.	*Calyptocarpus vialis*	+	+	+	–	+
14.	*Lagascea mollis*	+	+	+	–	+
15.	*Sclerocarpus africanus*	+	+	+	–	+
16.	*Helianthus rigidus*	–	–	–	–	+
17.	*Rudbeckia nitida*	+	+	+	–	+
18.	*Bidens pilosa* (normal ray)	+	+	+	–	+
19.	*Bidens pilosa* (Intermediate)	–	–	–	–	+
20.	*Bidens pilosa* (Rayless)	+	+	+	–	+
21.	*Bidens humilis*	–	+	–	–	+
22.	*Moonia heterophylla*	+	+	+	–	+
23.	*Galinsoga quadriradiata*	+	+	+	–	+
24.	*Galinsoga parviflora*	+	+	+	–	+
25.	*Galinsoga* (Intermediate)	–	+	+	–	+
26.	*Spilanthes calva* (small head)	+	+	+	+	+
27.	*Spilanthes calva* (large head)	+	+	+	–	+
28.	*Spilanthes paniculata*	+	+	+	+	+
29.	*Parthenium hysterophorus*	–	–	–	–	+
30.	*Xanthium indicum*	+	+	+	–	+
31.	*Senecio wightii*	–	+	+	–	+
32.	*Senecio lawsnoi*	–	+	–	–	+
33.	*Senecio belgaumensis*	–	–	–	–	+

1	2	3	4	5	6	7
34.	*Senecio neelgherryanus*	–	–	–	–	+
35.	*Senecio tenuifolius*	–	+	–	+	+
36.	*Senecio hohenackeri*	–	+	–	–	+
37.	*Senecio grahami*	–	–	–	–	+
38.	*Senecio edgeworthii*	–	–	–	–	+
39.	*Senecio edgeworthii* (small leaf)	–	–	–	–	+
40.	*Senecio ludens*	+	+	+	+	+
41.	*Senecio cineraria*	–	+	+	–	+
42.	*Senecio corymbosus*	–	+	–	–	+
43.	*Senecio walkeri*	–	+	+	–	+
44.	*Gynura nitida*	+	+	+	–	+
45.	*Gynura lycopersicifolia*	–	–	–	–	+
46.	*Gynura aurantiaca*	+	+	+	–	+
47.	*Notonia grandiflora*	–	+	+	–	+
48.	*Notonia walkeri*	–	–	–	–	+
49.	*Emilia sonchifolia*	+	+	+	–	+
50.	*Emilia scabra*	+	+	+	–	+
51.	*Emilia ramulosa*	–	–	–	–	+

7

APPENDIX II

(Tables - Nodal and Foliar Anatomy)

Table No.2.1 : Number of leaf trace bundles in leaf node continuum at different levels in members of Asteraceae.

(Level 1, node ; Level 2, base of the petiole; Level 3, middle of the petiole; Level 4, base of the lamina; Level 5, middle of the lamina.

	1	2	3	4	5
Ageratum conyzoides	3	5	3	3	1
Ageratum houstonianum	3	3	3	3	1
Adenostemma lavenia	3	3	3	3	1
Emilia scabra	3	5	3	5	3
Emilia sonchifolia	3	3	3	3	3
Gynura nitida	3	5	5	5	3
Gynura aurantiaca	3	many	many	many	1
Notonia grandiflora	3	5	5	5	3
Notonia walkeri	3	5	5	5	1
Senecio ludens	3	3	3	3	1
Senecio walkeri	3	3	3	5	1
Acanthospermum hispidum	3	3	3	3	1
Siegesbeckia orientalis	3	3	3	3	1
Bidens pilosa (normal ray)	3	5	5	5	3
Bidens pilosa (rayless)	3	5	5	5	3
Eclipta prostrata	3	5	5	5	3
Galinsoga quadriradiata	3	3	3	3	1
Galinsoga (intermediate)	3	3	3	3	1
Galinsoga parviflora	3	3	3	3	1
Lagascea mollis	3	3	3	3	1
Moonia heterophylla	3	5	5	5	3
Rudbeckia nitida	3	8	8	10	10

Sclerocarpus africanus	3	7	7	7	3
Spilanthes calva (small head)	3	5	5	5	3
Spilanthes calva (large head)	3	3	3	3	3
Spilanthes paniculata	3	3	3	3	1
Calyptocarpus vialis	3	3	3	3	1
Synedrella nodiflora	3	3	3	3	1
Xanthium indicum	3	7	many	many	5

Table No.2.2 : Venation type and Vein density in different members of Asteraceae

Name of the plant	Vein density per	Venation type
Ageratum conyzoides	16	p. sem.cr.
Ageratum houstonianum	31	- Do -
Adenostemma lavenia	21	- Do -
Emilia scabra	3.5	p. si. cr.
Emilia sonchifolia	2.3	- Do -
Gynura nitida	13	p. sem. cr.
Gynura aurantiaca	10	- Do -
Notonia grandiflora	12	hy.
Notonia walkeri	15	p. en. cam.
Senecio ludens	22	p. sem. cr.
Senecio walkeri	25	- Do -
Acanthospermum hispidum	37	- Do -
Bidens pilosa (normal ray)	24	p. si. cr.
Bidens pilosa (rayless)	38	- Do -
Eclipta prostrata	38	p. sem. cr.
Galinsoga quadriradiata	22	p. en. cam.
Galinsoga (intermediate)	22	- Do -
Galinsoga parviflora	27	p. sem. cr.
Lagascea mollis	27	p. si. cr.
Moonia heterophylla	18	p. sem. cr.
Rudbeckia nitida	52	- Do -
Sclerocarpus africanus	29	- Do -
Siegesbeckia orientalis	26	- Do -
Spilanthes calva (small head)	30	- Do -
Spilanthes calva (large head)	15	- Do -
Spilanthes paniculata	29	p.en. cam.
Calyptocarpus vialis	38	- Do -
Synedrella nodiflora	26	- Do -
Xanthium indicum	37	- Do -

p. sem. cr. = Pinnate semicraspidodromous
p. si. cr. = Pinnate simple craspidodromous
p. cn. cam. = Pinnate eucamptodromous
Hy = Hyphodromous.

Table 2.3 : Various parameters of stomatal structure
(Numbers in brackets indicate standard deviation)

Sl. No.	Taxon		No. of epidermal	No. of stomata	Stomatal frequency	Stomatal index	Length of stomata in mm	Breadth of stomata in mm	L/B ratio
1	2		3	4	5	6	7	8	9
1.	*Ageratum conyzoides*	U	–	–	–	–	–	–	–
		L	23	14	60.87	37.84	28.79	20.59 (0.523)	1.40 (0.125)
2.	*A. houstonianum*	U	17	4	23.53	19.05	45.78 (1.161)	18.45 (0.558)	2.48
		L	28	12	42.86	30.00	46.47 (1.289)	21.64 (0.728)	2.15
3.	*Adenostemma lavenia*	U	–	–	–	–	–	–	–
		L	74	24	32.43	24.40	32.34 (0.626)	19.13 (0.720)	1.19
4.	*Chromolaena odorata*	U	69	2	2.90	2.82	28.97 (0.144)	24.83 (0.172)	1.17
		L	39	27	69.23	40.91	33.21 (0.158)	28.43 (0.469)	1.17
5.	*Emilia scabra*	U	15	3	20.00	16.67	30.98 (0.584)	21.09 (0.773)	1.47
		L	28	18	64.29	39.13	27.01	19.32	1.40

							(0.606)	(0.560)	
6.	*E. sonchifolia*	U	24	11	45.83	31.43	34.39 (1.358)	18.45 (1.010)	1.86
		L	28	25	89.29	47.17	29.20 (1.056)	21.24 (0.465)	1.38
7.	*Gynura nitida*	U	15	2	13.33	11.76	57.54 (1.082)	32.66 (0.193)	1.76
		L	12	7	58.33	36.84	44.83 (0.802)	34.44 (0.709)	1.30
8.	*G. aurantiaca*	U	12		16.67	14.29	46.01 (1.020)	29.11 (0.515)	1.58
		L	12		50.00	33.33	40.95 (0.932)	31.25 (0.515)	1.31
9.	*Notonia grandiflora*	U	11	2	18.18	15.38	55.58 (1.136)	36.90 (1.289)	1.51
		L	10	1	10.10	9.09	51.02 (1.092)	37.81 (1.252)	1.35
10.	*N. walkeri*	U	–	–	–	–	–	–	–
		L	55	13	23.64	19.12	37.61 (0.892)	25.51 (0.727)	1.48
11.	*Senecio cineraria*	U	–	–	–	–	–	–	–
		L	17	10	58.82	37.04	35.99 (0.498)	29.02 (0.742)	1.24
12.	*S. corymbosus*	U	–	–	–	–	–	–	–

		L	22	5	22.73	18.52	32.75	29.29 (0.120)	1.42 (0.242)
13.	*S. honenackeri*	U	27	8	29.63	22.86	36.85 (0.187)	29.06 (0.277)	1.27
		L	29	18	62.07	38.30	37.17 (0.129)	24.69 (0.125)	1.51
14.	*S. lawsnoi*	U	19	9	47.37	32.14	34.12 (0.534)	21.41 (0.431)	1.59
		L	19	9	42.86	30.00	29.02 (0.168)	27.83 (0.475)	1.04
15.	*Senecio ludens*	U	96	32	33.33	25.00	31.07 (0.586)	22.78 (0.555)	1.36
		L	87	29	33.33	25.00	21.18 (0.232)	20.09 (0.144)	1.05
16.	*S. neelgherryanus*	U	17	4	23.53	19.05	51.66 (0.565)	40.13 (0.755)	1.29
		L	17	9	52.94	34.62	45.28 (0.961)	32.62 (0.259)	1.39
17.	*S. candicans*	U	–	–	–	–	–	–	–
		L	51	22	43.14	30.14	38.49	22.09	1.74
18.	*S. tenuifolius*	U	19	8	42.11	29.63	40.27	25.28	1.59
		L	23	9	39.13	28.13	42.96 (0.675)	32.20 (0.622)	1.33

19.	*S. walkeri*	U	–	–	–	–	–	–	–
		L	110	27	24.55	19.71	28.93 (0.789)	21.18 (0.322)	1.37
20.	*S. wightii*	U	21	1	4.76	4.55	35.72 (0.866)	20.36 (0.232)	1.75
21.	*Acanthospermum hispidum*	U	61	19	31.15	23.75	33.48 (0.852)	21.41 (0.859)	1.56
		L	59	26	44.07	30.59	35.31 (1.319)	20.73 (0.879)	1.73
22.	*Bidens humilis*	U	18	5	27.78	21.74	35.03 (0.543)	25.56 (0.872)	1.37
		L	28	8	28.57	22.22	33.44 (0.898)	21.09 (0.274)	1.59
23.	*B. Pilosa* (normal ray)	U	16	10	62.50	38.46	48.52 (1.019)	31.98 (0.587)	1.52
		L	18	13	72.22	41.94	48.11 (1.471)	31.11 (1.283)	1.55
24.	*B. pilosa* (rayless)	U	17	8	47.06	32.00	41.87 (0.808)	21.73 (1.177)	1.93
		L	22	18	81.82	45.00	40.00 (1.074)	26.60 (0.901)	1.50
25.	*Eclipta prostrata*	U	16	8	50.00	33.33	52.89 (1.063)	31.57 (0.511)	1.68
		L	20	14	70.00	41.18	34.30	22.41	1.53

							(0.565)	(0.553)	
26.	*Galinsoga quadriradita*	U	–	–	–	–	–	–	–
		L	14	8	57.14	36.36	32.57 (0.668)	20.23 (0.295)	1.61
27.	*Galinsoga* (Intermediate)	U	–	–	–	–	–	–	–
		L	57	13	22.81	18.57	28.70 (1.205)	16.86 (0.626)	1.70
28.	*G. parviflora*	U	–	–	–	–	–	–	–
		L	17	10	58.82	37.04	31.66 (0.459)	20.14 (0.208)	1.57
29.	*Lagascea mollis*	U	62	13	20.97	17.33	35.53 (0.852)	18.68 (0.752)	1.90
		L	90	19	21.11	17.43	34.39 (1.105)	20.27 (0.387)	1.70
30.	*Moonia heterophylla*	U	–	–	–	–	–	–	–
		L	16	6	37.50	27.27	33.35 (0.971)	20.86 1.172)	1.60
31.	*Rudbeckia nitida*	U	–	–	–	–	–	–	–
		L	101	18	17.32	15.13	35.75 (0.859)	20.50 (0.456)	1.74
32.	*Sclerocarpus africanus*	U	59	15	25.42	20.27	34.17 (0.852)	19.13 (0.558)	1.79
		L	60	17	28.33	22.08	37.13 (1.224)	19.13 (0.72)	1.94
33.	*Siegesbeckia orientalis*	U	69	8	11.59	10.39	38.49	19.59	1.96

							(0.774)	(0.340)	
		L	40	20	50.00	33.33	35.31 (0.505)	20.18 (0.230)	1.75
34.	*Spilanthes calva* (small head)	U	15	7	46.67	31.82	39.41 (1.105)	24.24 (0.974)	1.63
		L	12	8	66.67	40.00	39.45 (1.060)	27.33 (1.476)	1.44
35.	*S. calva* (large head)	U	12	1	8.33	7.69	42.91 (0.762)	22.09 (0.740)	1.94
		L	16	7	43.75	30.43	42.59 (1.280)	28.34 (0.785)	1.50
36.	*S. paniculata*	U	16	2	13.30	11.76	42.78 (0.550)	28.87 (0.896)	1.96
		L	15	9	60.00	37.50	42.37 (1.044)	21.68 (0.560)	1.95
37.	*Synedrella nodiflora*	U	–	–	–	–	–	–	–
		L	17	3	17.65	15.00	36.92 (0.663)	19.45 (0.877)	1.38
38.	*Calyptocarpus vialis*	U	–	–	–	–	–	–	–
		L	14	4	38.57	22.22	23.23 (1.116)	15.94 (0.628)	1.46
39.	*Xanthium indicum*	U	23	10	43.48	30.30	35.08 (1.136)	25.97 (0.911)	1.35
		L	16	9	56.25	36.00	37.2 (1.384)	27.70 (1.063)	1.34

8

APPENDIX III

(Tables - Embryology)

Table 3.1 : Embryological features of the tribe Heliantheae-Structure and development of Anther and Pollen
(Adopted from Embryology of Compositae - Pullaiah, (1984)

Species investigated and author	Archesporium	Wall layers	Tapetum	Sporagenous tissue	Cytokinesis	Tetrads	Pollen grains
(1)	(2)	(3)	(4)	(5)	(6)	(7)	(8)
Eclipta erecta (Bharagava, 1935	Single row	3 wall layers (Endothecium, middle layer, tapetum)	Periplasmodial, dial, nuclear divisions and fusions occur	Two rows	By furrowing	Tetrahedral or isobilaeral eral	3-celled with 3 germ pores, some times double triple & quardruple pollen grains occur, pollen degenerations are common
Tridas procumbens (Banherji, 1940a, c; Raghavan & Venkata Subban, 1941; Rogers, 1979a)	-	"	Periplasmodial 2 Nucleate	Sometimes PMCs occur	"	"	3-celled with spinous exine 4 germ pores

(1)	(2)	(3)	(4)	(5)	(6)	(7)	(8)
Rudbeckia bicolor (Maheshwari & Srinivasan, 1944)	-	-	Periplasmodial Divisions and fusions of nuclei occur	Moderately extensive	"	"	3-celled
Synedrella nodiflora (Banerji & Pal, 1959)	Single row	3 layers (endothecium, tapetum	Periplasmodial, cells 4 nucleate	2 rows	"	Mostly tetrahedral	3-nucleate echinate, ellipsocial and tetracolpate.
Parthenium hysterophorus (Deshpande, 1960b; Pullaiah, 1982a)	Single row	Wall 3 layers (Endothecium, middle layer and tapetum)	Periplasmodial cells 2 nucleate	4 rows	"	Tetrahedral and isobilateral lateral	3-nucleate with 3 germ pores
Acanthospermum hispidum (Vaidya, 1961, 1973; Pullaiah, 1976)	"	"	Periplasmodial	Moderate sized mass	"	2	3-celled at anther dehiscence
Glossocardia bosavallea (Deshpande, 1962d; Pullaiah, 1976	"	"	Periplasmodial 2 large, barrel shaped polyploid nuclei	2 rows	"	Tetrahedral or isobilateral	Tricoplate and 3 celled
Wedelia calendulaceae (Ghosh, 1962)	"	"	Periplasmodial cells 4 nucleate	Moderately extensive	"	"	Spherical, echinate and 3-nucleate, pollen dimorphism.
Coreopsis tinctoria (Maheswari devi, 1963)	"	"	Periplasmodial	"	"	"	3-celled and triporate at the time of shedding
Bidens pilosa (Maheswari Devi 1963)	"	"	"	Single row	"	"	"
Xanthium strumarium (Maheswari Devi 1963)	"	"	"	Moderately extensive	By furrowing	Tetrahedral or isobilateral	3 celled and triporate at the time of shedding

(1)	(2)	(3)	(4)	(5)	(6)	(7)	(8)
Zinna pauciflora (Maheswari Devi, 1963)	Single row	3 layers (Endothecium middle layer, tapegum)	Periplasmodial	Moderately extensive	By furrowing	Tetrahedral or isobilateral	3 celled and triporate at the time of shedding
Eclipta erecta (Maheswari Devi, 1963)	"	"	"	"	"	"	"
Cosmos sulphureus (Maheswari Devi, 1963)	"	"	Periplasmodial Divisions and fusions of nuclei result in large polyploid nuclei	"	"	"	"
Helianthus annuus (Maheswari Devi, 1963)	"	"	Periplasmodial	"	"	"	"
H. debilis	"	"	Periplasmodial, cells polyploid	"	"	"	"
Bidens biternata (Deshpande, 1964b)	"	"	Periplasmodial, 2 large polyploid nuclei are formed.	Two rows	"	"	Tricolpate and 3-celled
Siegesbeckia orientalis (Misra, 1965)	-	3 layers (Endothecium,	Periplasmodial, 2 large polyploid nuclei are formed	2 or 3	By furrowing	Tetrahedral or decussate or isobilateral	Spheroidal, tricolpate and 3-celled . Pollen degenaraions are common.
Lecocarpus pinnatifidus (Eliasson, 1971)	-	"	Periplasmodial, cells 2-4 nucleate	-	-	Tetrahedral	Spheroidal and triporate; may shed at 2 celled stage.
Melampodium paludosum (Eliasson, 1971)	-	"	Periplasmodial	-	-	-	Spheroidal and triporate.

(1)	(2)	(3)	(4)	(5)	(6)	(7)	(8)
Blainvillea rhomboidea (Sundara Rajan 1972)	Two rows	"	Periplasmodial, cells 2 nucleate	2 rows	-	Tetrahedral or isobilateral or decussate	3-celled
Macreae laricifolia (Elisaon, 1972)	-	"	Periplasmodial, Nuclear divisions occur	-	-	-	Spheroidal, triporate and 3 celled
Spilanthes acmella (Mukherjee & Kothare, 1973a; Sundara Rajan, 1947a)	Single row	"	"	2 rows	By furrowing	Isobilateral or tetrahedral	3-celled spherical, tri or tetracolpate
Scalesia spp. (Eliasson, 1974)	-	"	Periplasmodial 2 large synkaryons are formed	-	-	Tetrahedral	Spheroidal, triporate with spinous exine. 3-celled.
Eclipta alba (Sundara Rajan, 1974b)	-	3 layers (Endothecium, middle layer, tapetum)	Periplasmodial type cells become binucleate	4 rows	-	Tetrahedral	Spheroidal, triporate with spinous exine.
Guizotia abyssinica (Chopra & Singh 1974; Maheswari Devi & Pullaiah, 1976a)	Single row of 5-8 cells	"	Periplasmodial, celle become multinucleate and polyploid	"	By furrowing	Decussate and tetrahedral	3-celled with thick spinous exine
Melampodium divaricatum (Maheswari Devi & Pullaiah, 1976b)	"	"	"	"	"	"	3-celled and tricolpate with thick spinous exine, Pollen degenerations seen.
Tithonia rotundifolia (Pullaiah, 1978b)	-	"	Periplasmodial, celled become polyploid	"	"	"	3-celled with thick spinous exine and 3 germ pores

(1)	(2)	(3)	(4)	(5)	(6)	(7)	(8)
Melampodium cinereum (Maheswari Devi & Pullaiah, 1980)	Single row of 5-8 cells	"	Periplasmodial, cells become multinucleate andpolyploid	Moderate sized mass	"	Tetrahedral or isobilateral	3-celled with thick spinous exine
Lagascea mollis (Pullaiah, 1981a)	"	"	"	"	"	"	3-celled with thick spinous exine. Pollen polymorphism is seen
Zinnia angustifolia (Pullaiah, 1981a)	"	"	"	"	"	"	"
Galinsoga parviflora (Pullaiah, 1981a)	"	"	"	2 rows of cells	"	"	"
Coreopsis lanceolaria	"	"	"	Moderate sized mass	"	"	"

Table 3.2 : Embryological features of tribe Heliantheae-Structure and development of ovule and female gametophyte
(Adopted from Embryology of Compositae - Pullaiah, 1984).

Species investigated and author	Ovule	Archesporium	Megaspore tetrad	Type of embryo sac development	Synergids	Antipodals
(1)	(2)	(3)	(4)	(5)	(6)	(7)
Helianthus annus (Hofmeister, 1849; Carano place 195: a, 1921; Maheswari 1964; New Comb, 1973a)	Integumentary Vascular Traces present	Unicellular	Linear, chalazal functional	Polygonum type	Hooked synergids	Two antipodals, nuclear divisions and fusions take forming polyploid nuclei
Helopsis laeris (Hegelmaier, 1889)	-	-	-	"	"	-
Silphium laciniatum (Lund, 1900)	-	-	-	"	-	-
S. perfoliatum (Merrel, 1900)	-	-	-	"	-	-
S. integrifolium (Merrel, 1900)	-	-	-	"	-	-
S. terebinthaceum (Merrel, 1900)	-	-	-	"	-	-
S. trifoliatum (Merrel, 1900)	-	-	-	"	-	-
Parthenium aregentatum	-	-	Linear tetrad	"	-	-
P. inanum (Kirkwood, 1910)	-	-	Linear tetrad	Polygonum type	-	-
Dahlia coronata	-	1-celled •	Linear tetrad chalazal functional	"	-	3-uninucleate antipodals

(1)	(2)	(3)	(4)	(5)	(6)	(7)
Cosmidium burridgeanum (Tackholm 1916)	-	-	-	"	-	-
Bides tripartitus (Dahigren, 1920)	-	-	-	"	-	-
Zinnia grandiflora (Plam, 1931)	-	-	Linear tetrad, chalazal functional	"	-	Two antipodals, multinucleate and polyploid.
Eclipta erecta (Bhargava, 1953)	Antaropus unitegmic and tenuinucellate	Unicellular	Linear, chalazal functional	"	Pear shaped	Usually 3, two nucleate antipodals
Galinsoga ciliata (Popham, 1938)	Anatropous, unitegmic and tenuinucellate	Unicellular	Linear, chalazal functional	Polygonum type	Large and angular	Two to four antipodals, nuclear divisions observed.
Tridax procumbens (Banerji, 1940a; Maheswari & Roy 1952; robers, 1969b)	"	"	"	"	Pear shaped and possess prominent beak like structure	Two to three antipodals with two to three nuclei
Rudbeckia bicolor (Maheswari and Srinivasan, 1944)	Anatropous, unitegmic and tenuinucellate	Unicellular	Coenomegaspore	Fritillaria type	Hooked	Three antipodals resemble egg apparatus.
Enhydra flucturans (Mitra, 1947)	"	"	Linear, chalazal functional	Polygonum type	"	Three binucleate antipodals.
Tridax trilobata (Hjelmqvist, 1951)	"	"	3 cells are formed of which the middle contains nuclei and the cells towards poles contain one nucleus	Polygonum & Allium types occur in the same species 2	-	Three binucleate antipodals.

(1)	(2)	(3)	(4)	(5)	(6)	(7)
Synedrella nodiflora (Banerji & Pal, 1959; Sundara Rajan, 1969b)	"	Unicellular sometimes multicellular	Linear, chalazal functional	Polygonum type	Elongated smooth synergids	Usually 3 uninucleate, rarely multinucleate.
Parthenium hysterophorus (Deshpande, 1960b; Pullaiah, 1982b)	"	Unicellular	"	"	Pear shaped and possess beak like structure	3 antipodals upper 2 become binucleate.
Acanthospermum hispidum (Vaidya, 1961, 1973; Pullaiah, 1981b)	"	"	"	"	Hooked	2 or 3 antipodals
Rudbeckia missourensis (Cesca, 1961)	Anatropous, unitegmic and tenuinucellate	Unicellular	Coenomegaspore	Fritillaria type	Hooked	3 antipodals resemble egg apparatus
Savitalia procumbens (Hjelmqvist & Holmberg, 1961)	"	"	Linear, chalazal functional	Polygonum and Allium types occur in the same species	-	-
Wedelia calendulacea (Ghosh, 1962)	"	"	"	Polygonum type	Pear shaped with like processes	2 or 3 antipodals with variable number of nuclei
Glossocardia bosvallea (Deshpande, 1962d; Pullaiah, 1976a)	Anatropous, unitegmic and	Unicellular	"	"	Synergids project into the micropyle	2 or 3 antipodals, with variable number of nuclei
Coreopsis tinctoria (Maheswari Devi, 1963)	"	"	"	"	Hooked synergids	Antipodal haustoria are present
Bidens pilosa	"	"	"	"	"	"
Eclipta erecta (Maheswari Devi, 1963)	"	"	"	"	"	3 antipodals,become multinucleate.

(1)	(2)	(3)	(4)	(5)	(6)	(7)
Xanthium strumarium (Maheswari Devi, 1963;	Branched integumentary vascular traces present	"	"	"	"	3 antipodals become nucleate
Zinna pauciflora (Maheswari Devi, 1963)	Branched integumentary vascular traces present	Unicellular	Linear, chalazal functional	Polygonum type	Hooked synergids	3 antipodals become mutlinucleate
Cosmos sulphureus (Maheswari Devi, 1963)	Integumentary present	"	"	"	"	2 antipodals become multinucleate and haustorial
Helianthus debilis	"	"	"	"	"	2 antipodals, nuclear divisions and fusions take place forming polyploid nuclei.
Bidens biternata (Deshpande, 1946b)	Anatropous, unitegmic and tenuinucellate	"	"	"	Pear shaped with beak like processes	2 or 3 antipodals with variable number of nuclei.
Siegesbeckia oreintalis (Misra, 1965)	Integumentary	"	"	"	Hooked	Two antipodals
Guizotia abyssinica Chopra, 1968a; Chopra & Singh, 1976; Maheswari Devi & Pullaiah, 1976a; Pullaiah 1976)	"	"	"	"	"	2 or 3 antipodals persist till a heart-shaped embryo is formed.
Lecocarpus pinnatifidus (Eliasson, 1971)	Anatropous, unitegmic and tenuinucellate	Unicellular	Linear chalazal functional	Polygonum type	Pear shaped	2 or 3 antipodals with 1 or 2 nuclei

(1)	(2)	(3)	(4)	(5)	(6)	(7)
Melampodium paludosum (Eliasson, 1971)	"	"	-	-	"	"
Blainvillea rhomboidea (Sundara Rajan, 1972)	"	"	Linear chalazal functional	Polygonum type	-	2 antipodals with variable number of nuclei in them.
Macraea laricifolia (Eliasson, 1972)	"	"	"	"	-	2 or 3 antipodals synkaryons formed
Spilanthes acmella (Mukherjee & Kothare, 1973; Sundara Rajan, 1974)	"	Usually single celled, occasionally 2 celled	"	"	Pear shaped	Varying number of cells with varying number of nuclei in them.
Scalesia app. (Eliasson, 1974)	"	Usually single celled, sometimes multicelled	"	"	-	Usually 2 or 3 sometimes increase upto 6. Due to nuclear divisions and fusions synkaryons are formed.
Melampodium divaricatum (Maneswari Devi &	"	1-celled of multicelled	Linear or isobilateral tetrads, chalazal or rarely both chalazal and micropylar functional	"	Hooked	Antipodal cells increase upto 8 in number become polyploid
Coreposis lanceolaria (Pullaiah, 1976)	Antropous, unitegmic & tenuinucellate	1-celled	Linear tetrad, chalazal functional	Polygonum type	Beak shaped	2 cells. Haustorial in nature. Cells become multinucleate.
Tithonia rotundifolia (Pullaiah, 1987b)	"	"	Linear or isobilateral, chalazal functional. Sometimes more than 1 megaspore functional	"	Hooked, persist upto the 2-celled embryo stage	Antipodal cells increase in number Multinucleate and polyploid cells are also observed

(1)	(2)	(3)	(4)	(5)	(6)	(7)
Melampodium cinereum (Mehaswari Devi & Pullaiah, 1980)	"	1-celled or multicelled	Linear tetrads chalazal functional	"	Hooked	Antipodal cells increase in number & cells become multinucleate and polyploid.
Lagascea mollis (Pullaiah, 1981a)	"	"	"	"	"	"
Zinna angustifolia	"	"	"	"	"	2 or 3 antipodal cells, multinucleate.
Galinsoga parviflora (Pullaiah, 1981a)	"	1-celled	"	"	"	2 or 3 antipodal cells.

Table 3.3 : Embryological characters of Heliantheae - *Fertilization, Endosperm and Embryo*
(Adopted from Embryology of Compositae - Pullaiah, 1984).

Species investigated and author	Fertilization	Endosperm development	Embryo development	Apomixis	Polyembryony
(1)	(2)	(3)	(4)	(5)	(6)
Helianthus annus (Hofmeister, 1849; Carano, 1951a, 1921; Dahlgren, 1924; Maheswari Devi, 1963; New Comb, 1973b)	Normal	Nuclear	*Senecio* variation of the Asterad type	-	-
Dahlia coronata (Palm, 1915	-	"	-	-	-
Bidens tripartita (Dahlgren, 1920)	-	"	-	-	-
Dahlia merckii (Dahlgren, 1924)	-	"	-	-	-
Coreopsis bicolar (Gelin, 1934)	-	-	-	Hieracium type of somatic apospory	-
Eclipta erecta (Bhargava, 1935; Maheswari Devi, 1963)	Normal	Cellular	*Senecio* variation of the Asterad type	-	-
Galinsoga ciliata	"	Nuclear	Agrees with Asterad type	-	-
Tridax procumbens (Maheswari and Roy, 1952)	"	Nuclear, in 1 case primary endosperm nucleus (?) remains undivided	*Senecio* variation of Asterad type	- -	In one case additional embryo probably from integumentary tapetum is recorded

(1)	(2)	(3)	(4)	(5)	(6)
Parthenium argentatum (Kirkwood, 1910; Dianova et al., 1934, 1935; Easu, 1944, 1946)	Normal	Cellular	Agrees with Asterad type	-	-
P. argentatum (2n - 72) (Easu, 1944, 1946)	No fertilisation	Autonomous	"	Antennaria (alpina) type of gonial apospory is more frequent while Hieracium type of somatic apospory also occurs	-
Rudbeckia laciniata (Battaglia, 1945, 1946c. d, 1947a, b; Fagerlind, 1946; Movsesian, 1964; Solntseva, 1973)	No fertilisation	Autonomous	Unreduced partheno genetic. In some cases semigamy is observed	Various types of Apomixis like Taraxacum, Ixeris, Rudbeckia type I, Type II 8 and 4 nucleate, and Rudbeckia type III occur, Sometimes 2 types co-exist, Semgamy is observed	Synergid embryo is observed in addition to normal embryo
R.deami (Battaglia, 1952 b, 1955b)	-	-	-	Ixeris type of Aneuspory	-
R.triloba (Battaglia, 1952b, 1955b)	-	-	-	"	-
R.sullivanti (Battaglia, 1952b, 1955b)	-	-	Unreduced partheno genetic. In some cases semigamy is recorded. Asterad type	Ixeris type of Aneuspory, semigamy are observed	Semigamous antipodal embryo recorded
R.missouriensis (Cesca, 1961)	Normal	Nuclear, pentaploid	Asterad type	-	-
Glossocardia bosvallea (Deshpande, 1962d; Pullaiah, 1976)	"	Nuclear	"	-	-

(1)	(2)	(3)	(4)	(5)	(6)
Tridax trilobata (Kapil & Sethi, 1962b)	"	"	"	-	-
Wedelia calendulaceae (Ghosh, 1962)	"	"	"	-	-
Coreopsis tinctoria (Maheswari Devi, 1963)	"	"	Senecio variation of Asterad type	-	-
Bidens pilosa (Maheswari Devi, 1963)	"	Nuclear, in 1 case the primary endosperm nucleus remains undivided	"	-	-
Xanthium strumarium (Crete 1955; Maheswari Devi, 1963)	"	Nuclear	"	-	-
Zinna pauciflora (Maheswari Devi, 1963)	"	"	"	-	-
Helianthus debilis (Maheswari Devi, 1963)	"	"	"	-	-
Cosmos sulphureus (Maheswari Devi, 1963)	"	"	"	-	-
Bidens biternata (Deshpande, 1964b)	Normal	Nuclear, in 1 case the primary endosperm nucleus remains undivided	*Senecio*, variation of Asterad type	-	-
Siegesbeckia orientalis (Misra, 1965)	"	"	"	-	-
Blainvillea rhomboidea (Sundara Rajan, 1972)	"	Cellular	"	-	-

(1)	(2)	(3)	(4)	(5)	(6)
Chrysanthellus indicum (Deshpande & Kothare, 1973)	"	"	"	-	-
Spilanthes acmella (Mukherjee & Kothare, 1973; Sundara Rajan 1974a)	"	"	"	-	-
Guizotia abyssinica (Chopra & Singh, 1976; Maheswari Devi & Pullaiah, 1976a; Pullaiah, 1976)	Porogamous	Nuclear	"	-	-
Melampodium divaricatum (Maheswari Devi &	Normal fertilisa-tion Nuclear 10-15% of the ovules type only polyspermy in area cases	Nuclear	*Senecio* variation of Asterad type	Somatic apospory of the Hieracium type, adventive embryony has been observed	-
Tithonia rotundifolia (Pullaiah, 1978b)	Porogamous normal	Nuclear type	*Senecio* variation of Asterad type	Somatic apospry of the Hieracium type, adentive embryony has been observed	-
Melampodium cinereum (Maheswari Devi & Pullaiah, 1980)	"	Nuclear	"	-	-
Legascea mollis (Pullaiah, 1981a)	"	-	"	-	-
Zinna angustifolia (Pullaiah, 1981a)	"	-	"	-	-
Galinsoga parviflora (Pullaiah, 1981b)	"	Cellular	"	-	-

(1)	(2)	(3)	(4)	(5)	(6)
Acanthospermum hispidum (Pullaiah, 1981b)	"	Nuclear	"	-	-
Parthenium hysterophorus (Pullaiah, 1982b)	"	Cellular	"	-	-

Table 3.4 : Embryological features of the tribe Senecioneae Structure and development of Anther and Pollen
(Adopted from Embryology of Compositae - Pullaiah, (1984)

Species investigated and author	Archesporium	Wall layers	Tapetum	Sporagenous tissue	Cytokinesis	Tetrads	Pollen grains
(1)	(2)	(3)	(4)	(5)	(6)	(7)	(8)
Erechtites hieracifolia (G.O. Cooper, 1963b)	Single row	3 layers (endothecium, middle layer and tapetum	-	Single row	By furrowing	Tetrahedral	3-celled with thick exine and 3 germ pores
Emilia sanchifolia (Sundara Rajan, 1978a)	-	No fibrous thickenings in the endothecium	Periplasmodial, cells uninucleate	"	"	"	2-celled and triporate at the time of shedding
E. sonchifolia (Pullaiah, 1976)	Singlr row	3-wall layers (endothecium, middle layer and tapetum	Periplasmodial, binucleate	"	"	Tetrahedral and isolbilateral	3-celled and echinate with 3 germ pores
Gynura nitida (Pullaiah, 1976)	Single row of 6-8 cells	"	"	Moderate sized mass	"	"	"
Notonia grandiflora (Pullaiah, 1976)	"	"	Periplasmodial, cells become polyploid	"	"	"	Pollen grains in uninucleate condition.
Emilia flammea (Prakasa Rao *et al.*, 1979)	"	"	Secretory , tapetum cells become polyploid	"	"	Tetrahedral, isobilateral and decussate	2 or 3-celled and 3 or 4 zonicolpate

Table 3.5 : Embryological features of tribe Senecioneae-Structure and development of Ovule and Female gametophyte
(Adopted from Embryology of Compositae - Pullaiah, 1984).

Species investigated and author	Ovule	Archesporium	Megaspore tetrad	Type of embryo sac development	Synergids	Antipodals
(1)	(2)	(3)	(4)	(5)	(6)	(7)
Senecio vulgaris (Strassburger, 1879); Small, 1919; Carano, 1951b; 1921; Winge, 1913); Dahlgren, 1920; Afzelius, 1924)	Anatropous, unitegmic and tenuinucellate	Unicellular	Linear, chalazal functional	Polygonum type	-	3 antipodals. The chalazal later divides to form 4 cells of which Chalazal 1 is haustorial cells become multinucleate.
Senecio crassifolius (Guignard, 1882)	-	1-celled	Linear tetrad, chalazal functional	"	-	-
Tussilago farfara (Guignard, 1882; Schurhoff, 1920; Langlet, 1925)	-	-	-	"	-	-
Senecio aureus (Mottier, 1893)	Antropous, unitegmic and tenuinucellate	1-celled	Linear tetrad, rad, chalazal functional	"	Pear shaped	Number of cells upto 6; cells become multinucleate and polyploid.
Emilia saqittata (Palm, 1915; Afzelius, 1924)	-	"	"	"	-	"
E. amplexicaulis (Palm, 1915)	-	"	Linear tetrad	"	-	-
Senecio disciflolius (Afzelius, 1924)	Antropous, untegmic and tenuinucellate	1-celled	Linear tetrad, chalazal functional	Polygonum type	-	-
S. nebrodensis (Afzelius, 1924)	"	"	"	"	-	2 antipodal cells, both binucleate

(1)	(2)	(3)	(4)	(5)	(6)	(7)
S. abrotanifolius (Afzelius, 1924)	"	"	Linear or middle megaspore side by side	"	-	-
S. jacobaea (Afzelius, 1924)	"	"	Linear	"	-	-
S. erucifllius (Afzelius, 1924)	"	"	"	"	-	20-40 antipodals.
S. cineraria (Afzelius, 1924)	"	"	"	"	-	upto 20 antipodal cells, sometimes multinucleate
S. fuchsii (Afzelius, 1924)	"	"	"	"	-	-
S. subalpinus (Afzelius, 1924)	"	"	"	"	-	-
S. heriteri (Afzelius, 1924)	Anatropous, unitegmic & tenuinucellate	1-celled	Linear or T. shaped	Polygonum type	-	-
S. roberti-friesii (Afzelius, 1924)	"	"	Linear tetrad	"	-	-
Crassocephalum crepidodes (Afzelius, 1924)	"	"	"	"	-	-
C. rubens (Afzelius, 1924)	"	"	"	"	-	-
Culcitum reflexum (Afzelius, 1924)	"	"	"	"	-	-

(1)	(2)	(3)	(4)	(5)	(6)	(7)
Arnica alpina (Afzelius, 1924)	"	-	No reduction division	Apomictic antenaria type of Gonial apospory	-	-
Erechtites hieracifolia (G.O. Cooper, 1936)	"	Unicellular	Linear, chalazal functional	Polygonum type	Club shapped	3 antipodals, become binucleate
Senecio leucanthemum (Bachetti, 1961)	-	"	"	"	-	3 antipodals, become polyploid
Emilia sonchifolia (Sundara Rajan, 1978a)	Anatropous, unitegmic and	"	"	"	Spindle shaped without any hooks or breaks	3 uninucleate antipodals
E. sonchifolia (Pullaiah, 1976)	"	1-celled	"	"	Pear shaped	Number of cells increases upto 6. Cells become multinucleate and polyploid.
Gynura nitida (Pullaiah, 1976)	Anatropous, unitegmic & tenuinucellate	1-celled	Linear chalazal functional	Polygonum type	Hooked	Cells become multinucleate and polyploid
Notonia grandiflora (Pullaiah, 1976)	"	"	"	"	'	"

Table 3.6 : Embryological characters of Senecioneae- *Fertilization, Endosperm and Embryo*
(Adopted from Embryology of Compositae - Pullaiah, 1984).

Species investigated and author	Fertilization	Endosperm development	Embryo development	Apomixis	Polyembryony
(1)	(2)	(3)	(4)	(5)	(6)
Senecio aureus (Motteir, 1893	Porogamous normal	Nuclear	Resembles *Senecio* variation of the Asterad type	-	-
Senecio vulgaris (Dahlgren, 1920; Soueges, 1920a, b)	-	"	Asterad type	-	-
Erechtites hieracifolia (G.O. Copper, 1936)	Normal	-	Agrees with Asterad type	-	-
Arnica alpina ssp. *alpina* (Afzelius, 1936)	No fertilisation	Autonomous	Unreduced parthenogenetic	Antennaria (alpina) type of gonial apospory	-
A.chamissonia (Afzelius, 1936)	"	"	"	"	-
A. diversifolia (Afzelius, 1936)	"	"	"	"	-
A. alpina ssp. *angustifolia* (Engell, 1970)	"	Autonomous cellular	Unreduced parthenogenetic. *Senecio* varaition of Asterad type	"	-
Tussilago farfara	Normal	Cellular	*Senecio* variation of Asterad type		
Emilia sonchifolia (Pullaiah, 1976)	Porogamous, normal	"	"	-	-
Gunura nitida (Pullaiah, 1976)	"	"	"	"	-
Notonia grandiflora (Pullaiah, 1976)	No fertilization	No endosperm	No embryo development	-	-

9

APPENDIX IV

(Tables - Chemosystematics)

Table 4.1 Major groups of secondary compounds in Plants
(Adopted from, The Biochemistry of plants. Volume VII. Eds. Stumpf and Conn, 1981)

Group	Biosynthetic Origin
Amino Acids	Citric acid cycle, shikimic acid, others.
Alkaloids	Amino acids, aromatic amino acids, often with other pathways such as terpenoids, acetate etc.
Alkanes, Alkenes, aldehydes, alcohols, ketones, waxes (long chain)	Acetyl - malonyl CoA (through fatty acids)
Acetylenic compounds,	Acetyl malonyl CoA (through fatty acids)
Anthraquinones etc.	Polyketides or shikimic acid and meralonic acid and others.
Carbohydrates	Pentose cycle, photosynthesis.
Cinnamic acids	Shikimic acid.
Coumarins	Shikimic acid (through cinnamic acid)
Cynogenic glycosides and lipids	Amino acids and carbohydrates or fatty acids
Fatty acids	Acetyl malonyl CoA
Flavonoids	Aromatic amino acids and acetyl malonyl CoA
Glucosinolates	Amino acids and carbohydrates
Phenyl propanoids	Cinnamic acids
Phenols (Simple types)	Shikimic acids, acetylenic compounds, meralonic acid, acetyl malonyl CoA.
Polyketides	Acetyl malony CoA
Terpenes (steroids carotenoids, etc)	Acetyl malonyl CoA (through meralonic acid).

Table 4.2 : Properties of different flavonoid classes
(Adopted from Phytochemical methods second edition; Harborne J.B. 1984)

Flavonoid Classes	Distribution	Characteristic properties
Anthocyanins	Scarlet, red, mauve and blue flower pigments also in leaf and other tissues.	Water soluble
Protoanthocyanidins	Mainly colourless. In heartwoods and in leaves of woody plants.	Yield anthocyanidins
Flavonols	Mainly colourless co-pigments in both cyanic and acyanic flowers; widespeard in leaves	After acid hydrolysis bright yellow spots on florestal chromatograms
Flavones	As flavonols	After acid hydrolysis, dull brown spots on florestal chromatoram
Glycoflavones	As flavonols	Contain C-C linked sugar mobile in H_2O
Biflavonylis	Colourless, almost confined to gymnosperms	Dull absorbing spots with very high Rf values
Chalcones and aurones	Yellow flower pigments occasionally found in other tissues	Give red and colours with ammonia
Flavonones	Colourless, found in leaf and fruit (Specially in *Citrus*)	Give intense red colours with Mg/Hcl.
Isoflavones	Colourless, often in root common in family the Leguminosae	No specific colour test available

Table 4.3 : Colour properties of flavonoids in visible and UV light
(Adopted from Phytochemical methods. Second edition : Harborne J.B. 1984)

Visible colour	Alone	With Ammonia	Indication
Orange, Red, Mauve	Dull, oragne red or mauve flourecent yellow Blue cerise or pink	Blue	Anthocyanidin 3-glycosides. Most anthocyanidin 3,5 diglycosides
Bright Yellow	Dark brown or black	Dark brown or black	6-hydroxyloated flavonols and flavones, some chalcones

		Dark red or bright orange	Most chalcones and aurones
	Bright yellow or yellow green	Bright orange or red	Aurones
Very pale yellow	Dark brown	Bright yellow or yellow brown	Most flavonol glycosides
		Vivid yellow green	Most flavonol glycosides
		Dark brown	Biflavonyls and unusually substituted flavones.
None	Dark Mauve	Faint brown	Most isoflavones and flavonols.
	Faint blue	Intense blue	5 desoxy isoflavones and 7, 8 dehydrory flavonols
	Dark mauve	Pale yellow or yellow green	Flavonones and flavonanols and 7 glycosides

Table 4.4 : Phenolic spot pattern in *Ageratum conyzoides*

		Rf			
Sl. No..	Spot. No.	I	II	Colour	Area in mm^2
1.	1	0.662	0.218	Grey	165
2.	3	0.712	0.237	Grey	54
3.	4	0.743	0.487	Greenish yellow	276
4.	5	0.837	0.687	Greenish yellow	93
5.	9	0.887	0.875	Grey	136
6.	10	0.893	0.156	Light yellow	49
7.	12	0.943	0.031	Grey	44

Table 4.5 : Phenolic spot pattern in *Ageratum houstonianum*.

		Rf			
Sl. No..	Spot. No.	I	II	Colour	Area in mm^2
1.	2	0.687	0.0875	Light yellow	69
2.	4	0.743	0.625	Greenish yellow	74
3.	5	0.812	0.463	Greenish yellow	221

4.	6	0.812	0.462	Light yellow	152
5.	7	0.818	0.312	Light green	77
6.	8	0.843	0.2	Light green	65
7.	11	0.906	0.25	Light green	67
8.	13	0.95	0.1	Yellow	82
9.	14	0.95	0.275	Light purple	83

Table 4.6 : Phenolic spot pattern in ***Adenostemma lavenia.***

		Rf			
Sl. No..	Spot. No.	I	II	Colour	Area in mm²
1.	1	0.575	0.05	Light yellow	236
2.	2	0.662	0.325	Green	186
3.	3	0.668	0.506	Light yellow	192
4.	4	0.731	0.043	Light yellow	85
5.	5	0.862	0.068	Light yellow	178
6.	6	0.875	0.206	Light yellow	149
7.	7	0.875	0.906	Light yellow	63
8.	8	0.881	0.968	Light yellow	91

Table 4.7 : Phenolic spot pattern in ***Chromolama odorata.***

		Rf			
Sl. No..	Spot. No.	I	II	Colour	Area in mm²
1.	1	0.437	0.052	Grey	61
2.	2	0.55	0.306	Light yellow	103
3.	3	0.593	0.05	Light yellow	239
4.	4	0.606	0.318	Green	97
5.	5	0.643	0.237	Green	169
6.	6	0.656	0.356	Green	79
7.	7	0.65	0.537	Green	166
8.	8	0.693	0.187	Greenish yellow	134
9.	9	0.712	0.337	Light yellow	104
10.	10	0.712	0.512	Green	122
11.	11	0.775	0.043	Light yellow	251
12.	12	0.918	0.106	Light yellow	203
13.	13	0.956	0.837	Green	236

Table 4.8 : Phenolic spot pattern in *Acanthospermum hispidum*

Sl. No..	Spot. No.	Rf I	Rf II	Colour	Area in mm^2
1.	1	0.731	0.431	Light yellow	82
2.	2	0.831	0.693	Light yellow	244
3.	3	0.968	0.681	Light yellow	169
4.	4	0.975	0.818	Greenish yellow	137

Table 4.9 : Phenolic spot pattern in *Polymnia grandis*

Sl. No..	Spot. No.	Rf I	Rf II	Colour	Area in mm^2
1.	1	0.731	0.431	Light yellow	81
2.	2	0.806	0.731	Light yellow	113
3.	3	0.918	0.706	Grey	78
4.	4	0.931	0.037	Grey	72

Table 4.10 : Phenolic spot pattern in *Siegesbeckia orientalis*

Sl. No..	Spot. No.	Rf I	Rf II	Colour	Area in mm^2
1.	1	0.668	0.606	Light yellow	171
2.	2	0.743	0.593	Light yellow	47
3.	3	0.775	0.725	Light yellow	64
4.	4	0.8	0.686	Grey	61
5.	5	0.868	0.112	Grey	46
6.	6	0.943	0.143	Grey	51

Table 4.11 : Phenolic spot pattern in *Blainvellia acmella*

Sl. No..	Spot. No.	Rf I	Rf II	Colour	Area in mm^2
1.	1	0.037	0.918	Grey	152
2.	2	0.731	0.7	Greenish yellow	70
3.	3	0.708	0.775	Grey	53

4.	4	0.781	0.656	Greenish yellow	43
5.	5	0.812	0.743	Light yellow	41
6.	6	0.868	0.05	Grey	156
7.	7	0.987	0.762	Grreenish yellow	127

Table 4.12 : Phenolic spot pattern in *Eclipta prostrata* (normal leaf)

		Rf			
Sl. No..	Spot. No.	I	II	Colour	Area in mm^2
1.	1	0.681	0.062	Greenish yellow	24
2.	4	0.800	0.125	Light yellow	62
3.	5	0.837	0.137	Greenish yellow	33
4.	6	0.850	0.218	Grey	88
5.	8	0.925	0.156	Grey	113

Table 4.13 : Phenolic spot pattern in *Eckuota oristrata* (small leaf) *Eclipta prostrata*

		Rf			
Sl. No..	Spot. No.	I	II	Colour	Area in mm^2
1.	1	0.681	0.162	Greenish yellow	61
2.	2	0.731	0.175	Grey	72
3.	3	0.756	0.343	Light yellow	173
4.	7	0.881	0.756	Light yellow	137
5.	0	0.956	0.800	Light yellow	124
6	10	0.943	0.087	Violet	162

Table 4.14 : Phenolic spot pattern in *Synedrella nodiflora* (normal leaf)

		Rf			
Sl. No..	Spot. No.	I	II	Colour	Area in mm^2
1.	1	0.668	0.218	Greenish yellow	58
2.	2	0.656	0.300	Greenish yellow	53
3.	8	0.700	0.275	Green	45
4.	9	0.731	0.356	Greenish yellow	195
5.	10	0.750	0.531	Green	89
6.	14	0.781	0.025	Light yellow	122
7.	17	0.843	0.843	Light yellow	183
8.	19	0.918	0.062	Grey	140
9	20	0.968	0.781	Grey	122

Table 4.15 : Phenolic spot pattern in *Synedrella nodiflora* (small leaf)

Sl. No..	Spot. No.	Rf I	Rf II	Colour	Area in mm²
1.	3	0.700	0.587	Green	35
2.	4	0.643	0.531	Light violet	62
3.	5	0.625	0.668	Dark green	58
4.	6	0.637	0.737	Light green	30
5.	7	0.675	0.575	Dark green	90
6.	9	0.718	0.300	Greenish yellow	114
7.	10	0.750	0.537	Green	130
8.	19	0.906	0.112	Grey	150

Table 4.16 : Phenolic spot pattern in *Calyptocarpus vialis*

Sl. No..	Spot. No.	Rf I	Rf II	Colour	Area in mm²
1.	4	0.650	0.512	Light violet	89
2.	5	0.656	0.637	Dark green	76
3.	9	0.750	0.268	Greenish yellow	65
4.	11	0.781	0.136	Greenish yellow	37
5.	12	0.734	0.206	Grey	24
6.	13	0.825	0.218	Greenish yellow	76
7.	15	0.812	0.637	Greenish yellow	57
8.	16	0.831	0.750	Greenish yellow	100
9.	18	0.850	0.212	Grey	54
10.	19	0.900	0.062	Grey	105

Table 4.17 : Phenolic spot pattern in *Lagascea mollis*

Sl. No..	Spot. No.	Rf I	Rf II	Colour	Area in mm²
1.	1	0.637	0.268	Grey	54
2.	2	0.675	0.337	Green	33
3.	3	0.712	0.375	Grey	157
4.	4	0.812	0.968	Grey	114
5.	5	0.95	0.937	Green	87

Table 4.18 : Phenolic spot pattern in *Sclerocarpus africanus*

Sl. No..	Spot. No.	Rf I	Rf II	Colour	Area in mm^2
1.	1	0.706	0.356	Grey	24
2.	2	0.731	0.275	Grey	85
3.	3	0.762	0.012	Light green	89
4.	4	0.893	0.437	Green	93
5.	5	0.968	0.05	Green	58
6.	6	0.962	0.843	Green	348

Table 4.19 : Phenolic spot pattern in *Helianthus rigidus*

Sl. No..	Spot. No.	Rf I	Rf II	Colour	Area in mm^2
1.	1	0.762	0.3	Light green	196
2.	2	0.737	0.418	Light green	119
3.	3	0.75	0.515	Grey	228
4.	4	0.981	0.043	Light green	88
5.	5	0.931	0.225	Light yellow	53
6.	6	0.918	0.612	Light yellow	230
7.	7	0.968	0.881	Green	286

Table 4.20 : Phenolic spot pattern in *Rudbeckia nitida*

Sl. No..	Spot. No.	Rf I	Rf II	Colour	Area in mm^2
1.	1	0.812	0.237	Light yellow	42
2.	2	0.918	0.062	Light green	41
3.	3	0.893	0.543	Green	81
4.	4	0.887	0.15	Green	53
5.	5	0.881	0.212	Light green	59
6.	6	0.9	0.293	Light green	52

Table 4.21 : Phenolic spot pattern in *Bidens pilosa* (Normal ray)

Sl. No..	Spot. No.	Rf I	Rf II	Colour	Area in mm^2
1.	1	0.5	0.87	Yellow	166
2.	6	0.762	0.106	Light yellow	100
3.	9	0.818	0.325	Yellow	116
4.	11	0.843	0.431	Light yellow	90
5.	13	0.85	0.093	Light grey	102
6.	14	0.868	0.268	Greenish yellow	88
7.	15	0.893	0.343	Grey	63

Table 4.22 : Phenolic spot pattern in *Bidens pilosa* (intermediate)

Sl. No..	Spot. No.	Rf I	Rf II	Colour	Area in mm^2
1.	8	0.768	0.418	Light green	94
2.	12	0.843	0.368	Light green	68
3.	14	0.893	0.262	Greenish yellow	99
4.	16	0.837	0.587	Light yellow	93
5.	18	0.906	0.506	Light yellow	100
6.	17	0.875	0.768	Light green	53
7.	20	0.981	0.106	Light yellow	29

Table 4.23 : Phenolic spot pattern in *Bidens pilosa* (Rayless)

Sl. No..	Spot. No.	Rf I	Rf II	Colour	Area in mm^2
1.	1	0.462	0.056	Yellow	418
2.	2	0.775	0.068	Light yellow	181
3.	4	0.762	0.425	Light green	233
4.	5	0.787	0.243	Grey	136
5.	10	0.843	0.243	Grey	79
6.	14	0.893	0.25	Greenish yellow	108
7.	15	0.912	0.343	Grey	70
8.	17	0.881	0.706	Light green	298
9.	19	0.912	0.556	Greenish yellow	171
10.	20	0.962	0.093	Light yellow	146

Table 4.24 : Phenolic spot pattern in *Bidens humilis*

Sl. No..	Spot. No.	Rf I	Rf II	Colour	Area in mm^2
1.	1	0.437	0.1	Yellow	269
2.	3	0.743	0.412	Yellow	77
3.	7	0.787	0.268	Greenish yellow	78
4.	11	0.825	0.45	Light yellow	65
5.	20	0.937	0.1	Light yellow	122

Table 4.25 : Phenolic spot pattern in *Moonia heterophylla*

Sl. No..	Spot. No.	Rf I	Rf II	Colour	Area in mm^2
1.	1	0.468	0.175	Light yellow	169
2.	2	0.568	0.443	Grey	133
3.	3	0.618	0.268	Green	179
4.	4	0.637	0.15	Green	63
5.	5	0.706	0.206	Greenish yellow	425
6.	6	0.768	0.468	Grey	160
7.	7	0.831	0.375	Grey	140
8.	8	0.912	0.062	Light yellow	127
9.	9	0.962	0.724	Greenish yellow	142

Table 4.26 : Phenolic spot pattern in *Galinsoga parviflora*

Sl. No..	Spot. No.	Rf I	Rf II	Colour	Area in mm^2
1.	6	0.756	0.362	Light yellow	183
2.	7	0.8	0.181	Greenish yellow	59
3.	9	0.825	0.662	Grey	72
4.	10	0.831	0.162	Greenish yellow	25
5.	11	0.831	0.268	Yellow	99
6.	12	0.956	0.912	Grey	405

Table 4.27 : Phenolic spot pattern in *Galinsoga parviflora* and *Galinsoga quadriradiata* (intermediate)

Sl. No..	Spot. No.	Rf I	Rf II	Colour	Area in mm^2
1.	3	0.681	0.156	Greenish yellow	89
2.	4	0.725	0.181	Grey	44
3.	5	0.775	0.206	Greenish yellow	106
4.	12	0.925	0.906	Grey	227
5.	13	0.906	0.081	Grey	168
6.	14	0.975	0.806	Greenish yellow	134

Table 4.28 : Phenolic spot pattern in *Galinsoga quadriradiata*

Sl. No..	Spot. No.	Rf I	Rf II	Colour	Area in mm^2
1.	1	0.225	0.843	Grey	163
2.	2	0.368	0.918	Grey	197
3.	5	0.762	0.175	Greenish yellow	93
4.	6	0.756	0.331	Light yellow	86
5.	8	0.793	0.425	Light yellow	86
6.	12	0.925	0.931	Grey	118
7.	13	0.968	0.793	Greenish yellow	168
8	14	0.975	0.087	Grey	79

Table 4.29 : Phenolic spot pattern in *Spilanthes calva* (small head)

Sl. No..	Spot. No.	Rf I	Rf II	Colour	Area in mm^2
1.	1	0.487	0.625	Yellow	174
2.	2	0.468	0.443	Light yellow	158
3.	3	0.531	0.781	Light yellow	177
4.	4	0.662	0.462	Greenish yellow	515
5.	7	0.043	0.825	Grey	104
6.	8	0.7	0.393	Grey	86
7.	9	0.743	0.312	Grey	37
8.	11	0.781	0.306	Light yellow	44
9.	13	0.743	0.007	Light yellow	282

Table 4.30 : Phenolic spot pattern in *Spilanthes calva* (large head)

Sl. No..	Spot. No.	Rf I	Rf II	Colour	Area in mm^2
1.	1	0.45	0.118	Yellow	163
2.	4	0.668	0.325	Greenish yellow	229
3.	4A	0.675	0.431	Light greenish yellow	171
4.	5	0.662	0.575	Grey	183
5.	6	0.712	0.231	Greenish yellow	222
6.	13	0.737	0.096	Light yellow	357
7.	15	0.812	0.231	Light yellow	86
8.	16	0.825	0.818	Grey	143
9.	17	0.906	0.031	Greenish yellow	38
10.	19	0.95	0.75	Grey	109

Table 4.31 : Phenolic spot pattern in *Spilanthes paniculata*

Sl. No..	Spot. No.	Rf I	Rf II	Colour	Area in mm^2
1.	6	0.75	0.237	Grey	51
2.	8	0.731	0.45	Grey	78
3.	10	0.737	0.331	Green	107
4.	12	0.781	0.237	Green	18
5.	13	0.825	0.005	Light yellow	219
6.	14	0.806	0.237	Green	58
7.	18	0.95	0.098	Grey	79
8.	20	0.975	0.862	Greenish yellow	286

Table 4.32 : Phenolic spot pattern in *Parthenium hysterophorus*

Sl. No..	Spot. No.	Rf I	Rf II	Colour	Area in mm^2
1.	1	0.631	0.456	Light violet	109
2.	2	0.625	0.356	Violet	82
3.	3	0.656	0.281	Green	88
4.	4	0.662	0.193	Light yellow	193
5.	5	0.706	0.056	Light yellow	208
6.	6	0.762	0.25	Grey	86
7.	7	0.868	0.981	Grey	77
8.	8	0.837	0.956	Light yellow	125

Table 4.33 : Phenolic spot pattern in *Xanthium indicum*

Sl. No..	Spot. No.	Rf I	Rf II	Colour	Area in mm²
1.	1	0.6	0.287	Light yellow	105
2.	2	0.6	0.287	Light yellow	89
3.	3	0.737	0.525	Grey	117
4.	4	0.737	0.643	Light yellow	81
5.	5	0.793	0.587	Grey	51
6.	6	0.806	0.506	Grey	55

Table 4.34 : Phenolic spot pattern in *Senecio wightii*

Sl. No..	Spot. No.	Rf I	Rf II	Colour	Area in mm²
1.	8	0.718	0.55	Light green	454
2.	11	0.750	0.337	Grey	71
3.	27	0.893	0.181	Grey	68
4.	28	0.912	0.412	Violet	52
5.	35	0.987	0.893	Light yellow	336

Table 4.35 : Phenolic spot pattern in *Senecio lawsoni*

Sl. No..	Spot. No.	Rf I	Rf II	Colour	Area in mm²
1.	8	0.687	0.425	Light green	84
2.	11	0.718	0.312	Light green	95
3.	27	0.887	0.168	Greenish yellow	130
4.	31	0.943	0.343	Light yellow	129
5.	35	0.981	0.868	Light yellow	210

Table 4.36 : Phenolic spot pattern in *Senecio belgaumensis*

Sl. No..	Spot. No.	Rf I	Rf II	Colour	Area in mm²
1.	8	0.7	0.406	Light green	231
2.	13	0.75	0.768	Light yellow	88
3.	15	0.768	0.537	Grey	205
4.	27	0.893	0.168	Grey	72
5.	35	0.956	0.9	Light yellow	385

Table 4.37 : Phenolic spot pattern in *Senecio neelgherryanus*

Sl. No..	Spot. No.	Rf I	Rf II	Colour	Area in mm^2
1.	8	0.656	0.456	Light green	245
2.	13	0.762	0.768	Light yellow49	
3.	24	0.812	0.418	Violet	141
4.	27	0.9	0.206	Grey	56
5.	28	0.937	0.437	Violet	234
6.	35	0.887	0.9	Light yellow	88

Table 4.38 : Phenolic spot pattern in *Senecio tenuifolius*

Sl. No..	Spot. No.	Rf I	Rf II	Colour	Area in mm^2
1.	1	0.362	0.412	Light yellow	115
2.	6	0.587	0.700	Blue	90
3.	20	0.843	0.331	Grey	54
4.	21	0.856	0.218	Yellow	68
5.	25	0.868	0.212	Grey	53
6.	27	0.881	0.137	Grey	25
7.	29	0.918	0.81	Light yellow	63
8.	33	0.981	0.617	Grey	70

Table 4.39 : Phenolic spot pattern in *Senecio hohenackeri*

Sl. No..	Spot. No.	Rf I	Rf II	Colour	Area in mm^2
1.	3	0.637	0.293	Grey	445
2.	8	0.731	0.425	Light green	58
3.	13	0.768	0.618	Light yellow	112
4.	18	0.806	0.537	Grey	65
5.	21	0.806	0.218	Grey	74
6.	22	0.837	0.593	Grey	32
7.	25	0.862	0.250	Grey	44
8.	29	0.906	0.118	Light yellow	72

Table 4.40 : Phenolic spot pattern in *Senecio grahami*

Sl. No..	Spot. No.	Rf I	Rf II	Colour	Area in mm^2
1.	3	0.543	0.25	Light yellow	171
2.	10	0.718	0.093	Light yellow	207
3.	12	0.718	0.356	Dark green	125
4.	25	0.837	0.481	Violet	100
5.	27	0.931	0.181	Grey	121
6.	36	0.975	0.868	Yellow green	79

Table 4.41 : Phenolic spot pattern in *Senecio edgeworthii*

Sl. No..	Spot. No.	Rf I	Rf II	Colour	Area in mm^2
1.	3	0.5	0.275	Light yellow	107
2.	27	0.875	0.181	Grey	87
3.	29	0.9	0.481	Light yellow	90
4.	31	0.925	0.356	Light yellow	91
5.	33	0.893	0.731	Grey	61
6.	35	0.968	0.875	Light yellow	150

Table 4.42 : Phenolic spot pattern in *Senecio edgeworthii* (small leaf)

Sl. No..	Spot. No.	Rf I	Rf II	Colour	Area in mm^2
1.	3	0.5	0.262	Light yellow	107
2.	27	0.887	0.181	Grey	83
3.	29	0.912	0.475	Light yellow	97
4.	31	0.925	0.368	Light yellow	81
5.	35	0.968	0.875	Light yellow	169

Table 4.43 : Phenolic spot pattern in *Senecio ludens*

Sl. No..	Spot. No.	Rf I	Rf II	Colour	Area in mm^2
1.	3	0.606	0.275	Light yellow	83
2.	12	0.756	0.350	Dark green	44
3.	14	0.781	0.300	Grey	29
4.	18	0.812	0.300	Grey	48
5.	23	0.862	0.618	Light yellow	15
6.	26	0.875	0.162	Grey	46
8	32	0.937	0.643	Dark green	133

Table 4.44 : Phenolic spot pattern in *Senecio cineraria*

Sl. No..	Spot. No.	Rf I	Rf II	Colour	Area in mm^2
1.	3	0.537	0.256	Light yellow	98
2.	12	0.762	0.356	Dark green	88
3.	13	0.768	0.631	Light yellow	.285
4.	17	0.806	0.337	Grey	39
5.	20	0.837	0.331	Grey	30
6.	24	0.868	0.400	Violet	100
7.	27	0.887	0.156	Grey	36
8.	30	0.900	0.015	green	41

Table 4.45 : Phenolic spot pattern in *Senecio corymbosus*

Sl. No..	Spot. No.	Rf I	Rf II	Colour	Area in mm^2
1.	2	0.487	0.456	Light yellow	268
2.	4	0.562	0.687	Light yellow	203
3.	5	0.575	0.631	Light yellow	133
4.	7	0.643	0.531	Green	455
5.	8	0.706	0.35	Light green	86
6.	9	0.725	0.206	Green	157
7.	10	0.743	0.068	Light yellow	188
8.	14	0.768	0.906	Greenish yellow	118
9.	16	0.812	0.218	Yellow	63
10.	27	0.881	0.143	Grey	232
12.	35	0.975	0.831	Light yellow	203

Table 4.46 : Phenolic spot pattern in *Senecio walkeri*

Sl. No..	Spot. No.	Rf I	Rf II	Colour	Area in mm^2
1.	5	0.562	0.618	Light green	60
2.	9	0.706	0.218	Green	170
3.	11	0.737	0.343	Dark green	60
4.	12	0.756	0.4	Dark green	52
5.	15	0.775	0.493	Grey	43
6.	16	0.793	0.237	Light yellow	57
7.	17	0.80	0.337	Dark green	43
8.	20	0.831	0.356	Grey	82
9.	27	0.868	0.187	Grey	42
10.	34	0.943	0.75	Green	56

Table 4.47 : Phenolic spot pattern in *Gynura nitida*

Sl. No..	Spot. No.	Rf I	Rf II	Colour	Area in mm^2
1.	1	0.556	0.081	Light yellow	115
2.	3	0.681	0.15	Light yellow	145
3.	6	0.731	0.468	Light yellow	39
4.	7	0.775	0.362	Light yellow	114
5.	8	0.818	0.075	Light yellow	16
6.	9	0.831	0.275	Green	92
7.	10	0.812	0.493	Light green	114
8.	13	0.962	0.493	Green	27

Table 4.48 : Phenolic spot pattern in *Gynura lycopersicifolia*

Sl. No..	Spot. No.	Rf I	Rf II	Colour	Area in mm^2
1.	1	0.556	0.1	Light yellow	147
2.	2	0.65	0.237	Yellow	124
3.	3	0.718	0.137	Light yellow	156
4.	4	0.700	0.431	Greenish yellow	37
5.	5	0.750	0.256	Green	94
6.	7	0.793	0.343	Light yellow	134
7.	8	0.856	0.056	Light yellow	131
8.	9	0.850	0.243	Green	152
9.	11	0.850	0.493	Yellow	183

Table 4.49 : Phenolic spot pattern in *Gynura aurantiaca*

Sl. No..	Spot. No.	Rf I	Rf II	Colour	Area in mm^2
1.	5	0.768	0.219	Green	148
2.	9	0.843	0.331	Green	122
3.	12	0.837	0.900	Light green	190
4.	14	0.962	0.968	Green	83

Table 4.50 : Phenolic spot pattern in *Notnoia grandiflora*

Sl. No..	Spot. No.	Rf I	Rf II	Colour	Area in mm^2
1.	2	0.631	0.525	Grey	81
2.	4	0.675	0.387	Greenish yellow	97
3.	5	0.712	0.175	Greenish yelllow	56
4.	6	0.737	0.212	Green	37
5.	8	0.75	0.262	Light green	46
6.	12	0.925	0.837	Green	173

Table 4.51 : Phenolic spot pattern in *Notonia walkeri*

Sl. No..	Spot. No.	Rf I	Rf II	Colour	Area in mm^2
1.	1	0.593	0.031	Light yellow	179
2.	2	0.625	0.562	Grey	133
3.	3	0.656	0.250	Light yellow	148
4.	7	0.668	0.425	Light yellow	151
5.	8	0.731	0.25	Light green	99
6.	9	0.756	0.093	Yellow	376
7.	10	0.856	0.243	Light yellow	197
8.	11	0.856	0.857	Green	102
9.	12	0.918	0.862	Green	153
10.	13	0.887	0.081	Light green	224

Table 4.52 : Phenolic spot pattern in *Emilia sonchifolia*

Sl. No..	Spot. No.	Rf I	Rf II	Colour	Area in mm^2
1.	8	0.681	0.95	Grey	173
2.	11	0.843	0.193	Greenish yellow	52
3.	12	0.862	0.55	Light	151
4.	13	0.875	0.225	Greenish yellow	93
5.	16	0.975	0.231	Light yellow	104
6.	18	0.968	0.812	Greenish yellow	116
7.	19	0.968	0.931	Greenish yellow	82

Table 4.53 : Phenolic spot pattern in *Emilia scabra*

Sl. No..	Spot. No.	Rf I	Rf II	Colour	Area in mm^2
1.	1	0.562	0.325	Light yellow	71
2.	5	0.625	0.575	Grey	91
3.	6	0.643	0.675	Grey	67
4.	7	0.687	0.781	Grey	231
5.	8	0.725	0.95	Grey	194
6.	10	0.837	0.093	Light yellow	120

Table 4.54 : Phenolic spot pattern in *Emilia ramulosa*

Sl. No..	Spot. No.	Rf I	Rf II	Colour	Area in mm^2
1.	2	0.575	0.468	Light yellow	195
2.	3	0.612	0.662	Light yellow	120
3.	4	0.618	0.818	Greenish yellow	139
4.	8	0.737	0.962	Grey	166
5.	9	0.843	0.981	Light yellow	96
6.	14	0.925	0.131	Grey	173
7.	15	0.912	0.993	Green	47
8.	17	0.981	0.706	Light green	158

Table No. 4.55 : Co-efficient of similarity for species of Ageratum

	a	b
a	x	
b	0.25	x

(a - *A. conyxoides;* b = *A.houstonianum*)

Table No. 4.56 : Co-efficient of similarity for species of *Eclipta*

	a	b
a	x	
b	0.15	x

(a - *A. prostrata* (normal leaf); b = *E. prostrata* (small leaf)

Table No. 4.57 : Matching co-efficient for species of *Synedrella* and *Calyptocarpus*

	a	b	c
a	x		
b	0.17	x	
c	0.18	0.33	x

Table No. 4.58 : Co-efficient of similarity for species of *Synedrella* and *Calyptocarpus*

	a	b	c
a	x		
b	0.23	x	
c	0.21	0.44	x

(a = *S.nodiflora* (normal leaf) b= *S.nodiflora* (small leaf) c= *Calyptocarpus vialis*)

Table No. 4.59 : Matching Co-efficient for species of *Bidens*

	a	b	c	d
a	x			
b	0.26	x		
c	0.095	0.26	x	
d	0.18	0.18	0.095	x

Table No. 4.60 : Co-efficient for of similarity for species of *Bidens*

	a	b	c	d
a	x			
b	0.35	x		
c	0.14	0.35	x	
d	0.33	0.26	0.16	x

(a = *Bidens pilosa* (normal ray); b = *B.pilosa* (rayless) c = *B.pilosa* (intermediate ray); d = *B.humilis*)

Table No. 4.61: Matching Co-efficient for species of *Galinsoga*

	a	b	c
a	x		
b	0.25	x	
c	0.13	0.44	x

Table No. 4.62 : Co-efficient of similarity for species of *Galinsoga*

	a	b	c
a	x		
b	0.28	x	
c	0.16	0.57	x

(a = *G. parviflora*; b = *G. quadriradiata;* c = Possible hybrid between *G.parviflora* x *G. quadriradiata*)

Table No. 4.63 : Matching Co-efficient for species of *Bidens*

	a	b	c
a	x		
b	0.25	x	
c	0.17	0.17	x

Table No. 4.64: Co-efficient of similarity for species of *Spilanthes*

	a	b	c
a	x		
b	0.31	x	
c	0.23	0.22	x

(a = *S. calva* (small head); b = *S.calva* (large head); c = *S.paniculata*

Table No. 4.65 : Distribution of various flavonoid spots in 12 species of *Senecio*.
Spot nos. are shown in vertical rows. Species in horizontal rows (for names of species see table no.4.71)

	1	2	3	4	5	6	7	8	9	10	11	12
1.												X
2.									X			
3.	X	X	X	X								
4.									X			
5.									X	X		
6.											X	
7.									X			
8.					X	X	X	X	X			X
9.									X	X		
10.	X								X			
11.					X	X	X	X		X		
12.	X	X	X	X						X		
13.			X				X	X			X	
14.				X					X			
15.							X					
16.									X	X		
17.			X							X		
18.											X	
19.				X								
20.			X							X		X

21.											X	X
22.											X	
23.				X								
24.	X		X					X				
25.											X	X
26.				X								
27.	X	X	X	X	X	X	X	X	X	X		X
28.						X		X				
29.		X										
30.			X									
31.					X							
32.				X								X
34.										X		
35.		X			X	X	X	X	X			
36.	X											

Table No. 4.66 : Similarity matrix based on Paired Affinity indices for 12 species of Senecio
(for names of species see table no. 4.71)

	1	2	3	4	5	6	7	8	9	10	11	12
1.	X											
2.	50	X										
3.	57	43	X									
4.	43	43	38	X								
5.	18	36	15	15	X							
6.	18	36	15	15	80	X						
7.	17	33	29	14	73	73	X					
8.	31	31	40	13	67	83	77	X				

	1	2	3	4	5	6	7	8	9	10	11	12
9.	24	24	11	21	38	38	35	33	X			
10.	25	25	45	22	27	27	38	24	38	X		
11.	14	31	25	13	15	15	31	26	11	0	X	
12.	14	43	25	13	15	15	14	13	11	22	38	X

Table 4.67 : Similarity matrix based on matching coefficient for 12 species of *Senecio*
(for names of species see table no.76

	1	2	3	4	5	6	7	8	9	10	11	12
1.	X											
2.	85	X										
3.	95	79	X									
4.	79	79	74	X								
5.	75	70	70	70	X							
6.	75	80	70	70	94	X						
7.	71	78	73	62	92	92	X					
8.	76	76	76	64	90	94	92	X				
9.	66	65	54	60	74	74	71	69	X			
10.	68	70	71	63	71	71	74	66	67	X		
11.	65	73	68	62	70	70	73	71	54	50	X	
12.	67	79	68	62	70	70	67	64	54	63	70	X

Table No. 4.68 : Non similarity matrix based on Negative correlation for 12 species of *Senecio*
(for names of species see Table no.76

	1	2	3	4	5	6	7	8	9	10	11	12
1.	X											
2.	50	X										
3.	42	50	X									
4.	57	57	62	X								
5.	81	63	84	84	X							
6.	81	63	84	84	19	X						

7.	83	66	71	85	27	27	X					
8.	69	69	60	86	33	16	23	X				
9.	76	76	89	78	62	62	64	66	X			
10.	75	75	55	77	73	73	62	76	61	X		
11.	85	71	75	86	84	84	71	73	89	100	X	
12.	85	71	75	86	84	84	85	86	89	77	62	X

Table No. 4.69 : Similarity matrix based on biochemical distance for 12 species of *Senecio*
(for names of species see table no.76)

	1	2	3	4	5	6	7	8	9	10	11	12
1.	X											
2.	3.38	X										
3.	5.41	2.02	X									
4.	0.25	3.38	1.36	X								
5.	0.25	2.25	2.6	2.33	X							
6.	1.90	4.02	0.88	0.6	11.38	X						
7.	1.36	4.11	4.7	0.07	4.5	6.88	X					
8.	0	0.16	11.75	0.94	2.05	8.33	2.44	X				
9.	3.08	4.88	5.44	7.6	2.69	9.36	3.25	0.52	X			
10.	4.2	2.6	0.72	0.11	3.4	1.02	4.86	0.86	5.0	X		
11.	7.6	8.75	4.80	10.05	0.72	11.0	2.88	5.05	0.77	0	X	
12.	2.66	3.22	0.36	0.58	2.9	1.19	1.30	1.52	5.75	1.25	0.16	X

Table No. 4.70 : Group affinity values for 12 species of Senecio based on P.A. indices

1.	*S. grahami*	311
2.	*S. edgeworthii*	395
3.	*S. cineraria*	329
4.	*S. ludens*	250
5.	*S. lawsnoi*	399

6.	*S. wightii*	415
7.	*S. belgaumensis*	434
8.	*S. neelgherryanus*	438
9.	*S. corymbosus*	[illegible]
10.	*S. walkeri*	286
11.	*S. hohenackeri*	216
12.	*S. tenuifolius*	222

Table No. 4.71 : Group Non affinity values for 12 species of *Senecio,* based on negative correlation

1.	*S. grahami*	784
2.	*S. edgeworthii*	711
3.	*S. cineraria*	747
4.	*S. ludens*	894
5.	*S. lawsnoi*	694
6.	*S. wightii*	677
7.	*S. belgaumensis*	664
8.	*S. neelgherryanus*	657
9.	*S. corymbosus*	812
10.	*S. walkeri*	804
11.	*S. hohenackeri*	880
12.	*S. tenuifolius*	884

Table No. 4.72 : Group Non affinity values for 12 species of *Senecio,* based on matching coefficient

1.	*S. grahami*	822
2.	*S. edgeworthii*	845
3.	*S. cineraria*	798
4.	*S. ludens*	745

5.	*S. lawsnoi*	857
6.	*S. wightii*	860
7.	*S. belgaumensis*	845
8.	*S. neelgherryanus*	838
9.	*S. corymbosus*	708
10.	*S. walkeri*	734
11.	*S. hohenackeri*	719
12.	*S. tenuifolius*	734

Table No. 4.73 : Group Non affinity values for 12 species of *Senecio*, based on matching coefficient

Sl.No.	Taxon	Spot No. in the	No. of unique Composite key	Total spots spots	Iv_n	Iv_s
1.	*S. grahami*	36	1	6	2.77	16.66
2.	*S. edgeworthii*	-	0	6	0	0
3.	*S. cineraria*	30	1	8	2.77	12.5
4.	*S. ludens*	19, 23 26, 32	4	8	11.11	50
5.	*S. lawsnoi*	31	1	5	2.77	20
6.	*S. wightii*	-	0	5	0	0
7.	*S. belgaumensis*	-	0	6	0	0
8.	*S. neelgherryanus*	-	0	7	0	0
9.	*S. corymbosus*	2, 4, 7	3	11	8.33	27.27
10.	*S. walkeri*	34	1	10	2.77	10
11.	*S. hohenackeri*	18, 22	2	8	5.55	35
12.	*S. tenuifolius*	1, 6	2	8	5.5	25

Table No. 4.74: Matching coefficient for species of *Gynura*

	a	b	c
a	x		
b	0.13	x	
c	0.25	0.42	x

Table No. 4.75: Coefficient of similarity for species of *Gynura*

	a	b	c
a	x		
b	0.16	x	
c	0.30	0.47	x

(a = *G. aurantiaca;* b = *G.nitida;* c = *G. lycopersicifolia*)

Table No. 4.76 : Co efficient of similarity for species of *Notonia*

	a	b
a	x	
b	0.375	x

(a = *N. grandiflora,* b = *N.walkeri*)

Table No. 4.77 : Matching co efficient for species *of Emilia*

	a	b	c
a	x		
b	0.1	x	
c	0.1	0.1	x

Table No. 4.78 : Co efficient of similarity for species of *Emilia*

	a	b	c
a	x		
b	0.13	x	
c	0.15	0.14	x

(a = *E.sonchifolia;* b = *E. ramulosa;* c = *E.scabra*)